Minerals and Coal Process Calculations

Minerals and Coal Process Calculations

D.V. Subba Rao

*Formerly Head of the Department of Mineral Beneficiation,
S.D.S. Autonomous College, Andhra Pradesh, India*

CRC Press
Taylor & Francis Group
Boca Raton London New York

CRC Press is an imprint of the
Taylor & Francis Group, an **informa** business

A BALKEMA BOOK

Cover images credits: Flow diagram by Dr D.V. Subba Rao; Coal and Mineral images taken from Shutterstock.

CRC Press
Taylor & Francis Group
6000 Broken Sound Parkway NW, Suite 300
Boca Raton, FL 33487-2742

© 2016 by Taylor & Francis Group, LLC
CRC Press is an imprint of Taylor & Francis Group, an Informa business

Typeset by MPS Limited, Chennai, India

ISBN-13: 978-1-138-62662-1 (hbk)
ISBN-13: 978-0-367-88729-2 (pbk)

Library of Congress Cataloging-in-Publication Data

Visit the Taylor & Francis Web site at
http://www.taylorandfrancis.com

and the CRC Press Web site at
http://www.crcpress.com

Dedicated to

Prof. Dr. T.C. Rao
Who nurtured me professionally with love and affection

Table of contents

Preface		xi
Acknowledgements		xiii
List of tables		xv
List of figures		xix

1   Minerals and coal   1
  1.1   Types of minerals   1
  1.2   Mineral processing   7
  1.3   Coal   11
        1.3.1   Proximate analysis   14
                1.3.1.1   Determination of moisture   14
                1.3.1.2   Determination of volatile matter   14
                1.3.1.3   Determination of ash   15
                1.3.1.4   Determination of fixed carbon   15
        1.3.2   Expression of analytical results on different bases   17
                1.3.2.1   As received or As sampled   17
                1.3.2.2   Air-dried   18
                1.3.2.3   Dry or moisture free   18
                1.3.2.4   Dry, ash free (d.a.f)   18
                1.3.2.5   Dry, mineral matter free (d.m.m.f)   18
        1.3.3   Calculations on different bases   18
                1.3.3.1   As received or As sampled basis   18
                1.3.3.2   Air-dried basis   19
                1.3.3.3   Dry or moisture free basis   19
                1.3.3.4   Dry, ash free (d.a.f) basis   19
                1.3.3.5   Dry, mineral matter free (d.m.m.f) basis   20
  1.4   Varieties of coal   21
  1.5   Coal processing   22
  1.6   Problems for practice   23

2   Material (mass) balance   25

3   Sampling   27

4   Size analysis   29
  4.1   Sieve analysis   34
  4.2   Testing method   34

4.3   Presentation of particle size distribution data   37
4.4   Particle size distribution equations   42
4.5   Size assay analysis   48
4.6   Problems for practice   52

5   Screening   **55**
5.1   Purpose of screening   55
5.2   Screen   55
5.3   Screen action   57
    5.3.1   Factors affecting the rate of screening   58
5.4   Material balance   59
5.5   Screen efficiency   61
5.6   Tromp curve   68
5.7   Problems for practice   73

6   Density   **75**
6.1   Solids and pulp   78
6.2   Retention time   86
6.3   Miscible liquids   87
6.4   Problems for practice   88

7   Liberation   **91**

8   Comminution   **95**
8.1   Objectives of comminution   95
8.2   Laws of comminution   95
8.3   Types of comminution operations   97
8.4   Problems for practice   104

9   Crushing   **105**
9.1   Types of crushers   106
9.2   Crushing operation   113
9.3   Open and closed circuit crushing operations   115
    9.3.1   Open circuit crushing operation   116
    9.3.2   Closed circuit crushing operation   118
9.4   Problems for practice   124

10   Grinding   **127**
10.1   Grinding action   128
10.2   Wet and dry grinding   132
10.3   Grinding circuits   132
10.4   Problems for practice   141

11   Principles of settling   **143**
11.1   Laminar and turbulent flows   144
11.2   Fluid resistance   145
11.3   Terminal velocity   146

11.4 Free settling   155
11.5 Hindered settling   155
11.6 Equal settling particles   157
11.7 Settling ratio   158
11.8 Settling of large spheres in a suspension of fine spheres   161
11.9 Problems for practice   161

12 Classification   **163**
12.1 Classifiers   163
12.2 Efficiency of separation in hydrocyclone   180
12.3 Problems for practice   187

13 Beneficiation operations   **191**
13.1 Gravity concentration   191
13.2 Froth flotation   194
13.3 Magnetic separation   196
13.4 Electrical separation   197

14 Sink and float   **199**

15 Float and sink   **205**
15.1 Float and sink test   206
15.2 Near gravity materials   211
15.3 Yield reduction factor   212
15.4 Washability Index   213
15.5 Optimum degree of washability (ODW)   215
15.6 Washability number   216
15.7 Effect of sizing on washability characteristics of a coal   216
15.8 Mayer curve   227
       15.8.1 Construction of M-curve   227
       15.8.2 M-curve for a three product system   227
       15.8.3 M-curve for blended cleaned and un-cleaned coal   229
       15.8.4 M-curve for blending clean coal from two plants   230
15.9 Problems for practice   235

16 Metallurgical accounting   **237**
16.1 Two products beneficiation operations   237
       16.1.1 Ratio of concentration   238
       16.1.2 Ratio of recovery   238
       16.1.3 Ratio of enrichment   239
       16.1.4 Metallurgical efficiency   239
       16.1.5 Economic recovery or efficiency   239
16.2 Three products beneficiation operations   250
16.3 Separation efficiency   256
16.4 Economic efficiency   258
16.5 Problems for practice   263

17 Coal washing efficiency   **267**
  17.1 Dependent criteria   267
    17.1.1 Organic efficiency   267
    17.1.2 Anderson efficiency   268
    17.1.3 Ash error   268
    17.1.4 Yield error (or) yield loss   268
  17.2 Independent criteria   273
    17.2.1 Probable error (or) Ecart Probable Moyen (E_p)   278
    17.2.2 Error area or Tromp area   279
    17.2.3 Imperfection   279
  17.3 Problems for practice   284

18 Process plant circuits   **287**
  18.1 Circuits with complete material balance   288
  18.2 Flotation contact time   304
  18.3 Coal analysis and washability   309
  18.4 Additional problems for practice   315

Annexure: Procedure for determination of bonds work index   319
References   323
Further readings   325
Subject index   327
Index for calculations   331

Preface

This book Minerals and Coal Process Calculations has been conceived with an intention to provide a book exclusively on process calculations, the most vital part of mineral and coal processing operations. The aim of the process calculations is to evaluate the performance of mineral and coal processing operations in terms of the efficiency of the operation and recovery of the required constituents. The primary requirement to evaluate the performance is to measure different parameters like flow rates of the process streams and percent solids, size assay analysis, float and sink analysis of solids in each stream. The process calculations are also necessary to understand the principles of separation processes.

This book is designed to illustrate all the process calculations. The first chapter introduces the minerals and coal and grade calculations. The second chapter explains the material balancing method, the heart of the majority of process calculations. Three chapters sampling, liberation and beneficiation operations have been included for theoretical explanations necessary to understand process calculations easily. Size analysis, graphical representations, partition, washability, Mayer and Tromp curves, calculations of screen efficiency, density and percent solids, energy for comminution, open and closed circuit crushing and grinding operations, classification efficiency, float and sink analysis, and metallurgical accounting are illustrated in subsequent chapters. In the last chapter varieties of process plant circuits have been considered.

Majority of the calculations are performed by using two basic material balance equations without using derived formulae. However, calculations with derived formulae are also performed in few calculations simultaneously. Every effort has been put in to illustrate the calculations in simple way and self explanatory manner with an aim to familiarize the reader with different types of process calculations and to develop the abilities to evaluate the performance of the process with confidence. So far as I am aware, this is the first book entirely devoted to process calculations by way of worked out examples and problems for practice.

With in-depth detailed process calculations, this book is very useful for students, teachers, operating personnel, engineers, researchers, designers, equipment manufacturers and plant auditors concerned to mineral and coal processing.

D.V. Subba Rao
Formerly Head, Department of Mineral Beneficiation,
S D S Autonomous College, Garividi,
Vizianagaram District, Andhra Pradesh, India
dvsubbarao3@rediffmail.com
dvsubbarao3@gmail.com

Acknowledgements

At the outset I thank the management of S D S Autonomous college, particularly Sri R.K. Saraf, Chairman, for giving me the excellent opportunity to work as Head of the Department of Mineral Beneficiation for more than three decades and allowing me to bring many reforms in curriculum. Introducing the process calculations in the curriculum is one of the important reforms. The process calculations are class tested before they are introduced in the curriculum.

I am indebted to my students who raised questions in the class and expressed their difficulty to understand when I explain the process calculations. This made me to think of writing this book and present the process calculations in easy to understand way.

Discussion with the following of my students working in mineral and coal industries in India and abroad helped me to cover all process calculations in this book.

1   Dr. T. Gouricharan, Head, Coal Preparation, CIMFR, Dhanbad
2   Dr. C. Raghu Kumar, Head, Ore Beneficiation, Tata Steel Ltd., Jamshedpur
3   Dr. K. Srinivas, Section Head, Maaden Phosphate Company, Saudi Arabia
4   Dr. K. Udaya Bhaskar, Lead Engineer, ArcelorMittal Global R & D, USA
5   Mr. A. Srinivasulu, Sr.Manager, COB Plant, Tata Steel Ltd., Jamshedpur
6   Mr. P. Srinivasu, GM, BMM Ispat Ltd., Hospet, Karnataka
7   Dr. Y. Ramamurthy, Principal Researcher, Tata Steel Ltd., Jamshedpur
8   Dr. Murali Sekhar Jena, Scientist, IMMT, Bhubaneswar
9   Mr. T. Satyababu, Manager (Mineral Processing), NMDC, Kirandul
10 Mr. D.P. Chakravarthy, Sr.Manager, Rashi Steel & Power Ltd., Bilaspur
11 Mr. A. Jagga Rao, Asst. Manager, Trimex Sands Pvt. Ltd., Srikakulam, A.P
12 Dr. Sunil Kumar Tripathy, Principal Researcher, Tata Steel Ltd., Jamshedpur
13 Mr. Ch. Gopikrishna, Sr. Manager, Tata Steel Ltd., Jamshedpur
14 Mr. G. Satish Kumar, Dy.Manager, Weir Minerals India Pvt. Ltd., Visakhapatnam
15 Mr. Suryanarayan Bisoyi, Engineer-Minerals, IRE, Chavara, Kerala
16 Mr. R. Satyanarayana, Sr. Service Engineer, Metso India Pvt. Ltd., Visakhapatnam
17 Mr. A. Mohana Rao, Technical Assistant, NML, Jamshedpur
18 Mr. B. Ratnakar, Technical Assistant, JSW Steel Ltd., Vidyanagar, Karnataka
19 Mr. P. Srinivasa Rao, Assistant Engineer, Jindal Saw Ltd., Bhilwara, Rajasthan
20 Mr. L. Srinu, Graduate Engineer Trainee, Tenova India Private Limited, Bangalore

I thank Dr. C. Raghu Kumar, Head, Ore Beneficiation, Process Technology group, Tata Steel Ltd., Jamshedpur, Sri G.V. Rao, DGM, NMDC, Hyderabad, for the valuable discussions I had with them in identifying certain process calculations.

I do not have words to express my gratitude for the continuous support and help received from Dr. T.C. Rao, former Director, Regional Research Laboratory, Bhopal, and Professor and Head of the department of Fuel and Mineral Engineering, Indian School of Mines, Dhanbad. I had many useful discussions with him while conceptualizing and writing this book. His guidance in completing this book successfully is invaluable.

I am indebted to Sri A.L. Mohan, former Principal, S D S Autonomous College, Garividi, for his constant encouragement and blessings.

I acknowledge my gratitude to my colleagues Sri Y. Ramachandra Rao and Sri K. Satyanarayana for sharing their knowledge and experience which helped me in writing this book.

Much of the theoretical principles explained in this book have been taken from the books "Mineral Beneficiation – A Concise Basic Course" and "Coal Processing and Utilization" written by me.

I am thankful to Taylor & Francis group, its editorial and production staff for their excellent cooperation in bringing out this book.

Without the understanding and support of my wife Krishna Veni, and daughters Radha Rani and Lalitha Rani, writing this book would not have been possible.

List of tables

1.1.1	Non-metallic minerals	2
1.1.2	Metallic minerals	3
1.2.1	Mineral characterization methods	12
4.1	Comparison of Test Sieves of different Standards	33
4.2.1	Particle size distribution data from size analysis test	35
4.2.2	Calculated values for particle size distribution	36
4.2.1.1	Sieve analysis test data of a sample for example 4.2.1	36
4.2.1.2	Calculated values for example 4.2.1	37
4.3.1.3	Sieve analysis test data of a sample for example 4.3.1	39
4.3.1.4	Calculated values for example 4.3.1	39
4.3.2.1	Sieve analysis test data for example 4.3.2	40
4.3.2.2	Calculated values for example 4.3.2	41
4.4.1	Empirical particle size distribution equations	42
4.4.1.3	Screen analysis of a jaw crusher product	45
4.4.1.4	Calculated values for example 4.4.1	45
4.5.1	Size assay analysis of ROM Iron ore sample	48
4.5.2	Calculated values for Table 4.5.1	49
4.5.3	Size assay analyses of three products	49
4.5.4	Size assay analysis of ground copper ore	50
4.5.5	Copper distribution in ground copper ore of Table 4.5.4	50
4.5.6	Size wise ash analysis of ROM coking coal	51
4.5.7	Calculated values for Table 4.5.6	51
4.5.1.1	Size assay analysis of ground chrome ore	51
4.5.1.2	Calculated values for Table 4.5.1.1	52
4.6.1.1	Sieve analysis data for problem 4.6.1	52
4.6.2.1	Screen analysis for problem 4.6.2	53
4.6.3.1	Size analysis of a screen underflow for problem 4.6.3	53
4.6.4.1	Size assay analysis of ground lead ore	53
5.2.1	Types of screen surfaces	56
5.2.2	Principal types of industrial screens	57
5.5.3.1	Size analyses of feed, overflow and underflow for example 5.5.3	65
5.5.5.1	Analyses of vibrating screen's products for example 5.5.5	67
5.6.1	Size analyses of overflow and underflow streams of a screen	69
5.6.1.1	Calculated values to draw Tromp curve	70
5.6.1.2	Size analyses of Feed, overflow and underflow streams of a screen	72

5.6.1.3	Calculated values for example 5.6.1	72
5.7.3.1	Sieve analyses of feed and three products for problem 5.7.3	74
6.1	Aerated and packed densities of five materials	77
7.1	Liberation of the chrome ore at different particle size fractions	94
7.2	Liberation of the manganese ore at different particle size fractions	94
8.3.3.1	Screen analysis of gyratory crusher product	100
8.3.3.2	Calculated values for example 8.3.3	100
8.4.2.1	Screen analysis of crusher product	104
9.1.1	Types of Crushers	107
9.2.1.1	Size analysis of the jaw crusher product	113
9.2.1.2	Calculated values for Table 9.2.1.1	114
9.2.2.1	Crushing test data	114
9.2.2.2	Calculated values for Table 9.2.2.1	115
9.3.2	Closed circuit crushing calculations	118
9.4.5.1	Crushing Test Data for problem 9.4.5	124
10.3.2	Screen analyses of three samples for example 10.3.2	136
10.4.3	Screen analyses data for problem 10.4.3	141
12.1.2	Particle size distribution for example 12.1.2	167
12.1.7	Screen analyses data for example 12.1.7	172
12.1.8	Size analyses of hydrocyclone overflow and underflow	174
12.1.12	Analyses results of feed, overflow and underflow	179
12.2.1	Size analyses data of underflow and overflow of hydrocyclone	181
12.2.2	Partition coefficient calculations	182
12.2.1.1	Size distributions of underflow and overflow	185
12.2.1.2	Partition coefficient calculations	186
12.3.4	Size analyses for problem 12.3.4	188
12.3.5	Screen analyses for problem 12.3.5	188
12.3.8	Size analyses of underflow and overflow	189
12.3.9	Size analyses of underflow and overflow from hydrocyclone	189
12.3.10	Size distributions of feed, underflow and overflow of hydrocyclone	190
14.1	Results of sink and float studies of manganese ore	200
14.2	Results of sink and float studies of iron ore fines	200
14.3	Results of sink and float followed by microscopic study of Beach sands	201
14.4	Sink and float analysis results of chrome ore	201
14.5	Sink and float analysis results of manganese ore	201
14.6	Sequential sink and float analysis of $-10+2$ mm manganese ore	202
15.1.1	Laboratory observed values of float and sink analysis	207
15.1.2	Calculated values of float and sink analysis	208
15.1.3	Results of cumulative yields of floats and sinks	208
15.2.1	BIRD's classification	212
15.2.2	Values of ± 0.10 near gravity material	212
15.4	Calculated values to determine WI	214
15.5	Calculated values to determine ODW	215
15.7.1	Size-wise ash analysis of $-38+0.5$ mm coal	217
15.7.2	Size-wise float and sink data of raw coal crushed to -38 mm	217
15.7.3	Washability Characteristics of coal	218

15.7.1.1	Float and sink analysis data for example 15.7.1	218
15.7.1.2	Calculated values for example 15.7.1	218
15.7.2.1	Float and sink test data for example 15.7.2	219
15.7.2.2	Float and sink test results for example 15.7.2	219
15.7.2.3	Calculated values of NGM & DW for example 15.7.2	220
15.7.3.1	Float and sink analysis for example 15.7.3	222
15.7.3.2	Calculation of Weight% and Ash%	222
15.7.3.3	Calculation of Cumulative Weight%, Ash% and NGM	223
15.7.4.1	Float and sink analysis for example 15.7.4	224
15.7.4.2	Calculation of Weight% and Ash% for example 15.7.4	225
15.7.4.3	Calculation of Cumulative Weight% and Ash%	225
15.7.5.1	Washability test data for example 15.7.5	226
15.7.5.2	Calculated values for example 15.7.5	226
15.8.1.1	Washability test data for example 15.8.1	232
15.8.1.2	Cumulative percentages for example 15.8.1	232
15.9.1	Float and sink test data for problem 15.9.1	235
15.9.2	Float and sink analysis for problem 15.9.2	235
15.9.3	Float and sink data for problem 15.9.3	235
15.9.4	Washability test data for problem 15.9.4	236
15.9.5	Washability test data for problem 15.9.5	236
16.1.1	Quantities and assay values	240
16.1.9	Analyses of the streams of flotation circuit	245
16.2.1	Results of flotation test	251
16.2.1.1	Percent distribution of lead and zinc for example 16.2.1	252
16.3.2	Data of Iron ore concentration operation	257
16.5.1	Data of Iron ore concentration operation for problem 16.5.1	263
17.1.1.1	Float and sink test of raw coal	268
17.1.1.2	Calculated values for example 17.1.1	269
17.1.4.1	Float and sink experiment data for example 17.1.4	271
17.1.4.2	Calculated values for example 17.1.4	271
17.1.5.1	Float and sink analysis data for example 17.1.5	272
17.1.5.2	Calculated values for example 17.1.5	273
17.2.1	Float and sink analyses of clean coal and refuse	275
17.2.2	Calculated values for drawing Tromp Curve	276
17.2.3	Values of Independent criteria for different washing units	280
17.2.1.1	Float and sink analyses of clean coal and refuse for example 17.2.1	280
17.2.1.2	Calculated values for Tromp curve for example 17.2.1	281
17.2.2.1	Float and sink analyses of clean coal and refuse for example 17.2.2	282
17.2.2.2	Calculated values for Tromp curve for example 17.2.2	283
17.3.1	Float and sink analysis of raw coal for problem 17.3.1	284
17.3.2	Float and sink analysis for problem 17.3.2	284
17.3.5	Float and sink analyses of clean coal and refuse for problem 17.3.5	285
18.1.1	Results of tests for samples for example 18.1.1	288
18.1.2	Circuit sampling results for example 18.1.2	291
18.1.3	Analyses of samples for example 18.1.3	294
18.1.5	Results of tests for samples for example 18.1.5	300
18.2.3.1	Composition of each stream of flotation circuit	306

18.2.3.2 Solid water ratio and contact time   306
18.3.1.1 Ash product values   311
18.3.1.2 Calculated values for example 18.3.1   312
18.3.2.1 Float and sink analyses of two coals   312
18.3.2.2 Cumulative percentages for coal A   313
18.3.2.3 Cumulative percentages for coal B   313
18.3.2.4 Cumulative percentages for blended coal   314
18.3.2.5 Results from 6 curves   315
18.4.7.1 Float and sink analyses of floats and sinks of HMS   317

List of figures

1.3	Pictorial Presentation of proximate analysis on different bases	20
4.1	Test Sieve	31
4.2.1	Sieve analysis at the end of sieving	35
4.3.1.1	Graphical presentation of data tabulated in table 4.2.2	38
4.3.1.2	Linear scale and semi-log cumulative plots	38
4.3.1.3	Graphical presentation of data tabulated in table 4.3.1.4	40
4.3.1.4	Linear scale and semi-log cumulative plots for example 4.3.1	40
4.3.2.1	Linear scale cumulative graph for example 4.3.2	41
4.4.1.1	Gates-Gaudin-Schuhmann plot	43
4.4.1.2	Rosin Rammler plot	44
4.4.1.3	Gates-Gaudin-Schuhmann plot for example 4.4.1	46
4.4.1.4	Rosin-Rammler plot for example 4.4.1	46
4.4.2.1	Gates-Gaudin-Schuhmann plot for example 4.4.2	47
5.3	Simplified screen	58
5.6.1.1	Tromp curve	70
5.6.1.2	Tromp curve showing perfect separation and error area	71
5.6.1.3	Partition curve for the example 5.6.1	73
7.1	A particle of an Ore containing A, B, & C minerals	91
7.2	Liberation methods	92
7.3	Typical comminution product	93
9.1.1	Angle of nip of Roll Crusher	108
9.3.1	Open circuit and closed circuit crushing operations	116
9.3.1.1	Open circuit crushing details when screen efficiency is 100%	116
9.3.1.2	Open circuit crushing details when screen efficiency is 90%	116
9.3.1.3	Crushing circuit for example 9.3.1	117
9.3.2.1	Closed circuit crushing details when screen efficiency is 100%	119
9.3.2.2	Closed circuit crushing details when screen efficiency is 90%	119
9.3.2.3	Flow diagram for example 9.3.2	120
9.3.3	Crushing circuit for example 9.3.3	121
9.3.3.1	Closed circuit crushing operation	122
9.3.4	Crushing circuit for example 9.3.4	123
10.1.1	Motion of the charge in ball mill	128
10.1.2	Path of a ball	129
10.1.3	Forces on a ball	129
10.3.1.1	Open circuit grinding	133

10.3.1.2	Closed circuit grinding	133
10.3.1.3	Grinding action of rods	134
10.3.1.4	Closed circuit grinding for example 10.3.1	135
10.3.3	Circuit diagram for example 10.3.3	137
10.3.3.1	Rod mill–Ball mill primary grinding circuits	138
10.3.3.2	Primary grinding circuit with Autogenous mill	138
10.3.3.3	Regrinding circuit	138
10.3.5	Grinding circuit for example 10.3.5	139
10.3.6	Circuit diagram for example 10.3.6	140
11.3.1	Relationship of coefficient of resistance to Reynolds number	147
11.3.1.1	Determination of Reynolds number	150
11.3.2	Relationship of C_D to N_{Re} for different values of sphericity	151
11.3.5	Determination of N_{Re} for example 11.3.5	154
11.5	Settling of particles under (A) Free settling (B) Hindered settling	156
11.7	Free settling of (A) Fine particles (B) Coarse particles	159
12.1.1	(A) Free settling and (B) Hindered settling	164
12.1.2	Separation in sizing classifier	164
12.1.3	Separation in sorting classifier	165
12.1.4	Sizing in hydraulic classifier	166
12.1.2.1	Particle size versus cumulative weight fraction graph	168
12.1.5	Hydrocyclone	173
12.2.1.1	Efficiency curves	183
12.2.1.2	Reduced efficiency curve	184
12.2.1.3	Efficiency curves for example 12.2.1	187
13.2.1	Typical flotation circuit	196
14.6	Separability curves for Manganese ore	202
15.1	Washability Curves	210
15.4	Curve to determine Washability Index	214
15.5	Curve to determine Optimum Degree of Washability	216
15.7.2.1	Washability Curves for example 15.7.2	220
15.7.2.2	Graph to determine Washability Index for example 15.7.2	221
15.7.2.3	Graph to determine ODW for example 15.7.2	221
15.7.3.1	Washability curves for example 15.7.3	223
15.7.4.1	Total floats and yield gravity curves	225
15.8.1.1	M-curve for float-and-sink analysis data of Table 15.1.3	228
15.8.1.2	Example for prediction of cleaning properties in three product system	228
15.8.1.3	Predicting clean product by addition of un-cleaned coal to clean coal	230
15.8.1.4	Mixing of two coals at 50:50 ratio to obtain highest yield for a given ash	231
15.8.1.5	Yield gravity curve for example 15.8.1	232
15.8.1.6	M-curve for example 15.8.1	233
15.8.2.1	Prediction of clean coal product required for example 15.8.2	234
16.1	Flowsheet of three stage treatment	240
17.1.1	Total floats ash curve	269
17.1.4	Total floats ash curve & Yield gravity curve for example 17.1.4	272

17.2.1.1	Tromp Curve relates to Clean coal	277
17.2.1.2	Tromp Curve relates to Refuse	277
17.2.1.3	Tromp curve for example 17.2.1	281
17.2.2	Tromp curve for example 17.2.2	283
18.1.1	Closed circuit grinding diagram for example 18.1.1	288
18.1.1.1	Circuit diagram with complete material balance for example 18.1.1	290
18.1.2	Circuit diagram for example 18.1.2	291
18.1.2.1	Circuit diagram with complete material balance for example 18.1.2	293
18.1.3	Flotation circuit for example 18.1.3	293
18.1.3.1	Flotation circuit with complete material balance for example 18.1.3	296
18.1.4	Flow diagram for example 18.1.4	297
18.1.4.1	Flow diagram with complete material balance for example 18.1.4	299
18.1.5	Two stage water only cyclone circuit for example 18.1.5	300
18.1.5.1	Circuit with complete material balance for example 18.1.5	301
18.1.6	Process flow diagram for example 18.1.6	302
18.1.6.1	Process flow diagram with complete material balance for example 18.1.6	304
18.2.3	Flotation circuit diagram	307
18.3.1	Total floats ash curve and yield gravity curve for coal A	313
18.3.2	Total floats ash curve and yield gravity curve for coal B	314
18.3.3	Total floats ash curve and yield gravity curve for Blend coal	315
18.4.2	Closed circuit grinding diagram for problem 18.4.2	316
18.4.4	Circuit diagram for problem 18.4.4	316

Chapter 1

Minerals and coal

Minerals and Coal are non-renewable natural resources that occur in the earth's crust. As defined by Dana, a well known physicist, mineral is a substance having definite chemical composition and internal atomic structure and formed by the inorganic processes of nature [1].

1.1 TYPES OF MINERALS

All minerals contain one or more metals or metalloids. Minerals are broadly classified as metallic and non-metallic minerals.

Metallic minerals are the minerals from which a metal is extracted. Bauxite ($Al_2O_3 \cdot 2H_2O$), Hematite (Fe_2O_3), Ilmenite ($FeO\ TiO_2$), Pyrolusite (MnO_2), and Chromite ($FeO\ Cr_2O_3$) are few of the important metallic minerals.

Non-metallic minerals are the minerals used for industrial purposes for making cement, refractories, glass & ceramics, insulators, fertilizers etc. These minerals are also called **industrial minerals**. Metals are not extracted from these minerals. Some metallic minerals are also used for industrial purposes like Bauxite, Chromite and Zircon for refractory industry, Pyrolusite for dry battery cells and Ilmenite for pigment industry etc. Table 1.1.1 shows few non-metallic minerals, their chemical formulae and chief uses.

The following are the important terms used in describing the mineral deposits.

Rock is an aggregation of several minerals as occurred in the earth's crust. Granite, an aggregation of orthoclase, quartz and mica, is an example for the rock.

Ore is also an aggregation of several minerals from which one or more minerals can be exploited/separated at profit. Today's rock due to unavailability of technology to use it may become tomorrow's ore if technology is available to utilize it for our purpose. Similarly, at a place if the technology is available to separate required minerals profitably it is called ore. The same ore with the same characteristics in all respects is present in a place where it cannot be exploited at profit, it is called rock. Hence it should be understood that the term ore is used to represent its economic viability so that it can be attempted to exploit/separate for the required minerals.

Ore Minerals or **Valuable Minerals** (also called **Economic Minerals**) are those minerals of economic value and contain an economically exploitable quantity of some metal or non-metal.

Table 1.1.1 Non-metallic minerals.

Mineral	Chemical Formula	Main uses
Apatite	$Ca_4(CaF)(PO_4)_3$	in fertilizer, phosphorous chemicals, gems for jewelry
Asbestos (Crysotile)	$Mg_3Si_2O_5(OH)_4$	in fire proof fabrics, asbestos cement, friction products, roofing, flooring, pipe insulation
Baryte	$BaSO_4$	in well drilling fluids, barium chemicals, glass, paint, rubber
Bentonite	$(Ca\ Mg)O\ SiO_2$ $(Al\ Fe)_2O_3$	in oil well drilling, cosmetics, toothpaste, as binder for iron ore pellets
Calcite	$CaCO_3$	in cement & lime, as filler in paper, paint, adhesives and sealants
Dolomite	$CaMg(CO_3)_2$	in furnace linings, building material, flux in blast furnace
Feldspar	$(Na,K,Ca)\ AlSi_3O_8$	in glass and ceramics
Fluorite	CaF_2	in production of HF, as flux in steel making
Garnet (Almandine)	$3FeO\ Al_2O_3 3SiO_2$	in sand blasting, water filtration, wood polishing, as abrasive and gemstones
Gypsum	$CaSO_4 \cdot 2H_2O$	in cement & fertilizers
Kaolinite (China clay)	$H_4Al_2Si_2O_9$	in paper, paint, cosmetics, ceramics, rubber, fiberglass
Kyanite	Al_2SiO_5	in refractories, and glass
Limestone	Chiefly $CaCO_3$	in cement, glass, sculptures, building material, flux in blast furnace
Magnesite	$MgCO_3$	in refractories, magnesium chemicals and fertilizers
Phosphate rock	$Ca_3(PO_4)_2$	in fertilizers, phosphorous chemicals, food additives, detergents
Quartz	SiO_2	in glass, ferrosilicon, silicon metal, silicon carbide, and metallurgical fluxes
Sillimanite	$Al_2O_3\ SiO_2$	in refractories, and glass
Talc	$H_2Mg_3(SiO_3)_4$	in ceramics, paint, paper, roofing, plastics, cosmetics, and pharmaceuticals
Vermiculite	$3MgO(FeAl)_2O_3\ 3SiO_2$	in building industry, fireproofing, refractory insulation, acoustic panels, brake linings

Gangue Minerals are usually the non-metallic minerals associated with ore minerals which are worthless as a source for that metal or otherwise. These are usually unwanted or waste or useless minerals. These gangue minerals occasionally may find use as source of by-products. For example, pyrite present in Lead and Zinc ores is a gangue mineral but it is separated as by-product for extraction of sulphur after lead and zinc minerals are separated.

Ore Deposits are the natural deposits of ore minerals. These are also called economic mineral deposits.

Ore is an aggregation of valuable and gangue minerals.

Simple Ore is one from which a single metal can be extracted. For example, only Iron is extracted from Iron ore, Aluminium is extracted from Aluminium ore, Chromium is extracted from Chrome ore, etc. Such ores are called simple ores.

Complex Ore is one from which two or more metals can be extracted. Lead and zinc metals are extracted from lead zinc ore; Lead, Zinc and Copper metals are extracted from Lead-Zinc-Copper Ore. Such ores are called complex ores.

Table 1.1.2 Metallic minerals.

Mineral	Chemical Formula	Metal extracted	% metal
Hematite	Fe_2O_3	Iron	69.94
Magnetite	Fe_3O_4	Iron	72.36
Bauxite	$Al_2O_3 \cdot 2H_2O$	Aluminium	39.11
Braunite	$3Mn_2O_3\ MnSiO_3$	Manganese	63.60
Pyrolusite	MnO_2	Manganese	63.19
Chromite	$FeO\ Cr_2O_3$	Chromium	46.46
Galena	PbS	Lead	86.60
Sphalerite	ZnS	Zinc	67.10
Chalcopyrite	$CuFeS_2$	Copper	34.63
Ilmenite	$FeO\ TiO_2$	Titanium	31.57
Rutile	TiO_2	Titanium	59.95
Zircon	$ZrSiO_4$	Zirconium	49.76
Monazite	$(Ce,La,Th)PO_4$	Thorium	–

Metal Content of a mineral is generally expressed in percent of metal present in the mineral. It is calculated by taking the atomic weights of the elements present in the mineral as in example 1.1.1.

Example 1.1.1: *Calculate percent iron present in Hematite (Fe_2O_3). Atomic weights of Iron and Oxygen are 55.85 and 16.00 respectively.*

Solution:

Given

Atomic weight of Iron $= 55.85$
Atomic weight of Oxygen $= 16.00$
Chemical formula of Hematite is Fe_2O_3
Molecular weight of Hematite $= 55.85 \times 2 + 16 \times 3 = 159.7$

$$\text{Iron present in Hematite} = \frac{55.85 \times 2}{159.7} \times 100 = 69.94\%$$

It means that 69.94% Fe (Iron metal) by weight is present in mineral Hematite.

Similarly, percent metal present in any metallic mineral can be calculated. Some of the metallic minerals, their chemical formulae, metal extracted from them and the percent metal present in respective minerals by similar calculation are shown in Table 1.1.2.

Assay Value or **tenor** is the percent metal, percent valuable mineral, or ounces precious metal per ton of the ore depending upon the type of ore involved.

In case of ores of metallic minerals, percent metal present in the ore is the assay value. For an ore of precious metal like gold, ounces or gram precious metal per ton of the ore is the assay value as the precious metal is present in little quantities. An assay value of an ore of non-metallic minerals is represented by the percent valuable or required constituent of the ore. For example, lime (CaO) is the required constituent in the limestone (predominantly $CaCO_3$) for use in making cement. The assay value of the limestone is represented as %CaO. Example 1.1.2 illustrates this calculation.

Example 1.1.2:   *Determine percent lime (CaO) present in limestone of $CaCO_3$ composition. Atomic weights of Calcium, Carbon and Oxygen are 40.08, 12.01 and 16.00 respectively.*

Solution:

Given

Atomic weight of Calcium = 40.08
Atomic weight of Carbon = 12.01
Atomic weight of Oxygen = 16.00
Chemical formula of Limestone is $CaCO_3$

Molecular weight of Limestone $= 40.08 + 12.01 + 16 \times 3 = 100.09$

Lime (CaO) present in Limestone $= \dfrac{40.08 + 16.00}{100.09} \times 100 \ = 56\%$

It means that 56% lime (CaO) by weight is present in limestone.

$\%Al_2O_3$ and $\%SiO_2$ are normally determined as required constituents in China Clay $(Al_2O_3 \cdot 2SiO_2 \cdot 2H_2O)$ as illustrated in example 1.1.3.

Example 1.1.3:   *Calculate $\%Al_2O_3$ and $\%SiO_2$ present in china clay $(Al_2O_3 \cdot 2SiO_2 \cdot 2H_2O)$. Atomic weights of Aluminium, Silica, Hydrogen and Oxygen are 26.98, 28.09, 16.00 and 1.00 respectively.*

Solution:

Given

Atomic weight of Aluminium = 26.98
Atomic weight of Silicon   = 28.09
Atomic weight of Oxygen   = 16.00
Atomic weight of Hydrogen   = 1.00
Chemical formula of China Clay is $Al_2O_3 \cdot 2SiO_2 \cdot 2H_2O$ or $(Al_2Si_2H_4O_9)$

Molecular weight of china clay $= 26.98 \times 2 + 28.09 \times 2 + 1.00 \times 4 + 16.00 \times 9$

$$= 258.14$$

Molecular weight of $Al_2O_3 = 26.98 \times 2 + 16.00 \times 3 = 101.96$

Molecular weight of $2SiO_2 = 28.09 \times 2 + 16.00 \times 4 = 120.18$

Al_2O_3 present in china clay $= \dfrac{101.96}{258.14} \times 100 = 39.50\%$

SiO_2 present in china clay $= \dfrac{120.18}{258.14} \times 100 = 46.56\%$

Grade of an ore signifies the quality of the ore in general. An ore having high assay value is termed as **high grade ore** or **rich ore** and an ore of low assay value is termed as **low grade ore** or **lean ore**. The terms low grade and high grade are relative to the specific ore of acceptable tenor. As for example, Iron ore of around 60% Fe is acceptable for extraction of Iron whereas Lead ore of around 10% Pb is acceptable for extraction of Lead. Hence Iron ore of more than 60% Fe is rich ore and Lead ore of more than 10% Pb is rich ore.

The grade is expressed in terms of metal content in case of metallic ores like iron ore, manganese ore, base metal ores etc.; in terms of percentage of oxides in case of many metallic and non-metallic ores e.g. Cr_2O_3 in chrome ore, Al_2O_3 in Aluminium ore, CaO in limestone and P_2O_5 in apatite and rock phosphate. In some cases impurities also determine the grade, for example, the presence of sulphur and phosphorus in iron ore, manganese ore, and coal. Again in some cases, the strength of the material and colour are taken into account for grading purpose. Spinning and non-spinning types in case of asbestos; snow white, white and off colour in case of barites; and friable, compact or massive and crystalline in case of limestone are some of the examples. Rubber manufacturers specify absolutely copper free china clay as even small trace of copper results in early decay of rubber. Yet, in other case, the physical properties and the size are the only two factors considered for grading, the chemical composition does not come into the picture at all; as for example in case of mica and asbestos. Grading of Iron ore and coal are also done based on size as fines, lumps, etc.

Specification of an ore is intimately related with the grade. It pinpoints the tolerance limits of all constituents present in it. Individual consumers may prescribe different specifications for the same grade of mineral which is dependent upon two factors:

1   the technique of manufacturing process adopted by individual units, and
2   the grade of other raw material required to be used to obtain the end product

Different ores are classified commercially based on their use for particular purpose. For example manganese ore is classified as first grade, second grade, medium grade and low grade having manganese percent as 46–48%, 44–46%, 40–44% and <35% Mn respectively wherein manganese metal is extracted. Manganese ore of minimum 78% MnO_2 is classified as battery grade. Similarly Chrome ore of minimum 48% Cr_2O_3 is a metallurgical grade used for extraction of chromium metal. Chrome ore of 38–48% Cr_2O_3 is classified as refractory grade. If the chrome ore is used for extraction of chromium metal, %Cr in the ore is required to know whereas if it is used for making refractory bricks, %Cr_2O_3 in chrome ore is required to know. Hence in example 1.1.4, %Cr and %Cr_2O_3 are calculated in chromite mineral.

Example 1.1.4:   *Determine* %Cr_2O_3 *and* %*Cr in Chromite* $FeO \cdot Cr_2O_3$. *Atomic weights of Iron, chromium and Oxygen are 55.85, 52.00 and 16.00 respectively.*

Solution:

Given

Atomic weight of Iron   = 55.85
Atomic weight of Chromium   = 52.00
Atomic weight of Oxygen   = 16.00
Chemical formula of Chromite is $FeO \cdot Cr_2O_3$

$$\text{Molecular weight of Chromite} = 55.85 + 52.00 \times 2 + 16 \times 4 \quad = 223.85$$

$$\text{Chromium present in chromite} = \frac{52.00 \times 2}{223.85} \times 100 \quad\quad = 46.46\%\,Cr$$

$$Cr_2O_3 \text{ present in chromite} \quad = \frac{52.00 \times 2 + 16.00 \times 3}{223.85} \times 100 = 67.9\%$$

Assay value of an ore is determined usually by chemical analysis of a sample collected from the ore. Different minerals present in an ore are determined qualitatively by microscopic analysis of a sample. When once the assay value of an ore and the minerals present in that ore are determined, valuable and gangue mineral contents can be estimated by calculation. This aspect is well illustrated in example 1.1.5 and 1.1.6.

Example 1.1.5: *Lead ore of 2% Pb contains Galena (PbS) as only lead mineral. All other minerals are gangue minerals. What is the weight of gangue per ton of lead ore. Atomic weights of lead and sulphur are 207.19 and 32.06 respectively.*

Solution:

Given

Atomic weight of Lead $\quad = 207.19$
Atomic weight of Sulphur $\quad = 32.06$
Chemical Formula of Galena is PbS

$$\text{Molecular weight of Galena} = 207.19 + 32.06 = 239.25$$

$$\text{Lead present in galena} = \frac{207.19}{239.25} \times 100 \quad = 86.60\%$$

%Pb in Galena (PbS) $= 86.60\%$
%Pb in Lead Ore $= 2\%$

$\therefore$ % Galena (PbS) in Lead Ore $\quad = \dfrac{2}{86.60} \times 100 = 2.31\%$

% gangue in Lead Ore $\quad = 100 - 2.31 \quad = 97.69\%$

$\therefore$ Weight of gangue per ton of lead ore $= 97.69/100 \quad = 0.9769$ ton

Example 1.1.6: *A sample contains Ilmenite (FeO TiO$_2$), Rutile (TiO$_2$) and other minerals containing no Ti and Fe. Chemical analysis of the sample results the percent Fe and Ti as 28.57 and 20.87. Calculate the percent Ilmenite, rutile and other minerals in the sample. Atomic weights of Titanium, Iron, and Oxygen are 47.90, 55.85 and 16.00 respectively.*

Solution:

Given

Atomic weight of Titanium $= 47.90$
Atomic weight of Iron $\quad = 55.85$
Atomic weight of Oxygen $\; = 16.00$

%Fe in the sample $= 28.57\%$

%Ti in the sample $= 20.87\%$

Chemical formula of Ilmenite is $FeOTiO_2$

Chemical formula of Rutile is TiO_2

Molecular weight of Ilmenite $(FeOTiO_2) = 55.85 + 47.9 + 16 \times 3 = 151.75$

$$\%Fe \text{ in Ilmenite } (FeOTiO_2) = \frac{55.85}{151.75} \times 100 = 36.80\%$$

$$\% \text{ Ilmenite } (FeOTiO_2) \text{ in the sample} = \frac{20.87}{36.80} \times 100 = 56.71\%$$

$$\%Ti \text{ in Ilmenite } (FeOTiO_2) = \frac{47.9}{151.75} \times 100 = 31.57\%$$

%Ti in the sample associated with Ilmenite $(FeOTiO_2) = 56.71 \times (31.57/100)$
$= 17.90\%$

%Ti associated with Rutile $(TiO_2) = 20.87 - 17.90$ $\quad = 2.97\%$

Molecular weight of Rutile $(TiO_2) = 47.9 + 16 \times 2$ $\quad = 79.90$

$$\%Ti \text{ in Rutile } (TiO_2) = \frac{47.9}{79.9} \times 100 \qquad = 59.95\%$$

$$\% \text{ Rutile } (TiO_2) \text{ in the sample} = \frac{2.97}{59.95} \times 100 \qquad = 4.95\%$$

% other minerals in the sample $= 100 - 56.71 - 4.95 = 38.34\%$

1.2 MINERAL PROCESSING

Mineral Processing is a term which can be applied to a group of operations an ore is subjected to obtain a required end product. The ore is to be reduced in size to detach valuable mineral grains from the ore. Later detached valuable mineral grains are separated. Separation of valuable mineral from the aggregation of valuable and gangue minerals by physical methods is termed as Mineral Beneficiation. Hence size reduction and beneficiation are the two basic steps involved in Mineral Processing. The various operations in each step are as follows:

1   Liberation: Detachment or freeing of dissimilar particles from each other i.e., valuable mineral particles and gangue mineral particles.

  Operations: Crushing
            Grinding

2   Separation: Actual separation of liberated dissimilar particles i.e., valuable mineral particles and gangue mineral particles.

  Operations: Gravity concentration
            Heavy Medium Separation
            Jigging
            Spiraling
            Tabling
        Flotation
        Magnetic separation
        Electrical separation
        Miscellaneous operations like Hand Sorting

Before and after these size reduction and separation operations, several operations are carried out to obtain a required end product. These operations can be called supporting operations. Preliminary washing, Screening, Classification, Thickening, Filtration, Storage, Conveying, Feeding, Pumping, Pneumatic and Slurry transport etc. are the few supporting operations. Supporting operations (one or the other) are essential operations of any plant without which no plant can exists.

All these operations are known as unit operations which are defined as the operations conducted on any material and involve physical changes.

In practice, it is not possible to separate the pure valuable mineral product. It contains some gangue minerals. This product is called **concentrate**. Similarly gangue minerals product called **tailing** also contain little quantity of valuable mineral. This is due to either complex locking nature of the ore which leads to incomplete liberation or inefficiency of the equipment separating the valuable mineral.

Taking all the aspects into consideration, A.M.Gaudin has defined the Mineral Processing as follows:

Mineral Processing can be defined as processing of raw minerals to yield marketable products and waste by means of physical or mechanical methods in such a way that the physical and chemical identity of the minerals are not destroyed [2]

It follows that mineral processing is a process designed to meet the needs of the consumer of minerals. **Run-of-mine Ore** is an ore directly taken from the mine, as it is mined. It is a raw material for a Mineral Process Engineer from which much quantity of the valuable mineral is separated during its processing.

In a hypothetical situation, the concentrate from a beneficiation operation contains pure valuable mineral. Its assay value is equivalent to percent metal present in valuable mineral. Hence the maximum grade of the concentrate obtainable by beneficiation is equivalent to percent metal present in valuable mineral. This is illustrated in examples 1.2.1 and 1.2.2. Gangue content in the ore is also estimated in these examples.

Example 1.2.1: *What is the maximum grade of the copper concentrate obtainable from an ore of 2.1% Cu containing chalcopyrite (CuFeS₂)? Atomic weights of Cu, Fe, & S are 63.55, 55.85, & 32.06. Estimate the quantity of gangue per ton of ore.*

Solution:

Given

Atomic weight of Copper $= 63.55$
Atomic weight of Iron $\quad = 55.85$
Atomic weight of Sulphur $= 32.06$
%Cu in copper ore $\quad = 2.1\%$

Chemical formula of Chalcopyrite is $CuFeS_2$

Molecular weight of Chalcopyrite $= 63.55 + 55.85 + 32.06 \times 2$
$$= 183.52$$

$$\text{Copper present in Chalcopyrite} = \frac{63.55}{183.52} \times 100 = 34.63\%$$

$\therefore$ Maximum grade of concentrate obtainable is 34.63%Cu.

$$\% \text{ Chalcopyrite (CuFeS}_2\text{) in copper ore} = \frac{2.10}{34.63} \times 100 = 6.06\%$$

% gangue in copper ore $= 100 - 6.06 = 93.94\%$

Gangue in copper ore $= 0.9394$ ton per ton of ore

Example 1.2.2: *What maximum grade of the manganese concentrate can be obtained from an ore of 42% Mn containing braunite ($3Mn_2O_3 \cdot MnSiO_3$)? Atomic weights of Mn, Si and O are 54.94, 28.09, & 16.00. Estimate the quantity of gangue per ton of ore.*

Solution:

Given

Atomic weight of Manganese $= 54.94$
Atomic weight of Silicon $= 28.09$
Atomic weight of Oxygen $= 16.00$
%Mn in the ore $= 42\%$
Chemical formula of Braunite is $3Mn_2O_3 \cdot MnSiO_3$

Molecular weight of Braunite $= 54.94 \times 7 + 28.09 + 16.00 \times 12$

$$= 604.67$$

$$\text{Manganese present in Braunite} = \frac{384.58}{604.67} \times 100 = 63.60\%$$

Maximum grade of concentrate obtainable is 63.6%Mn.

$$\% \ 3Mn_2O_3 \cdot MnSiO_3 \text{ in manganese ore} = \frac{42.00}{63.60} \times 100 = 66.04\%$$

% gangue in manganese ore $= 100 - 66.04 = 33.96\%$

Gangue in manganese ore $= 0.3396$ ton per ton of ore

The following are the successive major steps involved to obtain the required end product.

Geological survey

⇩

Mining

⇩

MINERAL PROCESSING

⇩

Smelting or **Industrial use**

Geologists conduct geological survey and estimate the ore reserves, their quality and assay value. Mining engineers mine the ore and bring it to the surface of the earth. Mineral Processing Engineers beneficiate the ore to higher assay value. Thus beneficiated ore, if it is metallic ore, is smelted and the metal is extracted which is further utilized for the production of alloys. If the ore is non metallic, beneficiated ore is directly utilized for the production of various products like cement, refractories, fertilizers etc.

Smelting operation, for the extraction of a metal, requires

- Uniform quality of the ore
- Appropriate size of the ore
- Minimum tenor of the ore

Run-of-mine ore is processed to achieve the above. The primary object of Mineral Processing is to eliminate either unwanted chemical species or particles of unsuitable size or structure.

During processing, much of the gangue minerals, usually present in large quantities in many ores, are removed so that lean ores become suitable for extraction of metal. Mineral Processing is usually carried out at the mine site. The essential reason is to reduce the bulk of the ore which must be transported, thus saving the transport cost. The other benefits are reduction of cost of extraction of metal and loss of metal in slag.

Mineral Processing, previously called mineral dressing and ore dressing, is gaining importance chiefly as it is cheaper than metallurgical operations. The other reasons are the non-availability of good quality ore and use of bulk mining operations wherein processing is a must.

Ore/mineral characterization is the preliminary step to be performed before the ore is subjected to processing to obtain end product(s) from it. It involves identification of various minerals present in the ore, their chemical composition, their relative amounts, their texture, size and physical properties such as density, magnetic susceptibility, electrical conductivity and surface property. Ore characterization studies form an integral and often critical part of investigations. A comprehensive mineralogical investigation should encompass not only those minerals of economic value, but also harmful minerals, whose presence may negatively impact upon processing or salability of end products.

Furthermore, the chemical composition of the minerals, their size, morphology and association are all factors influencing the products produced from it, and therefore its success as a commercial venture. These and other factors, such as local variations in composition, grain size, mineralogy or the mineralogical distribution of harmful trace elements, can influence profitability or can affect the selection of processing routines to be employed. In short, a comprehensive ore characterization is an indispensable advantage, which can guide its processing for maximum profit.

The following are some of the physical properties of minerals through which minerals can be identified:

1   Characters dependent upon light

   a)   Colour
   b)   Streak

      c) Lustre
      d) Transparency
      e) Phosphorescence
      f) Fluorescence

2   State of aggregation

      a) Form
      b) Habit
      c) Pseudomorphism, Polymorphism and Polytipism
      d) Cleavage
      e) Fracture
      f) Hardness
      g) Tenacity

3   Specific gravity (density)
4   Magnetic susceptibility
5   Electrical conductivity
6   Surface property

The identification of minerals by their physical properties is termed as Megascopic Identification. Minerals are also identified by their optical properties under a microscope. Transparent minerals are identified under Petrological or Mineralogical microscope whereas opaque minerals are identified under Ore microscope. This microscopic examination, carried out for thorough understanding of the mineralogy of ore, for determining the mineral species present in the ore and their relative abundance.

Texture of mineral occurrences is an important property useful for separation of valuable minerals from their ores. Textures are mainly of three types:

Fine-grained	<1 mm
Medium-grained	1–5 mm
Coarse-grained	>5 mm

Minerals are analyzed by conventional chemical analysis. Two types of chemical analyses are:

1 **Qualitative Analysis**   in which elements present in the sample are identified
2 **Quantitative Analysis** in which quantity of elements or compounds present in the sample is estimated

Few mineral characterization methods and their use are given in Table 1.2.1.

1.3 COAL

Coal is a solid stratified rock, a natural fossil fuel, occurred in layers in the earth's crust, formed many millions of years ago from the remains of decaying trees and vegetation. Large trees, many resembling giant ferns, grew in dense forests on the low lying land or in the shallow waters of the lake and succeeding generations of trees as they died and accumulated on the floor of the lake, form a vegetable sludge.

Table 1.2.1 Mineral characterization methods.

Method	Use
1. Conventional wet chemical analysis 2. X-Ray Fluorescence (XRF) Spectrometry	Major elemental analysis
1. Atomic Absorption Spectroscopy (AAS)	Minor elemental analysis
1. Microwave Plasma – Atomic Emission Spectroscopy (MP-AES)	Minor and Trace elemental analysis
1. Inductively Coupled Plasma Optical Emission Spectrometry (ICP-OES) 2. Inductively Coupled Plasma Mass Spectrometry (ICP-MS)	Trace elemental Analysis
Optical microscopy 1. Reflected Light Microscopy 2. Transmitted Light Microscopy 3. Phase Contrast Microscopy (PCM)	Mineral identification and quantification
Electron microscopy 1. Scanning Electron Microscopy (SEM) with Energy Dispersive Spectroscopy (EDS) 2. Field Emission-Scanning Electron Microscopy (FE-SEM) 3. Quantitative Evaluation of Minerals by Scanning	Mineral structure, texture and elemental composition
Electron Microscopy (QEMSCAN) 4. X-ray Photoelectron Spectroscopy (XPS) 5. Electron Probe Micro-analyzer (EPMA) with Wavelength Dispersive Spectroscopy (WDS) 6. Transmission Electron Microscope (TEM)	
Proton microscopy 1. Proton Induced X-ray Emission (PIXE)	
1. Mineral Liberation analysis (MLA) with EDS/WDS	Mineral structure, texture and elemental composition with liberation characteristics
1. X-Ray Diffraction (XRD) 2. Raman spectroscopy 3. Infrared spectroscopy 4. Fourier transform infrared spectroscopy (FTIR)	Mineral phase identification
1. Atomic Force Microscopy (AFM)	Analysis of mineral surfaces
1. Differential Thermal Analysis (DTA)	Identification of minerals Semi quantitative mineral composition
1. Thermo Gravimetric Analysis (TGA)	Purity of mineral Distinguishes mineral compositions

From time to time sinking of the floor of the lake or swamp drowned the forests, and sediments (sand and mud) carried by rivers were deposited in layers above the vegetable sludge. Eventually, with a halt in the sinking, the water became shallow again, trees re-established themselves, and the whole cycle was repeated. This occurred many times over, so that there accumulated twenty or thirty or even more layers of vegetable sludge separated by layers of sand and mud. The last two materials, with little more change than hardening, eventually formed sandstone and shale, but the conversion of the plant material into coal involved a considerable change in composition. Each layer of vegetable sludge which forms further as coal bed is called **Coal Seam**. Formations of series of coal seams separated by layers of sand and mud (sedimentary rock) are known as **Coal Measure**.

Fallen trees and plants are attacked by oxygen from the surrounding air and slowly converted to carbon dioxide and water. The vegetation is water logged, however, air cannot penetrate to the cellulose and decay takes place anaerobically (i.e. in absence of air) by the action of bacteria. The products of decomposition have been buried to a considerable depth by the time the bacterial action ceases. The resulting product thus formed is called **peat**. It is further altered by the agencies of heat and pressure by the reactions named as coalification reactions to the varying degrees of completeness leading to formation of lignite, bituminous and anthracite coals.

There are two modes of origin of coal seams called **in situ** and **drift**. If coal seams occupy more or less the site on which the original plants grew and where their remains accumulated, it is called in situ origin. The coal of this origin is not polluted much with the extraneous dirts, and the seams thus formed are not very thick. In drift origin, plants were drifted from one place, where it actually grew and die, by flood or river transportation and accumulated in a lake or estuary. Under some earth quake or other geological sequence, everything had gone under ground and covered up with earth strata. The coal of drift origin is naturally contaminated by extraneous dirts.

Since coal is derived chiefly from vegetal matter, it consists mainly of the elements that go to compose plants, but it differs from plants in composition in as much as certain proportions changed during the formation of coal. As the plants accumulated under water, some silt also settled along with them. Thus at a later stage, some mineral matter got mixed with the coal as it formed. Hence the coal as obtained from the earth's crust is not strictly a coal but consist of **pure coal** (or **coal substance**), **mineral matter** and **moisture**.

Coal substance is that part of the plant's organic material (or cellulose) which has been later converted during the coalification reactions to form coal beds. It contains the elements carbon, hydrogen and oxygen with small amounts of nitrogen and sulphur and only traces of inorganic material which are not chemically combined with the organic material forming most of the coal.

Mineral matter refers to such impurities as they exist in the coal. The mineral matter is non-combustible. The residue from this mineral matter after coal has been burned is called ash. Ash forming mineral matter is of two types namely **inherent mineral matter** and **extraneous** or **adventitious mineral matter**. The inherent mineral matter represents inorganic elements present in plants giving raise to coal beds. It is very small in amount, about 2% or less of the total ash, different in composition and not possible to separate by usual beneficiation methods. The extraneous mineral matter is due to the substances which got associated with the decaying vegetable material during its conversion into coal, and also due to the rocks and dirt getting mixed up during mining and handling of coal. The former of extraneous mineral matter is in a fine state and intimately associated with the organic mass of coal. Hence difficulties are experienced in removing this from coal by mechanical methods. The second type of extraneous mineral matter is more amenable to coal cleaning methods. The major portion of mineral matter of commercial coal is extraneous.

The moisture in coal can be considered to occur in two parts. One part, called the free moisture, occurs on the surface of the coal and in its cracks and joints. The other part, called **inherent moisture**, is the amount of moisture in three forms as **Hygroscopic moisture** which is the water held inside the capillaries of the coal substance, **Decomposition moisture** which is the water incorporated in some of the coal's organic compounds and **Mineral moisture** which is the water that forms part of the crystal structure of clays and other minerals present in the coal.

There are two kinds of coal analysis. One is proximate analysis where the percentage of moisture, volatile matter, ash and fixed carbon are determined. The other is ultimate analysis where the elements carbon, hydrogen, nitrogen, sulphur, and oxygen are determined.

1.3.1   Proximate analysis

Analysis of coal is very important from a point of view for the selection of coal for different purposes like combustion, carbonization, gasification and liquefaction. Proximate analysis is the most often used type of analysis for characterizing coals in connection with their utilization. The proximate analysis determines the percentage of moisture, volatile matter, ash and fixed carbon. Proximate analysis is much more readily made and gives a preliminary indication of quality and suitability for various uses.

1.3.1.1   Determination of moisture

About 1 gm of −72 mesh B.S. Sieve coal sample is kept in a silica crucible and heated in a hot air oven at 105°–110°C for one hour. Thereafter, the crucible is taken out, cooled in a desiccator and weighed. The process of heating, cooling and weighing is repeated number of times till the constant weight of coal is achieved. The loss in weight is the weight of moisture.

$$\% \text{ Moisture} = \frac{\text{Weight of moisture}}{\text{Weight of original sample}} \times 100 \qquad (1.3.1.1)$$

1.3.1.2   Determination of volatile matter

About 1 gm of air-dried coal of −72 mesh B.S. Sieve is weighed in a standard cylindrical silica crucible with lid. The crucible is placed in a muffle furnace of specified dimensions at 925°C for a period of exactly seven minutes. Remove the crucible from the furnace and place on a cold iron plate to cool it rapidly. This prevents any oxidation of the contents in the crucible. Transfer the crucible to a desiccator while still warm. Allow it to cool and weigh. The loss of weight is taken as weight of volatile matter and air-dried moisture.

% Volatile matter and air-dried moisture

$$= \frac{\text{Weight of volatile matter and air-dried moisture}}{\text{Weight of original sample}} \times 100 \qquad (1.3.1.2)$$

% Volatile matter

$$= \% \text{ Volatile matter and air-dried moisture} - \% \text{ air-dried moisture}$$

$$(1.3.1.3)$$

1.3.1.3   Determination of ash

About 1 gm of air-dried coal of −72 mesh B.S. Sieve is taken in a clean dry silica crucible. The coal is distributed so that the thickness of the layer does not exceed 0.14 gm per square centimeter. The crucible is put into a muffle furnace at 500°C for 30 minutes and raised the temperature to 815 ± 10°C in another 30 to 60 minutes. The crucible is kept for one hour or more at this temperature until there is no loss in weight. The crucible is then taken out, cooled in a desiccator and weighed. The weight of the material left in the crucible is the weight of the ash.

$$\% \text{ Ash} = \frac{\text{Weight of ash}}{\text{Weight of original sample}} \times 100 \qquad (1.3.1.4)$$

It is also in practice to heat the coal sample directly at 750°C till a constant weight is obtained.

1.3.1.4   Determination of fixed carbon

Fixed carbon is the pure carbon present in the coal. It is the carbon available in the coal for combustion. Higher the fixed carbon content, higher will be its calorific value. **Total carbon** is the fixed carbon plus the carbon present in the volatile matters e.g. carbon monoxide, carbon dioxide, methane, hydrocarbons etc. Total carbon is always more than fixed carbon in any coal. High total carbon containing coal will have higher calorific value.

Fixed carbon is not determined directly. It is simply the difference between the sum of the other components and 100.

$$\% \text{ Fixed carbon} = 100 - (\% \text{Moisture} + \% \text{Volatile matter} + \% \text{Ash}) \qquad (1.3.1.5)$$

In examples 1.3.1.1 and 1.3.1.2, necessary calculations are made to determine proximate analysis of coal from the laboratory observed values.

Example 1.3.1.1:   *Three samples from medium coking coal were collected for the determination of proximate analysis. The following is the laboratory observed data obtained:*

Sample 1:

Weight of the empty crucible	*= 18.765 gm*
Weight of the crucible with coal sample	*= 19.649 gm*
Weight of the crucible with coal sample after heating at 105°C till constant weight	*= 19.570 gm*

Sample 2:

Weight of the empty crucible	*= 19.109 gm*
Weight of the crucible with coal sample	*= 20.012 gm*
Weight of the crucible with coal sample after heating at 925 ± 10°C for 7 minutes	*= 19.664 gm*

Sample 3:

$$
\begin{aligned}
&\text{Weight of the empty crucible} && = 18.567\,gm \\
&\text{Weight of the crucible with coal sample} && = 19.490\,gm \\
&\text{Weight of the crucible with coal sample} && \\
&\text{after heating at } 750°C \text{ till constant weight} && = 18.945\,gm
\end{aligned}
$$

Calculate the proximate analysis of the coal.

Solution:

Moisture determination (Sample 1)

$$
\begin{aligned}
&\text{Weight of the coal} = 19.649 - 18.765 = 0.884\,\text{gm} \\
&\text{Weight of moisture} = 19.649 - 19.570 = 0.079\,\text{gm}
\end{aligned}
$$

$$
\%\ \text{Moisture} = M \quad = \frac{0.079}{0.884} \times 100 \quad = 8.94\%
$$

Volatile matter determination (Sample 2)

$$
\begin{aligned}
&\text{Weight of the coal} = 20.012 - 19.109 = 0.903\,\text{gm} \\
&\text{Weight of volatile matter} + \text{moisture} = 20.012 - 19.664 = 0.348\,\text{gm}
\end{aligned}
$$

Since fresh sample is used

$$
\%\ (\text{Volatile matter} + \text{Moisture}) = \frac{0.348}{0.903} \times 100 \quad = 38.54\%
$$

$$
\%\ \text{Volatile matter} = V = 38.54 - 8.94 = 29.60\%
$$

Ash determination (Sample 3)

$$
\begin{aligned}
&\text{Weight of the coal} = 19.490 - 18.567 = 0.923\,\text{gm} \\
&\text{Weight of ash} = 18.945 - 18.567 = 0.378\,\text{gm}
\end{aligned}
$$

$$
\%\ \text{Ash} = A = \frac{0.378}{0.923} \times 100 = 40.95\%
$$

$$
\%\ \text{Fixed carbon} = FC = 100 - 8.94 - 29.60 - 40.95 = 20.51\%
$$

Proximate analysis of the coal taken is

$$
\begin{aligned}
\text{Moisture} = M &= 08.94\% \\
\text{Volatile matter} = V &= 29.60\% \\
\text{Ash} = A &= 40.95\% \\
\text{Fixed carbon} = FC &= 20.51\%
\end{aligned}
$$

Example 1.3.1.2: *Proximate analysis of a coal was carried out by taking three samples as follows:*

a) *First sample is taken in 25 ml silica crucible of 16.3256 gm and weighed as 17.1348 gm. It is heated at 105°C in a hot air oven till constant weight is obtained. Its weight is 17.1239 gm.*

b)   Second sample is taken in another 25 ml silica crucible of 17.0826 gm and weighed as 17.9301 gm. It is heated at 800°C in a muffle furnace till all the coal in it completely burns. Its weight is 17.3846 gm.

c)   Third sample is taken in a 18.5364 gm silica volatile matter crucible and weighed as 19.3579 gm. This is kept in a muffle furnace at 925°C for 7 minutes and then weighed as 19.1603 gm.

Calculate
   i) Percent moisture
   ii) Percent mineral matter
   iii) Percent coal substance

Solution:

(a) Weight of the coal sample $= 17.1348 - 16.3256 = 0.8092$ gm
Weight of the moisture $= 17.1348 - 17.1239 = 0.0109$ gm

$$\% \text{ Moisture} = M = \frac{0.0109}{0.8092} \times 100 = 1.35\%$$

(b) Weight of the coal sample $= 17.9301 - 17.0826 = 0.8475$ gm
Weight of the ash $= 17.3846 - 17.0826 = 0.3020$ gm

$$\% \text{ Ash} = A = \frac{0.3020}{0.8475} \times 100 = 35.63\%$$

(c) Weight of the coal sample $= 19.3579 - 18.5364 = 0.8215$ gm
Weight of volatile matter and moisture
$$= 19.3579 - 19.1603 = 0.1976 \text{ gm}$$

Since fresh sample is used

$$\% \text{ Volatile matter} + \text{Moisture} = \frac{0.1976}{0.8215} \times 100 = 24.05\%$$

$$\% \text{ Volatile matter} = V = 24.05 - 01.35 = 22.70\%$$

$$\% \text{ Fixed carbon} = 100 - (01.35 + 22.70 + 35.63) = 40.32\%$$
$$\% \text{ Mineral matter} = 1.1 \text{Ash} = 1.1 \times 35.63 = 39.19\%$$

$$\% \text{ Volatile matter from coal substance} = \% \text{ Volatile matter} - 0.1\% \text{Ash}$$
$$= 22.70 - 0.1 \times 35.63 = 19.14\%$$

$$\% \text{ Coal substance} = \% \text{ Fixed carbon} + \% \text{ Volatile matter from coal substance}$$
$$= 40.32 + 19.14 = 59.46\%$$

1.3.2   Expression of analytical results on different bases

The analytical results of proximate analysis can be modified by appropriate corrections to allow expressions on different bases. The most commonly used bases for reporting of analytical results are:

1.3.2.1   As received or As sampled

When the coal is received by a consumer from the mine, samples for analysis are collected, then its analysis is reported on **As received basis** or **As sampled basis**. Lot of physical and chemical changes occur during the transportation of coal from mine to the consumer and also during the processing such as size reduction, washing etc. Hence analysis of coal at consumers end is reported on **As received basis**.

1.3.2.2   Air-dried

Freshly mined coal looses its moisture due to exposure to atmospheric air, during transportation and storage. The data obtained by analyzing the coal at this stage is on **air dried basis**. The data on **As received** basis and **Air dried** basis may be the same because in both cases coal looses its moisture similarly depending on humidity and temperature of atmospheric air.

1.3.2.3   Dry or moisture free

When it is required to completely eliminate the effect of moisture on analytical data, the coal analysis is reported on **Dry** or **moisture free basis**. It is the data expressed as the percentages of the coal after all the moisture has been removed.

1.3.2.4   Dry, ash free (d.a.f)

If the effect of moisture and ash is to be eliminated, then the data is reported on **dry ash free basis**. The coal is considered to consist of volatile matter and fixed carbon on the basis of recalculation with moisture and ash removed. This does not allow for the volatile matter derived from minerals present in the air-dried coal. This is the simplest way to compare the organic fractions of coals without diluting the effects of inorganic components. This data is suitable for comparing low ash coals (ash <10%).

1.3.2.5   Dry, mineral matter free (d.m.m.f)

Here it is necessary that the total amount of mineral matter rather than ash is determined, so that the volatile matter content in the mineral matter can be removed. In case of high ash coals, the mineral matter content is around 10% more than its ash whereas mineral matter is almost equal to its ash in case of low ash coals. Hence the data expressed on **dry mineral matter free basis** is most suitable for comparing high ash coals (ash >10%).

1.3.3   Calculations on different bases

The details of calculations to express the proximate analysis on different bases are given in the following articles.

1.3.3.1   As received or As sampled basis

The results of proximate analysis are expressed as percentages of the coal including the total moisture content.

$$\% \text{ Moisture} = \frac{\text{Weight of moisture}}{\text{Weight of coal sample as received}} \times 100 \quad (1.3.3.1)$$

$$\% \text{ Volatile matter} = \frac{\text{Weight of volatile matter}}{\text{Weight of coal sample as received}} \times 100 \quad (1.3.3.2)$$

$$\% \text{ Ash} = \frac{\text{Weight of ash}}{\text{Weight of coal sample as received}} \times 100 \qquad (1.3.3.3)$$

$$\% \text{ Fixed carbon} = 100 - (\text{Moisture}\% + \text{Volatile matter}\% + \text{Ash}\%) \quad (1.3.3.4)$$

1.3.3.2 Air-dried basis

The results are expressed as percentages of the air-dried coal, including inherent but not surface or free moisture.

$$\% \text{ Moisture} = M = \frac{\text{Weight of moisture}}{\text{Weight of air-dried coal sample}} \times 100 \quad (1.3.3.5)$$

$$\% \text{ Volatile matter} = V = \frac{\text{Weight of volatile matter}}{\text{Weight of air-dried coal sample}} \times 100 \quad (1.3.3.6)$$

$$\% \text{ Ash} = A = \frac{\text{Weight of ash}}{\text{Weight of air-dried coal sample}} \times 100 \quad (1.3.3.7)$$

$$\% \text{ Fixed carbon} = FC = 100 - (M + V + A) \qquad (1.3.3.8)$$

1.3.3.3 Dry or moisture free basis

The results are expressed as percentages of the coal after the inherent moisture has been removed. By using the percentages on air-dried basis, the results of proximate analysis on dry basis can be calculated as follows:

On Dry basis

$$\% \text{ Ash} = \frac{A}{100 - M} \times 100 \qquad (1.3.3.9)$$

$$\% \text{ Volatile matter} = \frac{V}{100 - M} \times 100 \qquad (1.3.3.10)$$

$$\% \text{ Fixed carbon} = \frac{FC}{100 - M} \times 100 \qquad (1.3.3.11)$$

1.3.3.4 Dry, ash free (d.a.f) basis

The coal is considered, in proximate analysis, to consist of only volatile matter and fixed carbon, on the basis of recalculation with ash and moisture removed. The calculations are as follows:

On d.a.f basis

$$\% \text{ Volatile matter} = \frac{V}{100 - M - A} \times 100 \qquad (1.3.3.12)$$

$$\% \text{ Fixed carbon} = \frac{FC}{100 - M - A} \times 100 \qquad (1.3.3.13)$$

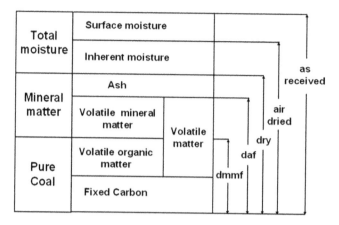

Figure 1.3 Pictorial Presentation of proximate analysis on different bases.

1.3.3.5 Dry, mineral matter free (d.m.m.f) basis

The coal is also considered to consist of solely volatile matter and fixed carbon, but it is necessary that the total amount of mineral matter rather than ash be determined.

Allowance is also made in dry, mineral matter free data, for the contribution to the air-dried volatile matter that comes from the mineral components. This may be done directly or indirectly. As it is difficult to determine volatile matter that comes from the mineral matter, it is suggested and agreed that the mineral matter of coal contributes to the volatile matter by an amount approximately equal to 10% of the ash.

Accordingly

$$\% \text{ Mineral Matter} = MM = A + 0.1A = 1.1A \tag{1.3.3.14}$$

The calculations are as follows:

On d.m.m.f basis

$$\% \text{ Volatile matter} = \frac{V - 0.1A}{100 - M - 1.1A} \times 100 \tag{1.3.3.15}$$

$$\% \text{ Fixed carbon} = \frac{FC}{100 - M - 1.1A} \times 100 \tag{1.3.3.16}$$

All the above bases are indicated pictorially in Fig. 1.3.

Detailed calculations on different bases are illustrated in example 1.3.3.

Example 1.3.3: *A coal has 2.34% moisture, 23.45% volatile matter and 45.67% ash. Calculate ash% on dry basis, volatile matter on d.a.f basis and fixed carbon on d.m.m.f basis.*

Solution:

$$\% \text{ Fixed carbon} = 100 - 2.34 - 23.45 - 45.67 = 28.54\%$$

Proximate Analysis of Coal $\quad M = 02.34\%$
$$V = 23.45\%$$
$$A = 45.67\%$$
$$FC = 28.54\%$$

$$\% \text{ Ash on dry basis} = \frac{A}{100 - M} \times 100 = \frac{45.67}{100 - 02.34} \times 100 = 46.76\%$$

$\%$ Volatile matter on d.a.f basis

$$= \frac{V}{100 - M - A} \times 100 = \frac{23.45}{100 - 02.34 - 45.67} \times 100 = 45.10\%$$

$\%$ Fixed carbon on d.m.m.f basis

$$= \frac{FC}{100 - M - 1.1A} \times 100 = \frac{28.54}{100 - 02.34 - 1.1 \times 45.67} \times 100 = 60.18\%$$

Alternately, fixed carbon can also be calculated through volatile matter

$\%$ Volatile matter on d.m.m.f basis

$$= \frac{V - 0.1A}{100 - M - 1.1A} \times 100 = \frac{23.45 - 0.1 \times 45.67}{100 - 02.34 - 1.1 \times 45.67} \times 100 = 39.82\%$$

$\%$ Fixed carbon on d.m.m.f basis $= 100 - 39.82 = 60.18\%$

1.4 VARIETIES OF COAL

The various classifications of coal have been given by different geologists. Most classifications are based upon some property of coal which varies with increasing maturity or rank of coal. The maturity or stage of coalification reached by a coal is referred to as its **rank**. Rank is not measured in percentages or other units, though it is judged by composition particularly the carbon or the oxygen content of the coal. The term is used only in a comparative sense, in such expressions as high rank, medium rank and low rank. It does not imply any superiority or inferiority of quality; low rank, for example, does not mean poor quality.

   For the classification to be used and developed for commercial purposes, the properties used are the proximate analysis, sulphur in coal and calorific value of coal in cal/gm. **Calorific value** of coal is the quantity of heat generated by complete combustion of unit weight of coal.

   Units for the calorific value are:

Calories/gm	(In C.G.S system)
Kilo Calories/kg	(In M.K.S system)
Btu/lb	(In F.P.S system)

   Various classifications proposed are less familiar. Certain varieties are recognized almost universally in science and commerce. These varieties are classified by ranks i.e., according to the degree of metamorphism in the series from lignite to anthracite. Those varieties are **Peat, Lignite** or **Brown coal, Sub-bituminous coal** or **Black lignite, Bituminous coal, Semi-anthracite** and **Anthracite**. They are not sharply separated and they grade into one another. Two coals with a certain percentage of fixed carbon may have very different calorific value.

From commercial point of view coal may be broadly classified as **coking coals** and **non-coking coals**. Coking coals are the coals when heated out of contact with air, volatile matter gets removed and carbon particles join each other to form a porous cellular mass with sufficient strength, called **coke**. These coals are also called **metallurgical coals**. Coking coals form a part of the bituminous group. Peat, Lignite, Sub-bituminous, Semi-anthracite and Anthracite are all non-coking coals. So also are some of the bituminous coals. Non-coking coals may resemble coking coal in all outward appearance. When they are heated in absence of air, they leave powdery residue. Many non-coking coals also leave a solid coherent residue but not possess the physical and chemical properties of the coke and not suitable for manufacture of coke. Non-coking coals are mainly used for power generation.

1.5 COAL PROCESSING

Coal Beneficiation means the separation of coal particles of low ash from the run-of-mine coal by physical and/or physico-chemical treatments. As it comes from the mine, coal is known as **Run-of-mine coal** (**ROM coal**) and consists of a range of sizes from chunks to small particles mixed with some dirt and rocks. In most cases, this ROM coal is subjected to various operations principally size reduction and screening, beneficiation (separation), and dewatering (if wet beneficiation operations are used) to meet certain market requirements as to sizes, ash, sulfur, moisture, and heating values. Coal Processing is a term that is used to designate the various operations performed on the ROM coal to prepare it for specific end uses such as feed to a coke oven or a coal-fired boiler or to a coal conversion process without destroying the physical identity of the coal. The terms coal preparation, coal dressing, coal cleaning, coal washing, coal upgradation and coal concentration are also applied to the same field. All of them are synonymous terms; different terms are used in different countries and different locations. When wet gravity methods are used for beneficiation of coal, those plants are called **Coal Washeries** and the low ash coal product is called **washed coal**.

The mineral matter, which gives ash after burning the coal, present in the coal is intimately associated with the coal substance in ROM coal. When ROM coal is reduced in size, all the coal particles produced will never contain the same quantity of mineral matter. Every coal particle may differ from other coal particles in its mineral matter (or ash) content. These coal particles are separated or beneficiated to obtain low ash coal particles and high ash coal particles as two products. The product of low ash coal particles is called **clean coal** or **combustibles**. The other product of high ash coal particles is called **refuse** or **rejects**. Some times another product, medium ash coal particles called **middling**, is also separated depending on the requirement.

The methods used in the beneficiation of coal are analogous in many ways to those used in the beneficiation of ores. The principles involved in coal beneficiation are similar to that of ore beneficiation. In some instances, the same machines could be used in both fields perhaps with slight modifications. An essential feature of coal beneficiation processes is that the treatment must be rapid and inexpensive, because coal is a cheap product. Accordingly, simple plants consisting of a few units, each capable of handling a large tonnage, are required.

Another factor important in coal processing is that the coal, during all stages of its handling, from mining, through beneficiation, transportation, and finally to its delivery to the consumer, must undergo a minimum of degradation because the value of coal depends on its particle size, and in general, the larger the particle size is, the better price the coal commands on the market.

Gravity separation principles form the basis of most of the coal beneficiation processes. In case of Coal Beneficiation, the valuable part, the low ash coal, is light and the impurities to be removed by washing or cleaning are heavier whereas in beneficiation of ores, the valuable mineral is heavy and impurities (gangue) is light.

Coal Processing forms a link between coal mining and coal utilization. Coal Processing literally means increasing the commercial value of coal by suitable preparation. The function of the modern coal processing plant is to produce the maximum yield of clean coal of suitable quality for the consumer at an economic cost.

1.6   PROBLEMS FOR PRACTICE

1.6.1:   *Calculate the Aluminium percent in Bauxite ($Al_2O_3 \cdot 2H_2O$). Atomic weights of Aluminium, Oxygen and Hydrogen are 26.98, 16.00 & 1.00 respectively.*
[39.11%]

1.6.2:   *What is the maximum grade of the Zinc concentrate obtainable from an ore of 8% Zn containing sphalerite (ZnS)? Atomic weights of Zn and S are 65.37 and 32.06 respectively.*
[67.10%]

1.6.3:   *An Iron Ore of 60% Fe contains only one Iron Mineral Hematite. What is the percent gangue in Iron Ore? Atomic weights of Iron and Oxygen are 55.85 and 16.00 respectively.*
[14.21%]

1.6.4:   *A sample contains Chalcopyrite ($CuFeS_2$), Pyrite (FeS_2), and non-sulfides (containing no Cu or Fe). Chemical analysis of the sample results the percent Cu and Fe as 22.5% and 25.6% respectively. Calculate the percent chalcopyrite, pyrite and non-sulfides in the sample. Atomic weights of Cu, Fe and S are 63.55, 55.85 and 32.06 respectively.*
[64.97%, 12.46%, 22.57%]

1.6.5:   *Proximate analysis of a coal was carried out by taking three samples as follows:*

a)   *First sample is taken in 25 ml silica crucible of 17.395 gm and weighed as 18.313 gm. It is heated at 105°C in a hot air oven till constant weight is obtained. Its weight is 18.221 gm.*

b) *Second sample is taken in another 25 ml silica crucible of 18.305 gm and weighed as 19.217 gm. It is heated at 800°C in a muffle furnace till all the coal in it completely burns. Its weight is 18.534 gm.*

c) *Third sample is taken in a 18.232 gm silica volatile matter crucible and weighed as 19.055 gm. This is kept in a muffle furnace at 925°C for 7 minutes and then weighed as 18.726 gm.*

Calculate
  i) Proximate analysis
  ii) Percent volatile matter on d.m.m.f basis
              [10.02%, 29.96%, 25.11%, 34.91%, 44.02%]

1.6.6: *A high volatile bituminous coal has 8% moisture, 34% volatile matter and 46% ash. Calculate fixed carbon percent on dry, d.a.f and d.m.m.f bases.*
              [13.04%, 26.09%, 28.99%]

Chapter 2

Material (mass) balance

The law of conservation of mass states that "the rate of mass entering the system equals that leaving as mass can be neither accumulated nor depleted within a system under steady state conditions". This is the basis for balancing the material over a system to analyze any process or operation. When a material is fed to a beneficiation operation, it yields usually two products. One product contains valuables called concentrate and another product contains waste called tailing.

The material balance equation for total material is written for a continuous operation as

Rate of material fed to the beneficiation operation (F)

$=$ Rate of material obtained as concentrate (C)

$+$ Rate of material obtained as tailing (T)

$$\Rightarrow \qquad F = C + T \qquad (2.1)$$

Let f, c and t are the fraction (or percent) of metal present in feed, concentrate and tailing respectively.

The material balance equation for the metal is written as

Metal in the feed to the beneficiation operation

$=$ Metal in the concentrate $+$ Metal in the tailing

$$\Rightarrow \qquad Ff = Cc + Tt \qquad (2.2)$$

These two are the fundamental equations required to estimate the recovery of the metal values, to evaluate the performance of the process (or) operation, to locate the process bottlenecks, to build a process model and to simulate the process.

Similar equations can be written for any process or operation. Majority of the beneficiation operations are wet operations where the ore is fed to an operation as a mixture of ore particles and water (called pulp or slurry). The beneficiation operation yields two pulp or slurry streams. In such a case F, C and T can be considered as feed slurry, concentrate slurry and tailing slurry respectively. Each stream is required to have a definite proportion of solids and water which normally represented as fraction or percent solids. f, c and t denote percent solids in feed, concentrate and tailing respectively. Then equation 2.1 is a total material (solids and water) balance equation and equation 2.2 is a material balance equation for solids.

Only solids without water are fed to a dry operation. These solids contain different sizes of the particles. In this case F, C and T are total solids in feed, concentrate and tailing respectively. f, c and t are the fraction or percent of a particular size of particles in feed, concentrate and tailing. Then equation 2.1 represents total solids balance and equation 2.2 represents balance for considered size particles.

Screening of different sizes of the particles is an example of dry operation. In this case, the equations are written as

$$F = P + U \qquad (2.3)$$

$$Ff = Pp + Uu \qquad (2.4)$$

where F, P and U are the material fed to the screen, overflow and underflow material obtained from the screen respectively. f, p and u are the fraction or percent of particular size of the material in feed, overflow and underflow. In this case of screening operation overflow contains oversize material and underflow contains undersize material.

Classification of fines is an example of wet operation where the given solid particles are fed to a classifier in the form of a slurry stream and it yields two products as slurry streams of overflow and underflow. In such a case, the equations 2.3 and 2.4 represent total solids balance and balance of particular size of solids considering only solids without water. The same equations can also be represented as slurry balance and solids balance if f, p and u are taken as fraction or percent solids in feed, overflow and underflow. In case of classification operation, overflow contains undersize material and underflow contains oversize material.

It is to be noted that equation 2.4 represents water balance if f, p and u are taken as fraction or percent water in feed, overflow and underflow. In case of dewatering operations like thickening and filtration, water balance is of vital importance to know or assess how much water is removed from the slurry.

Various expressions like recovery of metal, solids, particular size particles etc. can be obtained from these two material balance equations to evaluate the performance of an operation. Similarly, these equations are also useful to evaluate the efficiency of an operation. All these are dealt in subsequent chapters in this book with suitable numerical examples.

Chapter 3

Sampling

A Mineral/Coal Processing plant costs thousands of dollars to build and operate. The success of the plant relies on the assays of a few small samples. Representing large ore bodies truly and accurately by a small sample that can be handled in a laboratory is a difficult task. The difficulties arise chiefly in collecting such small samples from the bulk of the material.

The method or operation of taking the small amount of material from the bulk is called **Sampling**. It is the art of cutting a small portion of material from a large lot. The small amount of material is called **Sample** and it should be representative of the bulk in all respects (in its physical and chemical properties). Sampling was well defined by Taggart [3] as "the operation of removing a part, convenient in quantity for analysis, from a whole which is of much greater in bulk, in such a way that the proportion and distribution of the quality to be tested are the same in both the whole and the part removed (SAMPLE)". The conditions of the more stringent definition, that the sample shall be completely representative of the whole as regards all aspects, are practically never fulfilled when heterogeneous mineral mixtures are sampled.

Sampling is a statistical technique based on the theory of probability. The first and most obvious reason for sampling is to acquire information about the ore entering the plant for treatment. The second is to inspect its condition at selected points during its progress through the plant so that comparison can be made between the optimum requirements for efficient treatment and those actually existing, should these not coincide. The third is to disclose recovery and reduce losses.

The prerequisite for the development of a satisfactory flowsheet is the acquisition of a fully representative ore sample, even though in respect of a new ore-body, this sample may have to be something of a compromise. A bad sample will result in wastage of all test work and can lead to a completely wrongly designed mill.

A sample can be taken from any type of material dry, wet or pulp. But, in each case, the method of sampling and the apparatus necessary for them are different.

There are essentially three stages in obtaining a representative sample.

1   Collection of large quantity of material from a huge lot
2   Reduction of large quantity to a conveniently handled quantity
3   Final reduction to a required quantity

Large quantity of material collected in first stage is known as **gross sample** or **primary sample** or **composite sample**. The two methods used to obtain gross sample are **Random sampling** and **Systematic** or **stratified sampling**. Grab sampling and stream sampling are the two sampling methods used to collect the gross sample.

Grab sampling method is used to collect samples from stationary bulk material such as stockpile. Grab sampling is the simplest form of hand sampling. A small quantity of material is taken (or cut) from a place (or point) in the bulk. This small quantity of material taken in one cut is called **increment**. Grab sampling consists in taking small, equal portions of material from several places by scoop or shovel selected at random (random sampling) or at regular intervals (systematic or stratified sampling) from the bulk of the material to be sampled. The material thus collected is mixed together to form a primary or gross sample which is a base for the final sample. The various hand-tool samplers used are Drill, Shovel, Scoop, and Pipe samplers (slot, pocket and auger samplers). Pipe samplers are especially useful to collect the sample from inside the stockpile where it is not accessible for other tools.

Hand sample cutters and Automatic samplers are used for stream sampling in all plants. Automatic samplers are typically used in any systematic program and consist of a cutter that goes across the material stream to obtain a slice of the flow. A common falling ore sampling device is the Vezin sampler. Sampling devices called poppet valves are used for pulp sampling in pipes. In case of wet ore/coal or pulp, the gross sample is to be dried before it is subjected to second stage of sampling.

In the second stage, the gross sample is reduced to a quantity that can be handled with ease by **shovel sampling** also called **alternate shoveling** or **fractional shoveling** in stages depending upon the quantity of gross sample. Such reduced samples are called **secondary sample** and **ternary sample** depending upon the number of stages used.

Reduction of thus reduced sample to a quantity necessary for analysis, known as **final sample** or **test sample** or **laboratory sample**, is called **sample preparation**. It is the process of reducing the quantity by splitting. It is essential that the gross sample be thoroughly mixed before reduction in order to obtain a representative sub-sample or laboratory sample.

The two broad sampling methods are hand sampling and machine sampling. Coning and quartering is the oldest standard sample splitting method of hand sampling for sample preparation. **Jones riffle sampler, Table sampler, paper cone splitter, rotary cone splitter, rotary table splitter** or **micro splitter** are few of the machine samplers used for sample preparation. The sampler's knowledge, experience, judgment and ability are of greater value because instructions cannot cover every point or combination of circumstances encountered on each preparation.

When the sample is taken and subject to analysis, chances of error in single sample will exist. One must take the number of samples to reduce the error and to keep the overall error within the tolerable working limits.

Chapter 4

Size analysis

Size of the particle is most important consideration in mineral and coal processing as the energy consumed for reducing the size of the particles depends on size. Size of the particles determines the type of size reduction equipment, beneficiation equipment and other equipment to be employed.

The size of the particle of standard configuration like sphere and cube can easily be specified. For example, the size of a spherical particle is its diameter and that of a cubical particle is the length of its side. As the mineral particles are of irregular in shape, it is difficult to define and determine their size. The size of a particle is usually defined by comparison with a standard configuration, normally spherical particle.

An **equivalent size** or **equivalent diameter** of an irregular particle is defined as the diameter of a spherical particle having the same controlling characteristics as the particle under consideration.

Volume diameter, d_v, is defined as the diameter of a spherical particle having the same volume as the irregular particle

$$d_v = \left(\frac{6V_p}{\pi}\right)^{1/3} = 1.241 V_p^{1/3} \tag{4.1}$$

where V_p is the volume of the irregular particle.

Surface diameter, d_s, is defined as a diameter of a spherical particle having the same surface area as the irregular particle.

$$d_s = \left(\frac{A_p}{\pi}\right)^{1/2} = 0.564 A_p^{1/2} \tag{4.2}$$

where A_p is the surface area of the irregular particle.

Stokes' diameter is the diameter of a sphere having the same settling velocity as the particle under Stokisian conditions. **Newton's diameter** is the diameter of a sphere having the same settling velocity as the particle under Newtonian conditions.

It is obvious that each definition has its limitations.

Size of the irregular particle is the decisive factor in determining its surface area. For example, let us consider one gram of pure crystalline silica spheres of diameter d cm having density ρ_p gm/cm^3.

$$\text{The surface area of an individual particle} = \pi d^2 \text{ cm}^2$$

$$\text{The mass of an individual particle} = \frac{1}{6}\pi d^3 \rho_p \text{ gm}$$

$$\text{The number of particles per gram} = \frac{6}{\pi d^3 \rho_p}$$

$$\therefore \text{ The surface area of the particles per gram} = \frac{6}{\pi d^3 \rho_p}\pi d^2 = \frac{6}{d\rho_p} \text{ cm}^2 \quad (4.3)$$

This surface area per unit mass of the material is called **Specific surface** and in powder industry it is known as Blaine number.

When once the particle size is defined, the shape of the particle is also to be defined as the response of a particle during beneficiation is influenced to a large degree by its shape or by the ratio of its surface to its volume. For a given volume, a sphere has the minimum surface area while a flat plate has a maximum surface area for the same volume. The larger is the surface, the greater is the surface energy. The shape, more correctly the surface developed by the particle shape, has an impact on beneficiation.

The exact shape of an irregular particle is difficult to specify. One of the methods of defining particle shape is by using the term **Sphericity** Ψ_S. It is defined as the ratio of the surface area of the sphere having the same volume as that of the particle to the actual surface area of the particle. If d_v is the volume diameter of the sphere, A_p is the surface area of the particle, the sphericity is defined as

$$\Psi_S = \frac{\pi d_v^2}{A_p} \quad (4.4)$$

It must be noted that since sphericity compares the surface area of the particle to that of the equivalent spherical particle, it defines only the particle shape and is independent of particle size. Sphericity of a spherical particle is obviously 1.0. The sphericity of a cubical particle of size 'a' can be determined as follows:

The volume diameter of the cubical particle

$$d_v = \left(\frac{6V_p}{\pi}\right)^{1/3} = \left(\frac{6a^3}{\pi}\right)^{1/3} = 1.24a$$

$$\therefore \qquad \Psi_S = \frac{\pi(1.24a)^2}{6a^2} = 0.806$$

Similarly, the value of sphericity for a cylindrical particle of length equal to its diameter will be 0.874. Every mineral has a given crystallographic structure, thus a given shape or configuration and therefore a specific sphericity. The value of sphericity for an irregular particle is always less than unity. For any crushed material it is 0.6–0.7.

The reciprocal of sphericity is commonly called the shape factor or more precisely, the **surface shape factor** (λ_S).

Thus $$\text{Surface shape factor} = \lambda_S = \frac{1}{\Psi_S} = \frac{A_p}{\pi d_v^2} \tag{4.5}$$

Sphericity Ψ_V is defined as the ratio of the volume of the sphere having the same surface area as that of the particle to the actual volume of the particle. If d_S is the surface diameter of the sphere, V_p is the volume of the particle, the sphericity is defined as

$$\Psi_V = \frac{\frac{\pi}{6} d_S^3}{V_p} \tag{4.6}$$

Then $$\textbf{Volume shape factor} = \lambda_V = \frac{1}{\Psi_V} = \frac{6 V_p}{\pi d_S^3} \tag{4.7}$$

The volume shape factor (λ_V), is sometimes used for calculating the volume of an irregular particle.

In mineral and coal industry, Standard Test Sieves are used to measure the size of the small and the fine particles usually down to 74 microns. The side of a square aperture of a test sieve through which a particle just passes is taken as the size of the particle eventhough little or no importance is given to its shape.

Following are the definitions of important nomenclature used for size analysis.

Test Sieve is a circular shell of brass having 8 inch diameter and about 2 inch high as shown in Fig. 4.1.

Sieve cloth is made of wire, woven to produce nominally uniform cloth apertures (openings) of square shape. The sieve cloth of square opening is placed in the bottom of the shell so that material can be held on the sieve.

Figure 4.1 Test Sieve.

Aperture (or **Opening**) is a distance between two parallel wires.

Mesh number is the number of apertures per linear inch. Sieves are designated by Mesh number.

Mesh size is the size of an aperture i.e. the distance between two parallel wires. As mesh number increases, mesh size decreases.

Sieve Scale is the list of successive sieve sizes used in any laboratory, taken in order from coarsest to finest.

Standard Sieve Scale is the sieve scale adopted for size analyses and general testing work to facilitate the interchangeability of results and data. In this standard sieve scale, the sizes of successive sieves in series form a geometric progression.

For a standard sieve scale, the reference point is 74 microns, which is the aperture of a 200 mesh woven wire sieve. The ratio of the successive sizes of the sieves in the standard sieve scale is $\sqrt{2}$, which means that the area of the opening of any sieve in the series is twice that of the sieve just below it and one half of the area of the sieve next above it in the series.

In general, mesh number × mesh size in microns ≈ 15,000.

For closer sizing work the sieve ratio of $\sqrt[4]{2}$ is common.

The different standards in use are:

American Tyler Series
American Standards for Testing and Materials, ASTM E-11-01
British Standard Sieves, BSS 410-2000
French Series, AFNOR (Association Francaise de Normalisation) NFX 11-501
German Standard, DIN (Deutsches Institut fur Normung) 3310-1: 2000

The Indian Standard (IS) sieves, however, follow a different type of designation. For an IS sieve, the mesh number is equal to its aperture size expressed to the nearest deca-micron (0.01 mm). Thus an IS sieve of mesh number 50 will have an aperture width of approximately 500 microns. Such a method of designation has the simplicity that the aperture width is readily indicated from the mesh number.

For most size analyses it is usually impracticable and unnecessary to use all the sieves in a particular series. For most purposes, alternative sieves are quite adequate. For accurate work over certain size ranges of particular interest, consecutive sieves may be used. Intermediate sieves should never be chosen at random, as the data obtained will be difficult to interpret. In general, the sieve range should be chosen such that no more than about 5% of the sample material it retained on the coarsest sieve, or passes the finest sieve. These limits may be lowered for more accurate work.

Table 4.1 shows the comparison of test sieves of different standards.

Table 4.1 Comparison of Test Sieves of different Standards.

Tyler Sieve designation mesh no.	Tyler Width of aperture mm	Mesh double Tyler Series	U.S.A ASTM E-11-01 Sieve designation mesh no.	U.S.A ASTM Width of aperture mm	British B.S. 410-2000 Sieve designation mesh no.	British Width of aperture mm	Indian I.S. 460-1962 Sieve designation mesh no.	Indian Width of aperture mm	French AFNOR NFX-11-501 Sieve designation mesh no.	French Width of aperture mm	German DIN 3310-1:2000 Sieve designation mesh no.	German Width of aperture mm
–	–	–	–	–	–	–	–	–	38	5.00	–	5.00
4	4.75	–	4	4.75	3½	4.75	480	4.75	–	–	–	4.50
–	4.00	5	5	4.00	4	4.00	400	4.00	37	4.00	2E	4.00
6	3.35	–	6	3.35	5	3.35	340	3.35	–	–	–	–
–	–	–	–	–	–	3.15	320	3.18	36	3.15	–	3.15
–	2.80	7	7	2.80	6	2.80	280	2.80	–	–	–	2.80
8	2.36	–	8	2.36	7	2.36	240	2.39	35	2.50	–	2.50
–	2.00	9	10	2.00	8	2.00	200	2.00	34	2.00	3E	2.00
10	1.70	–	12	1.70	10	1.70	170	1.70	33	1.60	–	1.60
–	1.40	12	14	1.40	12	1.40	140	1.40	–	1.40	–	1.40
–	–	–	–	–	–	1.25	–	–	32	1.25	–	1.25
14	1.18	–	16	1.18	14	1.18	120	1.20	–	–	5	1.20
–	1.00	16	18	1.00	16	1.00	100	1.00	31	1.00	6	1.00
20	0.85	–	20	0.850	18	0.850	85	0.850	–	–	–	–
–	–	–	–	–	–	0.800	80	0.79	30	0.800	–	0.800
–	0.710	24	25	0.710	22	0.710	70	0.710	–	0.710	–	0.710
–	–	–	–	–	–	0.630	–	–	29	0.630	–	0.630
28	0.600	–	30	0.600	25	0.600	60	0.600	–	–	10	0.600
–	0.500	32	35	0.500	30	0.500	50	0.500	28	0.500	12	0.500
35	0.425	–	40	0.425	36	0.425	40	0.425	–	–	–	–
–	–	–	–	–	–	0.400	–	–	27	0.400	16	0.400
–	0.355	42	45	0.355	44	0.355	35	0.355	–	0.355	–	0.355
–	–	–	–	–	–	0.315	–	–	26	0.315	–	0.315
48	0.300	–	50	0.300	52	0.300	30	0.300	–	–	20	0.300
–	0.250	60	60	0.250	60	0.250	25	0.250	25	0.250	24	0.250
65	0.212	–	70	0.212	72	0.212	20	0.212	–	–	–	–
–	–	–	–	–	–	0.200	–	–	24	0.200	30	0.200
–	0.180	80	80	0.180	85	0.180	18	0.180	–	0.180	–	0.180
–	–	–	–	–	–	0.160	–	–	23	0.160	–	0.160
100	0.150	–	100	0.150	100	0.150	15	0.150	–	–	40	0.150
–	0.125	115	120	0.125	120	0.125	12	0.125	22	0.125	50	0.125
150	0.106	–	140	0.106	150	0.106	10	0.106	–	–	–	–
–	–	–	–	–	–	0.100	–	–	21	0.100	60	0.100
–	0.90	170	170	0.090	170	0.090	9	0.090	–	0.090	70	0.090
–	–	–	–	–	–	0.800	–	–	20	0.80	–	0.080
200	0.075	–	200	0.075	200	0.075	8	0.075	–	–	80	0.075
–	–	–	–	–	–	0.710	–	–	–	0.071	–	0.071
–	0.063	250	230	0.063	240	0.063	6	0.063	19	0.063	–	0.063
–	–	–	–	–	–	0.056	–	–	–	0.056	110	0.056
270	0.053	–	270	0.053	300	0.053	5	0.053	–	–	–	–
–	–	–	–	–	–	0.050	–	–	18	0.050	120	0.050
–	0.045	325	325	0.045	350	0.045	4	0.045	–	0.045	–	0.045
–	–	–	–	–	–	0.040	–	–	17	0.40	–	0.040
400	0.038	–	400	0.038	400	0.038	3	0.038	–	–	130	–

4.1 SIEVE ANALYSIS

Sieve analysis is the heart of the mineral/coal processing as the separation is achieved primarily based on the size of the particles. Sieve analysis is a method of size analysis using test sieves at smaller and fine sizes. It is performed to separate the sample into number of closely sized fractions by allowing the sample of material to pass through a series of test sieves and then determine the weight percentages of every fraction. **Closely sized material** (or **closed size material**) is the material in which the difference between maximum and minimum sizes of the particles in the material is less.

Sieving can be done by hand or by machine. Hand sieving method is considered more effective as it allows the particles to present in all possible orientations on to the sieve surface. However, machine sieving is preferred for routine analysis as hand sieving is long and tedious.

Owing to irregular shapes, particles cannot pass through the sieve unless they are presented in a favourable orientation particularly with the fine particles. Hence there is no end point for sieving. For all practical purposes, the end point is considered to have been reached when there is little amount of material passing through after a certain length of sieving.

Sieving is generally done dry. Wet sieving is used when the material is in the form of slurry. When little moisture is present, a combination of wet and dry sieving is performed by adding water initially.

4.2 TESTING METHOD

The sieves chosen for the test are arranged in a **stack**, or **nest**, starting from the coarsest sieve at the top and the finest at the bottom. A pan or receiver is placed below the bottom sieve to receive final undersize, and a lid is placed on top of the coarsest sieve to prevent escape of the sample.

The material to be tested is placed on uppermost coarsest sieve and close with lid. The nest is then placed in a Sieve Shaker and sieved for certain time. After sieving by sieve shaker, hand sieving is usually performed for better accuracy by taking individual sieves. Fig. 4.2.1 shows the sieve analysis at the end of the sieving.

The nest of sieves is taken out from the sieve shaker. The material collected on each sieve is removed and weighed. For better accuracy, after the nest is taken out from the sieve shaker, each individual sieve is taken by hand with lid and receiver and sieved for more length of time starting from the top sieve. The material collected in receiver is added to the next bottom sieve. The procedure is continued till the bottom most sieve. The complete set of values is known as **Particle Size Distribution** data. Particle size distribution data for an ore is given in a tabular form as shown in Table 4.2.1 for better explanation.

Another way of representing the material with mesh numbers is 20/28 which means the material passing through 20 mesh and retained on 28 mesh.

The weight percentages of the material retained on each sieve are to be calculated to form differential analysis. Cumulative weight percentage retained is obtained from differential analysis by adding, cumulatively, the individual differential weight percentages from the top of the table. Cumulative weight percentage passing is obtained

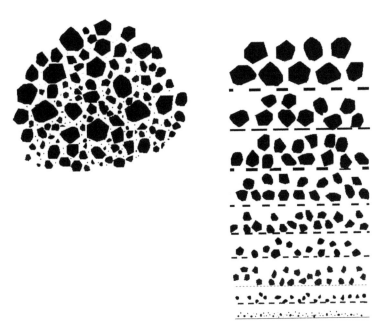

Figure 4.2.1 Sieve analysis at the end of sieving.

Table 4.2.1 Particle size distribution data from size analysis test.

Mesh number	Retained mesh size in microns	Weight of material gm
+20	840	7.7
−20 + 28	595	13.2
−28 + 35	420	29.7
−35 + 48	297	40.7
−48 + 65	210	28.6
−65 + 100	149	22.0
−100 + 150	105	16.5
−150 + 200	74	8.8
−200		52.8
		220.0

+ sign designates particles retained on that sieve
− sign designates particles passed through that sieve

by adding, cumulatively, the individual weight percentages from the bottom of the table.

All the fractions are fairly closely sized (except first fraction). Hence the size of the particles in each fraction may be calculated as arithmetic mean of the limiting sizes.

For example, the size of −20 + 28 mesh fraction is $\dfrac{840 + 595}{2} = 717.5$ microns.

It means, the particles which pass through 20 mesh and retain on 28 mesh are having the mean size of 717.5 microns. Similarly the mean sizes of each fraction are to be calculated. Table 4.2.2 shows all values.

Table 4.2.2 Calculated values for particle size distribution.

Mesh number	Retained mesh size microns	Mean size d_i microns	Weight gm	wt % retained w_i	Cum wt % retained	Cum wt% passing W
						100.0
+20	840		7.7	3.5	3.5	96.5
−20 + 28	595	717.5	13.2	6.0	9.5	90.5
−28 + 35	420	507.5	29.7	13.5	23.0	77.0
−35 + 48	297	358.5	40.7	18.5	41.5	58.5
−48 + 65	210	253.5	28.6	13.0	54.5	45.5
−65 + 100	149	179.5	22.0	10.0	64.5	35.5
−100 + 150	105	127.0	16.5	7.5	72.0	28.0
−150 + 200	74	89.5	8.8	4.0	76.0	24.0
−200		37.5	52.8	24.0	100.0	
			220.0	100.0		

Average size of the material is determined by using the following simple arithmetic formula

$$\therefore \quad \text{Average size} = \frac{100}{\Sigma \dfrac{w_i}{d_i}} \tag{4.2.1}$$

where w is the weight percent of the material retained by the sieve
d is the mean size of the material retained by the same sieve

$$\text{Specific surface of the sample of particles} = A = \frac{6\lambda_S}{\rho_p} \Sigma \frac{w_i}{d_i} \tag{4.2.2}$$

where w is the weight fraction of the material of size 'd'.

Calculation of average size and specific surface is illustrated in Example 4.2.1.

Example 4.2.1: *Sieve analysis test data of a sample of sand of specific gravity 2.65 is shown in Table 4.2.1.1. Calculate the average size and specific surface of the sample.*

Table 4.2.1.1 Sieve analysis test data of a sample for example 4.2.1.

Mesh number	Mesh size microns	Direct wt% retained
+18	853	7.0
−18 + 25	599	10.4
−25 + 36	422	14.2
−36 + 52	295	13.6
−52 + 72	211	9.2
−72 + 100	152	8.1
−100 + 150	104	8.2
−150 + 200	74	5.1
−200		24.2

Solution:

Average sizes for each fraction (d_i) and (w_i/d_i) are calculated and tabulated in Table 4.2.1.2.

Table 4.2.1.2 Calculated values for example 4.2.1.

Size of each fraction (d_i)	d_i (cm)	Wt % (w_i)	w_i/d_i
853.0	0.0853	7.0	82.06
$(853 + 599)/2 = 726.0$	0.0726	10.4	143.25
$(599 + 422)/2 = 510.5$	0.0511	14.2	277.89
$(422 + 295)/2 = 358.5$	0.0359	13.6	378.83
$(295 + 211)/2 = 253.0$	0.0253	9.2	363.64
$(211 + 152)/2 = 181.5$	0.0182	8.1	445.05
$(152 + 104)/2 = 128.0$	0.0128	8.2	640.63
$(104 + 74)/2 = 89.0$	0.0089	5.1	573.03
$(74 + 0)/2 = 37.0$	0.0037	24.2	6540.54
		100.0	9444.92

$$\therefore \quad \text{Average size of the sample} = \frac{100}{\sum \dfrac{w_i}{d_i}} = \frac{100}{9444.92} = 0.0106 \, \text{cm} = 106 \, \mu\text{m}$$

Assuming spherical particles $\lambda_S = 1$,

$$\text{Specific surface} = \frac{6\lambda_S}{\rho_p} \sum \frac{w_i}{d_i} = \frac{6 \times 1}{2.65} \times \frac{9444.92}{100} = 213.9 \, \text{cm}^2/\text{gm}$$

4.3 PRESENTATION OF PARTICLE SIZE DISTRIBUTION DATA

Particle size distribution refers to the manner in which particles are quantitatively distributed among various sizes in a sample of material; in other words a statistical relation between quantity and size. This data is best presented for use in the form of graphs. The simplest method is to plot a histogram of the weight percent of the material in the size interval against the size interval. When the size intervals are small enough, the histogram can be presented as a continuous curve taking the middle points of histogram. In other words, a graph is plotted between the weight percent of the material as ordinate and the arithmetic mean size as abscissa. It is called **linear scale frequency plot**. It gives the quantitative picture of the relative distribution of the material over the entire size range. In this plot, the size scale at the fine end is compressed and the data points are congested because the size intervals are in geometric series. Hence a logarithmic scale is used for abscissa of mean sizes. As the distances separating the points corresponding to consecutive mean sizes are equal, the data points at the finer end will spread. This plot is called **semi-log frequency plot**. In this plot, the area under the curve is in all cases equal to unity or 100 percent. This is often useful.

In studies of size distribution in crushed products, often a graph known as **log-log frequency plot** is obtained by plotting logarithm of the weight percent of the material against the logarithm of the mean size. All the three graphs for the tabulated data of Table 4.2.2 are shown in Fig. 4.3.1.1.

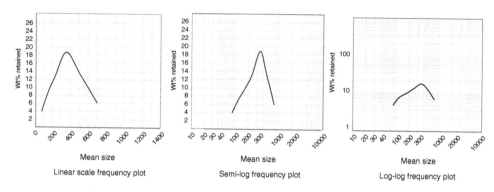

Figure 4.3.1.1 Graphical presentation of data tabulated in table 4.2.2.

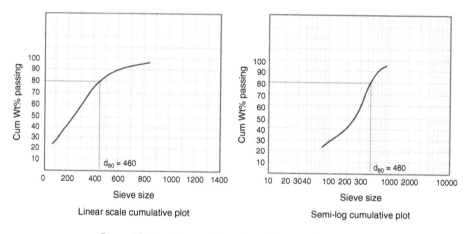

Figure 4.3.1.2 Linear scale and semi-log cumulative plots.

In many cases, the data is more commonly plotted as cumulative wt% passing versus actual size of opening. It is called **linear scale cumulative plot**. d_{80}, the size of the sieve at which 80% of the particles pass through (simply called 80% passing size) may be found from this plot. Crowding of data points at finer end is not a drawback in this plot since plotting of all points is not necessary. However, logarithmic scale of abscissa for actual size of opening is recommended in which case it is called **semi-log cumulative plot**. Although this plot appears to be a satisfactory presentation of the data, it is not widely used because the S shape of the curve is difficult to express mathematically. These two plots are shown in Fig. 4.3.1.2. wherein 80% passing size is also determined from the graph as 460 microns.

> **80% passing size (D80)** is the size of the sieve at which 80% of the particles pass through that sieve. 80% passing size can be determined from the plot of cumulative weight percent passing versus sieve size.

F80 is the 80% passing size of the feed material.
P80 is the 80% passing size of the product material.

80% passing size is used in all calculations to determine energy requirements for reducing the size of the particles by comminution equipment. In example 4.3.1, frequency and cumulative plots are drawn and 80% passing size is determined.

Example 4.3.1:   *For the sieve analysis test data of a sample shown in Table 4.3.1.3, draw linear scale frequency plot, semi-log frequency plot, log-log frequency plot, linear scale cumulative plot and semi-log cumulative plot. Determine 80% passing size.*

Table 4.3.1.3 Sieve analysis test data of a sample for example 4.3.1.

Mesh number	Retained mesh size in microns	Weight of material gm
+14	1200	2.5
−14 + 22	710	18.0
−22 + 30	500	18.5
−30 + 44	355	21.0
−44 + 60	250	27.5
−60 + 72	210	36.0
−72 + 100	150	31.5
−100 + 150	105	26.0
−150 + 200	74	18.5
−200		50.5
		250.0

Solution:

All necessary calculations for drawing frequency and cumulative plots are done and shown in Table 4.3.1.4.

Table 4.3.1.4 Calculated values for example 4.3.1.

Mesh number	Retained mesh size microns	Mean size microns	Weight gm	wt% retained	Cum wt% retained	Cum wt% passing W
						100.0
+14	1200	1200	02.5	1.0	1.0	99.0
−14 + 22	710	955	18.0	7.2	8.2	91.8
−22 + 30	500	605	18.5	7.4	15.6	84.4
−30 + 44	355	427.5	21.0	8.4	24.0	76.0
−44 + 60	250	302.5	27.5	11.0	35.0	65.0
−60 + 72	210	230	36.0	14.4	49.4	50.6
−72 + 100	150	180	31.5	12.6	62.0	38.0
−100 + 150	105	127.5	26.0	10.4	72.4	27.6
−150 + 200	74	89.5	18.5	7.4	79.8	20.2
−200		37	50.5	20.2	100.0	
			250.0	100.0		

Linear scale frequency plot, semi-log frequency plot and log-log frequency plot are drawn with the values of Table 4.3.1.4 as shown in Fig. 4.3.1.3.

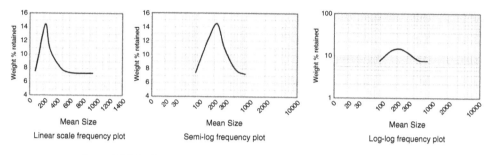

Linear scale frequency plot   Semi-log frequency plot   Log-log frequency plot

Figure 4.3.1.3 Graphical presentation of data tabulated in table 4.3.1.4.

Linear scale cumulative plot and semi-log cumulative plot are drawn with the values of Table 4.3.1.4 as shown in Fig. 4.3.1.4.

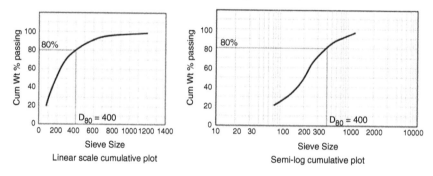

Linear scale cumulative plot   Semi-log cumulative plot

Figure 4.3.1.4 Linear scale and semi-log cumulative plots for example 4.3.1.

80% passing size is determined from these two plots and it is 400 microns.

Quantities of particles of any size outside the range of sieves taken can also be determined by the extrapolation of the curve obtained. Example 4.3.2 shows the determination of fine size particles from the sieve analysis of the sample of material.

Example 4.3.2:   *From the sieve analysis test data given in Table 4.3.2.1, determine percentage of $-75 + 37\,\mu m$ size particles present in the sample.*

Table 4.3.2.1 Sieve analysis test data for example 4.3.2.

Size Microns	Weight gm
+425	17
−425 + 300	31
−300 + 212	24
−212 + 150	19
−150 + 106	15
−106 + 75	12
−75	42

Solution:

Wt% and Cum. Wt% passing are calculated and shown in Table 4.3.2.2.

Table 4.3.2.2 Calculated values for example 4.3.2.

Size range Microns	Sieve Size microns	Weight gm	Wt%	Cumulative wt% passing
+ 425		17	10.6	100.0
−425 + 300	425	31	19.4	89.4
−300 + 212	300	24	15.0	70.0
−212 + 150	212	19	11.9	55.0
−150 + 106	150	15	09.4	43.1
−106 + 75	106	12	07.5	33.7
−75	75	42	26.2	26.2
		160	100.0	

Graph between cum. wt% passing and sieve size are drawn as shown in Fig. 4.3.2.1.

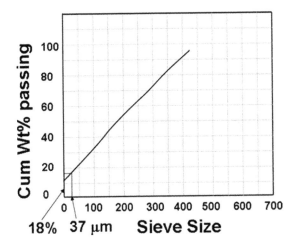

Figure 4.3.2.1 Linear scale cumulative graph for example 4.3.2.

From the graph Cumulative wt% passing corresponding to 37 microns is 18% obtained by extrapolation.

Cumulative wt% passing corresponding to 75 microns is 26.2% from the table.

$$\therefore \text{Wt\% of } -75 + 37 \text{ micron particles} = 26.2 - 18.0 = 8.2\%$$

Extrapolation of the curve is difficult many times and can not give good results.

4.4 PARTICLE SIZE DISTRIBUTION EQUATIONS

The particle size distribution equation defines quantitatively how the different size particles are distributed among the particles in the entire material. In practical applications, the entire particles are classified into classes each one of which is defined by its upper and lower boundary. A representative size is associated with each particle size class and it is assumed that all particles in the class will behave in processing systems as if it had a size equal to the representative size.

The weight fraction or percent W is measured experimentally at a number of fixed sizes that correspond to the mesh sizes of the set of sieves. This data is usually presented in tabular form as in Table 4.2.1. Graphical presentations are useful and are often preferred because it is generally easier to assess and compare particle size distributions when the entire distribution is visible. A variety of different graphical coordinate systems have become popular with a view to make the particle size distribution plot as or close to a straight line.

A number of empirical particle size distribution equations as given in Table 4.4.1 have been found to represent the particle size distribution of comminution products quite accurately in practice and these are useful in a number of situations.

Of the empirical particle size distribution equations as given in Table 4.4.1, Gates-Gaudin-Schuhmann and Rosin-Rammler distributions are most common.

Gates-Gaudin-Schuhmann distribution equation [4] is represented in mathematical form as follows:

$$W = 100 \left(\frac{d}{d_{100}} \right)^m \tag{4.4.1}$$

where $W =$ the cumulative weight percent passing
$d =$ particle size in microns

Table 4.4.1 Empirical particle size distribution equations.

Name	$W =$ Cumulative weight fraction passing d	Meaning of d*
Gates-Gaudin-Schuhmann	$\left(\dfrac{d}{d^*} \right)^m$	Maximum particle size
Rosin-Rammler or Weibull	$1 - e^{-\left(\frac{d}{d^*} \right)^n}$	Size at which 63.2% of material pass through
Broadbent-Calcott	$\dfrac{1 - e^{-\frac{d}{d^*}}}{1 - e^{-1}}$	Maximum particle size
Gaudin-Meloy	$1 - \left(1 - \dfrac{d}{d^*} \right)^n$	Maximum particle size
Log-Probability	$\text{erf} \left(\dfrac{\ln \frac{d}{d^*}}{\sigma} \right)$ erf – error function, σ – geometric standard deviation	Median particle size
Harris	$1 - \left[1 - \left(\dfrac{d}{d^*} \right)^s \right]^n$	Maximum particle size

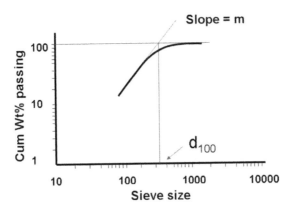

Figure 4.4.1.1 Gates-Gaudin-Schuhmann plot.

d_{100} = Gaudin-Schuhmann modulus or Size modulus or Maximum theoretical size or 100% passing size

m = distribution modulus (slope of the line)

Size distribution analyses of crushed and ground products are commonly plotted on log-log paper with W the cumulative wt% passing as ordinate and the particle diameter d in microns as abscissa. This plot, frequently referred as **Schuhmann, Gaudin-Schuhmann,** or **Gates-Gaudin-Schuhmann plot,** is shown in Fig. 4.4.1.1 and is convenient and most commonly used in the mineral industry to describe the size distributions and is represented in mathematical form as Gates-Gaudin-Schuhmann distribution.

Such plots as shown in Fig. 4.4.1.1 usually show a fairly straight line for the finer particle size range which begins to curve in the coarser sizes and often approaches tangency with the 100 percent passing line at the top of the plot. The slope of the straight line is the value of m. When the straight lower portion of the plotted line is extended at its slope m, it intercepts the 100 percent passing line at d_{100} microns. Interpolation is much easier from a straight line. Extrapolation of straight line, gives the size distribution data at finer sizes down to 5 microns. The burden of routine analysis can be greatly reduced as relatively few sieves will be needed to check essential features of the size distribution.

This Gates-Gaudin-Schuhmann distribution relationship is applicable for homogeneous brittle materials which have no natural weaknesses, particularly cleavage planes, cracks, and weak grain boundaries. It can be applied to as much as 80% of many materials with useful accuracy.

Rosin-Rammler distribution equation [5] (developed in 1933) is often used for representing the size distribution of broken coal and relatively fine crushed coal. Such products are found to obey the following relationship called Rosin-Rammler distribution.

$$W = 100 - 100e^{-\left(\frac{d}{d*}\right)^{n}} \qquad (4.4.2)$$

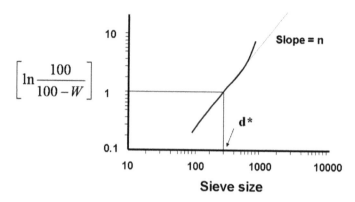

Figure 4.4.1.2 Rosin Rammler plot.

If $d = d^*$, $W = 100 - \dfrac{100}{e} = 100 - \dfrac{100}{2.71828} = 100 - 36.788 = 63.212$

Therefore d^* is the size at which 63.212% of material pass through.
Rosin-Rammler distribution equation can be rewritten as

$$e^{-(\frac{d}{d^*})^n} = \frac{100 - W}{100}$$

$$\Rightarrow \quad e^{(\frac{d}{d^*})^n} = \frac{100}{100 - W}$$

$$\Rightarrow \quad \ln \frac{100}{100 - W} = \left(\frac{d}{d^*}\right)^n$$

Taking log on both sides

$$\log\left[\ln \frac{100}{100 - W}\right] = n \log\left(\frac{d}{d^*}\right)$$

A plot of $\ln \dfrac{100}{100 - W}$ versus 'd' on log-log axes is to be drawn and it is shown in Fig. 4.4.1.2.

The graph is a fairly straight line for the finer particle size range and begins to curve upwards at the coarser end as shown in Fig. 4.4.1.2. The slope of the straight line is the value of the 'n'.

As the d^* is the size at which 63.212% of material passes,

$$\ln \frac{100}{100 - W} = \ln \frac{100}{100 - 63.212} = 1.0$$

The value of 'd' corresponding to $\ln \dfrac{100}{100 - W} = 1.0$ is to be read from the straight line which is the value of d^*.

The Gates-Gaudin-Schuhmann method is often preferred in mineral processing applications and Rosin-Rammler method is more often used in coal-preparation studies. The method chosen for application depends on how accurately the data can be fitted in a straight line. For a jaw crusher product, these two methods are applied, drawn the plots and obtained the equations in example 4.4.1.

Example 4.4.1: For the *screen analysis of a jaw crusher product given in Table 4.4.1.3, obtain Gates-Gaudin-Schuhmann equation and Rosin-Rammler equation.*

Table 4.4.1.3 Screen analysis of a jaw crusher product.

Mesh Number	Mesh Size microns	Weight% retained
2	9420	1.90
3	6730	15.58
4	4760	18.98
5	3360	14.88
8	2380	9.30
10	1680	7.22
14	1190	5.80
20	840	5.75
28	595	4.45
35	420	3.90
48	297	3.00
65	210	2.10
−65		7.14

Solution:

Required values are calculated and shown in Table 4.4.1.4.

Table 4.4.1.4 Calculated values for example 4.4.1.

Mesh Number	Mesh Size microns	Weight% retained	Cum. Wt% passing W	100 − W	$\frac{100}{100-W}$	$\ln\frac{100}{100-W}$
2	9420	1.90	98.10	1.90	52.632	3.963
3	6730	15.58	82.52	17.48	5.721	1.744
4	4760	18.98	63.54	36.46	2.743	1.009
5	3360	14.88	48.66	51.34	1.948	0.667
8	2380	9.30	39.36	60.64	1.649	0.500
10	1680	7.22	32.14	67.86	1.474	0.388
14	1190	5.80	26.34	73.66	1.358	0.306
20	840	5.75	20.59	79.41	1.259	0.230
28	595	4.45	16.14	83.86	1.192	0.176
35	420	3.90	12.24	87.76	1.139	0.130
48	297	3.00	9.24	90.76	1.102	0.097
65	210	2.10	7.14	92.86	1.077	0.074
−65		7.14				
		100.00				

Gates-Gaudin-Schuhmann Plot is drawn as in Fig. 4.4.1.3.

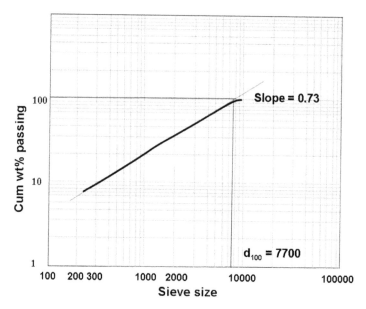

Figure 4.4.1.3 Gates-Gaudin-Schuhmann plot for example 4.4.1.

From the Gates-Gaudin-Schuhmann plot

$$\text{Slope} = m = 0.73 \quad \text{and} \quad d_{100} = 7700$$

Gates-Gaudin-Schuhmann equation is

$$W = 100 \left(\frac{d}{7700} \right)^{0.73}$$

Rosin-Rammler Plot is drawn as in Fig. 4.4.1.4.

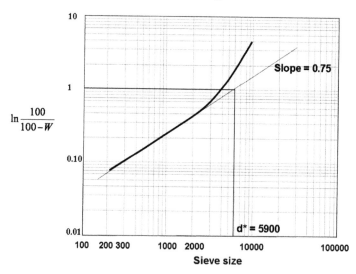

Figure 4.4.1.4 Rosin-Rammler plot for example 4.4.1.

From the Rosin-Rammler plot

$$\text{Slope} = n = 0.75 \quad \text{and} \quad d^* = 5900$$

Rosin-Rammler equation is

$$W = 100 - 100e^{-\left(\frac{d}{5900}\right)^{0.75}}$$

Example 4.4.2: *Obtain Gates-Gaudin-Schuhmann equation for the sieve analysis data given in Table 4.3.2.1 and determine percentage of $-75 + 37\,\mu m$ size particles present in the sample. Compare with the result obtained in example 4.3.2.*

Solution:

For the calculated values of example 4.3.2 as shown in Table 4.3.2.2, **Schuhmann plot is** drawn and is shown in Fig. 4.4.2.1.

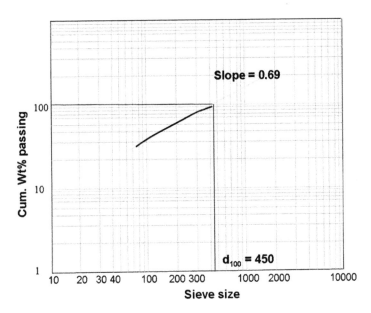

Figure 4.4.2.1 Gates-Gaudin-Schuhmann plot for example 4.4.2.

From the graph, slope $= 0.69$ and $d_{100} = 450$ microns.

Gates-Gaudin-Schumann equation is $W = 100\left(\frac{d}{450}\right)^{0.69}$

When $d = 37$ microns, is substituted in Schuhmann equation, $W = 17.9\%$

Cumulative wt% passing corresponding to 75 microns is 26.2%

$\therefore$ Wt% of $-75 + 37$ micron particles $= 26.2 - 17.9 = 8.3\%$

This value is almost same to that of the value obtained from graph in example 4.3.2.

4.5   SIZE ASSAY ANALYSIS

The assay value of the sample of material is determined by chemical analysis. Similarly, assay value of each size fraction of the material is also determined. This is usually called size assay analysis. Size assay analysis of an ore and coal and their products is of prime importance through which much information can be obtained. For example, distribution of the values among various sizes will be known by size assay analysis. Size assay analysis with or without certain physical properties of particles in each size fraction such as specific gravity is the basis for selection of the processes to be employed to obtain required product particles. It also gives the average assay value of the total material through which price of the material is decided based on the market value of the required product.

Size assay analysis of ROM Iron ore of -150 mm size is given in Table 4.5.1. Size assay analysis also includes percentages of silica and alumina in addition to iron as a minimum ratio of silica to alumina is required in the ore for extraction of iron.

From this data, assay of the sample, called head sample or simply head, can be calculated. Table 4.5.2 shows calculated values for determining head assay. Iron distribution in terms of percent of total iron among different sizes is also shown in the last column.

From Table 4.5.2, Head assay $= 6341.957/100 = 63.42\%$ Fe.

Size assay analysis of Table 4.5.1 shows percent iron decreases with decrease in size of the particles except in -10 mm $+ 100$ μm. Calculated iron distribution values in Table 4.5.2 reveals that 37.04% iron is distributed in -10 mm $+ 100$ μm size fraction and 20.21% is distributed in -100 μm size fraction.

Table 4.5.1   Size assay analysis of ROM Iron ore sample.

Size	wt%	Fe%	SiO$_2$%	Al$_2$O$_3$%
-150 mm $+ 50$ mm	14.5	65.81	1.13	0.68
-50 mm $+ 40$ mm	10.1	64.62	1.27	0.71
-40 mm $+ 30$ mm	7.6	63.90	1.81	0.84
-30 mm $+ 20$ mm	5.8	63.27	2.15	1.09
-20 mm $+ 10$ mm	4.0	62.84	2.68	1.28
-10 mm $+ 100$ μm	37.2	63.15	3.84	1.82
-100 μm	20.8	61.63	4.23	2.16
	100.0			

Table 4.5.2 Calculated values for Table 4.5.1.

Size A	wt% B	Fe % C	Fe D = (B × C)	% Fe distribution (D/6341.957)100
−150 mm + 50 mm	14.5	65.81	954.245	15.05
−50 mm + 40 mm	10.1	64.62	652.662	10.29
−40 mm + 30 mm	7.6	63.90	485.640	7.66
−30 mm + 20 mm	5.8	63.27	366.966	5.79
−20 mm + 10 mm	4.0	62.84	251.360	3.96
−10 mm + 100 μm	37.2	63.15	2349.180	37.04
−100 μm	20.8	61.63	1281.904	20.21
	100.0		6341.957	100.00

Table 4.5.3 Size assay analyses of three products.

−150 + 40 mm (22.5%)			−40 mm (77.5%)			−40 + 10 mm (12.3%)		
Size mm	Wt%	%Fe	Size mm	Wt%	%Fe	Size mm	Wt%	%Fe
−150 + 60	24.23	66.75	−40 + 30	8.45	65.23	−40 + 30	6.84	66.8
−60 + 50	39.84	65.84	−30 + 20	5.76	64.18	−30 + 20	34.19	65.1
−50 + 40	30.55	66.28	−20 + 10	21.75	63.36	−20 + 10	49.58	63.5
−40	5.38	64.59	−10	64.04	62.08	−10	9.39	63.1
	100.00	**66.19**		100.00			100.00	**64.2**

This ROM Iron ore is screened on 40 mm aperture primary screen. 22.5% of −150 + 40 mm fraction is obtained. 77.5% of −40 mm fraction from primary screen is fed to secondary screen with 10 mm aperture which yields 12.3% of −40 + 10 mm product. The size assay analysis of three products are given in Table 4.5.3.

In above table weight percentages shown in 2nd, 5th, and 8th columns are with reference to respective fractions.

%Fe in −150 + 40 mm screened fraction

$$= \frac{24.33 \times 66.75 + 39.84 \times 65.84 + 30.55 \times 66.28 + 5.38 \times 64.59}{100} = 66.19\%\,Fe$$

%Fe in −40 + 10 mm screened fraction

$$= \frac{6.84 \times 66.8 + 34.19 \times 65.1 + 49.58 \times 63.5 + 9.39 \times 63.1}{100} = 64.2\%\,Fe$$

Wt% of −150 + 10 mm fraction = 22.5% + 9.5% (12.3% of 77.5%) = 32%

$$\text{Average }\%Fe \text{ in } -150 + 10 \text{ mm fraction} = \frac{22.5 \times 66.19 + 9.5 \times 64.2}{22.5 + 9.5} = 65.6\%\,Fe$$

It is to be noted that −40 mm material in −150 + 40 mm screened fraction is 1.21% (5.38% of 22.5%). This material may contain very little quantity of −10 mm material

Table 4.5.4   Size assay analysis of ground copper ore.

Size range (μm)	Wt%	Assay (%Cu)
+500	2.0	0.05
−500 + 355	5.5	0.08
−355 + 250	10.5	0.35
−250 + 180	12.0	0.80
−180 + 125	15.5	1.20
−125 + 90	20.5	1.55
−90 + 63	22.0	1.80
−63	12.0	2.89

Table 4.5.5   Copper distribution in ground copper ore of Table 4.5.4.

Size range (μm) A	Wt% B	Assay (%Cu) C	Cu D = B × C	% Cu distribution (D/138.47)100
+500	2.0	0.05	00.100	00.07
−500 + 355	5.5	0.08	00.440	00.32
−355 + 250	10.5	0.35	03.675	02.65
−250 + 180	12.0	0.80	09.600	06.93
−180 + 125	15.5	1.20	18.600	13.43
−125 + 90	20.5	1.55	31.775	22.95
−90 + 63	22.0	1.80	39.600	28.60
−63	12.0	2.89	34.680	25.05
	100.0		138.470	100.00

which cannot be avoided from −150 + 10 mm fraction. Similarly, 0.89% (9.39% of 9.5%) of −10 mm material also included in −150 + 10 mm fraction.

This −150 + 10 mm fraction can directly send to the metallurgical plant for extraction of iron. The remaining 68% ore of −10 mm will not be suitable for extraction of iron with reference to its size. Hence it needs to be grounded to fine to prepare the pellets of +10 mm size to make suitable for extraction of iron.

Another example of size assay analysis of ground copper ore sample is given in Table 4.5.4.

Copper distribution among various sizes of the ground copper ore sample is calculated and shown in Table 4.5.5.

Head grade of the sample (%Cu in sample) = 138.47/100 = 1.38% Cu

Percent copper increases with decrease in size of the particles.

Calculated copper distribution values in Table 4.5.5 reveals that 90.03% copper is distributed in fine size fraction of −180 μm and only 9.97% copper is distributed in coarse size fraction of +180 μm.

An example of size assay analysis of ROM coking coal is shown in Table 4.5.6. Here, ash% of coal which is unwanted or waste is determined instead of valuable or wanted part. It is a size wise ash analysis. The coal, if required to use for any purpose, must contain as less ash% as possible. Coal of high ash% is inferior in quality and unsuitable for any use.

As in case of iron and copper ores considered previously, ash% of the sample of coal can be calculated. Table 4.5.7 shows the calculated values.

Table 4.5.6 Size wise ash analysis of ROM coking coal.

Size	Wt%	Ash%
+20 mm	4.09	41.35
−20 + 13 mm	10.35	37.49
−13 + 10 mm	6.01	34.83
−10 + 6 mm	17.29	33.28
−6 + 3 mm	14.53	29.74
−3 + 1 mm	17.29	27.99
−1 + 0.5 mm	6.49	25.06
−0.5 mm	23.95	23.98
	100.00	

Table 4.5.7 Calculated values for Table 4.5.6.

Size A	Wt% B	Ash% C	Ash product $D = B \times C$	Ash distribution $E = D/2986.91$
+20 mm	4.09	41.35	161.12	5.39
−20 + 13 mm	10.35	37.49	388.02	12.99
−13 + 10 mm	6.01	34.83	209.33	7.01
−10 + 6 mm	17.29	33.28	575.41	19.26
−6 + 3 mm	14.53	29.74	432.12	14.47
−3 + 1 mm	17.29	27.99	483.95	16.20
−1 + 0.5 mm	6.49	25.06	162.64	5.45
−0.5 mm	23.95	23.98	574.32	19.23
	100.00	29.87	2986.91	100.00

Column 4 of Table 4.5.7 is called ash product which is the ash content in respective size fractions of the coal if the values are divided by 100. Column 5 is the ash distribution in respective size fractions.

Ash % of the ROM coal sample $= 2986.91/100 = 29.87\%$

Example 4.5.1 *Size assay analysis of ground chrome ore sample is given in Table 4.5.1.1.*

Table 4.5.1.1 Size assay analysis of ground chrome ore.

Size fractions (μm)	Wt%	Assay %Cr_2O_3
−1000 + 850	2.5	19.8
−850 + 600	5.9	18.2
−600 + 425	11.6	19.7
−425 + 300	16.5	22.9
−300 + 212	16.0	21.5
−212 + 150	17.1	22.2
−150 + 75	19.4	21.3
−75 + 53	5.7	16.0
−53	5.3	15.6

Calculate the head assay of the sample and distribution of Cr_2O_3 in each size fractions.

Solution:

All necessary calculations are done and shown in Table 4.5.1.2.

Table 4.5.1.2 Calculated values for Table 4.5.1.1.

Size fractions (μm) A	Wt% B	Assay %Cr_2O_3 C	Cr_2O_3 $D = B \times C$	% Cr_2O_3 distribution $(D/2073.97)100$
$-1000 + 850$	2.5	19.8	49.50	2.38
$-850 + 600$	5.9	18.2	107.38	5.18
$-600 + 425$	11.6	19.7	228.52	11.02
$-425 + 300$	16.5	22.9	377.85	18.23
$-300 + 212$	16.0	21.5	344.00	16.59
$-212 + 150$	17.1	22.2	379.62	18.31
$-150 + 75$	19.4	21.3	413.22	19.93
$-75 + 53$	5.7	16.0	91.20	4.40
-53	5.3	15.6	82.68	3.99
	100.0		2073.97	

Head assay of the sample $(\% Cr_2O_3 \text{ in sample}) = 2073.97/100$
$$= 20.74\% \ Cr_2O_3$$

The distribution of Cr_2O_3 is calculated and shown in last column.

4.6   PROBLEMS FOR PRACTICE

4.6.1: *From the sieve analysis data of a sample shown in Table 4.6.1.1, determine 80% passing size.*

Table 4.6.1.1 Sieve analysis data for problem 4.6.1.

Mesh number	Mesh size microns	Direct wt% retained
$+18$	853	7.0
$-18 + 25$	599	10.4
$-25 + 36$	422	14.2
$-36 + 52$	295	13.6
$-52 + 72$	211	9.2
$-72 + 100$	152	8.1
$-100 + 150$	104	8.2
$-150 + 200$	74	5.1
-200		24.2

[580 μm]

4.6.2: *A sample of pyrite particles (Sp.gr. = 5) was screened. The screen analysis is given in Table 4.6.2.1. Calculate the mean surface diameter and specific surface.*

Table 4.6.2.1 Screen analysis for problem 4.6.2.

Mesh	% Retained	Size (cm)
8/10	0	1.651
10/14	21.2	1.168
14/20	19.6	0.833
20/28	17.4	0.589
28/35	14.0	0.417
35/48	15.8	0.295
48/65	12.0	0.208

[0.4629 mm; 2.593 cm²/gm]

4.6.3:   *For the size analysis of a screen underflow shown in Table 4.6.3.1, determine Gates-Gaudin-Schuhmann equation and Rosin-Rammler equation.*

Table 4.6.3.1 Size analysis of a screen underflow for problem 4.6.3.

Mesh Number	Mesh Size microns	Weight% retained
20	840	3.6
28	595	6.0
35	420	8.3
48	297	8.7
65	210	8.1
100	149	8.4
150	105	7.6
200	74	8.1
−200		41.2

$$[W = 100\left(\frac{d}{760}\right)^{0.35}; \; W = 100 - 100e^{-\left(\frac{d}{205}\right)^{0.63}}]$$

4.6.4:   *Size and assay analysis of ground lead ore sample is given in Table 4.6.4.1. Calculate the grade of the sample and distribution of lead among various sizes.*

Table 4.6.4.1 Size assay analysis of ground lead ore.

Mesh (Tyler)	Weight%	%Pb
+65	20.0	2.34
−65 + 100	14.0	5.34
−100 + 150	32.0	6.10
−150	34.0	5.06

[4.89%]

Chapter 5

Screening

The method of separating the particles according to their sizes is called screening. In Industrial screening, the particles of various sizes are fed to the screen surface. The material passing through the screen aperture is called **undersize** or **fines** or **underflow** while the material retained on the screen surface is called **oversize** or **coarse** or **overflow**.

5.1 PURPOSE OF SCREENING

Industrial screening is used

1. To remove oversize material before it is sent to next unit operation as in closed circuit crushing operations.
2. To remove undersize material before it is sent to next unit operation which is set to treat material larger than this size.
3. To grade materials into a specific series of sized (finished) products.
4. To prepare a closely sized (the upper and lower size limits are very close to each other) feed to any other unit operation.

Screening is generally used for dry treatment of coarse material. Dry screening can be done down to 10 mesh with reasonable efficiency. Wet screening is usually applied to materials from 10 mesh down to 30 mesh (0.5 mm) but recent developments of the Sieve Bend Screen have made possible wet screening at the 50 micron size.

5.2 SCREEN

Screen is a machine used to separate particles by size. It consists of a main frame, generally rectangular in shape, in which the screen surface that contains apertures for the passage of undersize particles is fixed. A support frame is installed directly below the screen surface to provide necessary strength. The main frame is fixed on to the stationary base frame or structure. All other mechanical parts necessary to give required motion to the screen is attached to the main frame. The three types of screen surfaces in common use today with simple description and applications are given in Table 5.2.1. Screens are classified as stationary and dynamic screens as shown below:

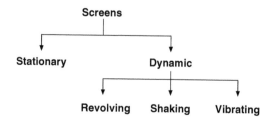

Principal types of industrial screens with simple description and applications are given in Table 5.2.2.

Table 5.2.1 Types of screen surfaces.

Type of screen surface	Description	Applications
Parallel rods or Profile bars	Rod/bar **Cross sections** Circular, Triangular, Wedge etc.	used for lumpy and coarser size particles
	Openings Circular, In-line and Staggered openings	used for coarser and small sizes
	Square, In-line and Staggered openings	
Punched or perforated plates	Slot-like Staggered openings	slotted openings are sometimes used for fine particles
	Openings Square	used for fairly coarse particles
	Rectangle	used for fine particles
Woven wires	Triple shute elongated	used for fine particles

Table 5.2.2 Principal types of industrial screens.

Type	Screen	Description	Applications
Stationary	Grizzly	Equally spaced parallel rods or bars running in flow direction	Lumpy or coarse
			Scalping before crushing
		Sloped to allow gravity transport	Dry separation
	Divergator	Parallel rods running in flow direction	400 to 25 mm
		Fixed at one end Gap increases from fixed to free end	Self cleaning and blockage free
		Alternate rods diverge at 5°–6°	Dry separation
	Sieve Bend	Curved slotted screen of wedge wire Feed slurry enters tangentially Imparts centrifugal action	2 mm to 45 μm wet separation
Revolving	Trommel	Rotating, punched or woven cylindrical shell	10 to 60 mm if dry finer if wet; used also for scrubbing
	Roller screen	Series of parallel driven rolls or discs	3–300 mm
Vibrating	Vibrating with different amplitudes	Vibrating: High speed motion Shaking: Slow linear motion	200 mm to 250 μm dry if coarse wet if fine
	Vibrating grizzly	Similar to stationary grizzly; and vibrates	Coarse Dry separation
Other types Reciprocating screen Gyratory screen Rotating probability screen Resonance screen Mogensen sizer		Similar to vibrating screens but differ in the type of motion given to the screen deck Used for varied purposes	

5.3 SCREEN ACTION

Consider a simplified screen having all apertures are of the same size as shown in Fig. 5.3.

The material is fed at one end of the screen. Screening is affected by continuously presenting the material to the screen surface which provides a relative motion with respect to the feed. The screen surface can be fixed or moveable. Agitation of the bed of material must be sufficient to expose all particles to the screen apertures several times during the travel of the material from feed end to the discharge end of the screen.

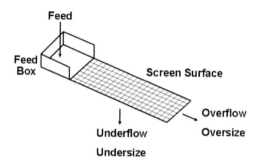

Figure 5.3 Simplified screen.

At the same time the screen must act as a transporter for moving retained particles from the feed end to the discharge end.

The industrial screens are arranged as single-deck and multi-deck screens. The screen having one screening surface is called Single-deck screen and if a screen has two or more screen surfaces, it is called Multi-deck screen.

Screening is performed either dry or wet. Wet screening is superior, adhering fines are easily washed off, and avoids the dust problem. Wet screens tend to have lower capacities than dry screens perhaps as much as 50% less. The cost of dewatering and drying of the products from wet screening becomes more.

The following are the terms used for screens in mineral industry according to their purpose:

1   Sizing screen: used to separate the particles according to their sizes
2   Feed screen: used to prepare the feed to any unit operation
3   Trash screen: used to remove the trash material
4   Scalping screen: used to remove small amounts of either oversizes or undersizes
5   Dewatering screen: used to remove water from mixture of solids and water
6   Desliming screen: used to remove slimes from the coarse material
7   Dedusting screen: used to remove extremely fine particles from dry material
8   Medium recovery screen: used to remove medium solids from coarse material

5.3.1   Factors affecting the rate of screening

A number of factors determine the rate at which particles pass through a screen surface and they can be divided into two groups: those related to particle properties and those dependent on the machine and its operation. Some important factors are:

Material factors

1   Bulk density of the material
2   Size and size distribution of the particles
3   Size of the particle relative to the aperture
4   Shape of the particle
5   Presence of clay material
6   Moisture content of the material

Machine factors

1   Size of the aperture
2   Shape of the aperture
3   Size of the screen surface
4   Percent opening area
5   Amount of near size material in the feed
6   Method of feeding the particles
7   Angle of incidence of the particle on the screen surface
8   Rate of feed – rate of flow over the screen
9   Speed at which the particle approaches the screen surface
10 Thickness of the material on the screen surface
11 Blinding of the screen surface
12 Type of screening, i.e., wet or dry screening
13 Type of motion given to the screen surface
14 Slope of the screen deck
15 Mechanical design for supporting and tightening the screen deck

5.4 MATERIAL BALANCE

Material balance is a method of balancing the material over a screen to analyze the process of screening. When a material containing different sizes of particles are fed to a screen, the screen separates the material into two fractions as overflow and underflow. The material balance equation for total material is

Rate of material fed to the screen (F)

$=$ Rate of material discharged as overflow (P)

$+$ Rate of material discharged as underflow (U)

$$\Rightarrow \qquad F = P + U \qquad\qquad (5.4.1)$$

The feed, overflow and underflow contain different sizes of particles. Let f, p and u be the fraction (or percent) of particular size particles in feed, overflow and underflow respectively. The material balance equation for the considered size particles is

Rate of considered size particles fed to the screen (F)

$=$ Rate of considered size particles discharged with overflow (P)

$+$ Rate of considered size particles discharged with underflow (U)

$$\Rightarrow \qquad Ff = Pp + Uu \qquad\qquad (5.4.2)$$

These two are the fundamental equations required to estimate the recovery of considered size material, to evaluate the performance of the screen, and to locate the process bottlenecks during operation. Example 5.4.1 and 5.4.2 illustrates the application of these material balance equations for a screening process.

Example 5.4.1: *Anthracite coal from a pulverization unit contains 80% by weight of fine material. In order to remove these fines, it is screened at the rate of 250 tons/hr using 1.8 mm screen. Oversize and undersize products are sampled and analysed. It is found that +1.8 mm material in oversize and undersize products are 40% and 10% respectively. Estimate the tonnage of fines removed in underflow, and the tonnage of fines left in overflow.*

Solution:

Given

Fraction of +1.8 mm material in the feed	$= f = 1 - 0.8 = 0.2$
Fraction of +1.8 mm material in the overflow product	$= p = 0.4$
Fraction of +1.8 mm material in the underflow product	$= u = 0.1$
Flow rate of Feed material	$= F = 250$ tons/hr

Total material balance $F = P + U$

+1.8 mm material balance $Ff = Pp + Uu$

On substitution of the given values the above equations becomes

$$250 = P + U$$

$$250 \times 0.2 = P \times 0.4 + U \times 0.1$$

Solving the above two equations $\Rightarrow$ $U = 166.67$ tons/hr

Tonnage of -1.8 mm fines removed in underflow

$$= U(1 - u) = 166.67 \times (1.0 - 0.1) = 150 \text{ tons/hr}$$

Tonnage of overflow $= F - U$ $= 250 - 166.67$ $= 83.33$ tons/hr

Tonnage of -1.8 mm fines left in overflow

$$= P(1 - p) = 83.33 \times (1.0 - 0.4) = 50 \text{ tons/hr}$$

Example 5.4.2: *1000 tons/day of run of mine iron ore assaying 60% Fe is separated into coarse and fine fractions. The amount of fines is 400 tons/day. The value of iron present in fines is estimated as 200 tons/day. What is the assay value of coarse fraction?*

Solution:

Given

Tonnage of iron ore feed $= F = 1000$ tons/day

Tonnage of fines separated $= U = 400$ tons/day

Tonnage of Fe in fines $= 200$ tons/day

Tonnage of coarse iron ore $\quad = P = 1000 - 400 = 600$ tons/day

Tonnage of Fe in feed ore $\quad = F \times f = 1000 \times 0.60 = 600$ tons/day

Tonnage of Fe in fines fraction $\quad = U \times u = 200$ tons/day (given)

Tonnage of Fe in coarse fraction $= Ff - Uu$

$$= 600 - 200 = 400 \text{ tons/day}$$

$$\% \text{ Fe in coarse fraction} = \frac{\text{Tonnage of Fe in coarse fraction}}{\text{Tonnage of coarse fraction}} \times 100$$

$$= \frac{400}{600} \times 100 = 66.67\%$$

This example 5.4.2 shows that much of the iron values are distributed in coarse fraction which is usual case in most of the iron ores. The grade of the coarse fraction is sufficient to extract iron from coarse fraction. Hence iron ores are separated by size and coarse fraction is sent for extraction and only fine fraction is beneficiated.

5.5 SCREEN EFFICIENCY

The efficiency or effectiveness of the screen and the capacity of the screen depends on the rate of screening. However, efficiency and capacity are opposite to each other in the sense that capacity can be increased at the expense of efficiency and vice versa. The efficiency of a screen may be improved by increasing the number of times that particle is presented at an aperture. This can be done in several different ways. The following are the two ways:

1   the length of the screen may be increased
2   the rate of flow may be decreased by decreasing the angle of inclination of the screen and amplitude of vibration so that the particles spend more time on the screen surface i.e., screening time is increased

A screen is said to behave perfectly if, in a mixture of different sizes of particles, all particles of a particular size less than the screen aperture are separated from the mix. In general, absolute separation of different sized particles using a screen is difficult as it involves probabilities of movement of particles at different stages that may be difficult to determine. Hence it is necessary to express the efficiency of the screening process.

Screen efficiency (often called the effectiveness of a screen), designated as η, is a measure of the success of a screen in closely separating oversize and undersize particles. There is no standard method for defining the screen efficiency. Depending on whether one is interested in removing oversize or undersize particles, screening efficiencies may be defined in a number of ways. In an industrial screen, if there are no broken or deformed apertures and screen is perfectly made, no single coarse particle coarser than the size of aperture pass through. In example 5.5.1, the screening is of this type.

Example 5.5.1: *An ore containing 55% of −1 mm material is screened on a 1 mm screen at the rate of 500 tons per day. If 250 tons per day of undersize is obtained from the screen and it contains no oversize, determine the efficiency of the screen.*

Solution:

Given

Quantity of the feed material	= 500 tons/day
Quantity of the undersize material obtained	= 250 tons/day
Percent −1 mm material in the feed	= 55%

Quantity of the undersize material present in the feed

$$= 500 \times \frac{55}{100} = 275 \text{ tons per day}$$

$$\text{Screen efficiency} = \eta = \frac{\text{Undersize material obtained}}{\text{Undersize material present in feed ore}} \times 100$$

$$= \frac{250}{275} \times 100 = 90.9\%$$

In reality, some coarse particles, may be less in quantity, will report to the underflow fraction. Under such cases screen efficiency is defined in different ways. Different screen efficiencies are hereby defined by taking an example for ease of understanding.

Let an ore containing 60% of −1 mm material is screened on 1 mm screen at the rate of 600 tons/day to obtain −1 mm fraction. 350 tons/day of underflow is obtained from the screen. Overflow contains 18% undersize and underflow contains 10% oversize.

If obtaining underflow is the sole criteria,

$$\text{Screen efficiency} = \eta = \frac{\text{Underflow particles obtained by screening}}{\text{Undersize particles present in feed ore}} \times 100$$

$$= \frac{350}{600 \times 0.60} \times 100 = 97.2\%$$

Here oversizes in underflow are not taken in to consideration or neglected. If obtaining overflow is the sole criteria,

$$\text{Screen efficiency} = \eta = \frac{\text{Oversize particles present in the feed ore}}{\text{Overflow particles obtained by screening}} \times 100$$

$$= \frac{600 \times 0.40}{600 - 350} \times 100 = 96.0\%$$

Here undersizes in overflow are not taken in to consideration or neglected.

To know how efficiently the screen could separate undersize particles

$$\text{Screen efficiency} = \eta = \frac{\text{Undersize particles obtained in underflow}}{\text{Undersize particles present in feed ore}} \times 100$$

$$= \frac{350 \times 0.90}{600 \times 0.60} \times 100 = 87.5\%$$

To know how efficiently the screen could separate oversize particles

$$\text{Screen efficiency} = \eta = \frac{\text{Oversize particles obtained in overflow}}{\text{Oversize particles present in feed ore}} \times 100$$

$$= \frac{250 \times 0.82}{600 \times 0.40} \times 100 = 85.4\%$$

If no oversize particle pass through the screen, the screen efficiency will be 100%. To know how efficiently the screen could separate both undersize and oversize particles

$$\text{Screen efficiency} = \eta$$

$$= \frac{\begin{array}{c}\text{Undersize particles}\\ \text{obtained in underflow}\end{array}}{\begin{array}{c}\text{Undersize particles}\\ \text{present in feed ore}\end{array}} \times \frac{\begin{array}{c}\text{Oversize particles}\\ \text{obtained in overflow}\end{array}}{\begin{array}{c}\text{Oversize particles}\\ \text{present in feed ore}\end{array}} \times 100$$

$$= \frac{350 \times 0.90}{600 \times 0.60} \times \frac{250 \times 0.82}{600 \times 0.40} \times 100 = 74.7\%$$

It should be noted that screen efficiency is defined with respect to the required particles to separate.

In processing plants, samples from feed ore, overflow and underflow fractions are collected and subjected to size analysis to know the different sizes present in feed and two product fractions to evaluate the performance or efficiency of the screen. Under such cases equations for efficiency can be derived by writing material balance equations on the screen as follows:

Total material balance	$F = P + U$	(5.5.1)
Oversize material balance	$Ff = Pp + Uu$	(5.5.2)
Undersize material balance	$F(1 - f) = P(1 - p) + U(1 - u)$	(5.5.3)

where F = Rate of feed material
P = Rate of overflow material obtained from the screen
U = Rate of underflow material obtained from the screen
f = fraction of oversize material in the feed
p = fraction of oversize material in the overflow obtained from the screen
u = fraction of oversize material in the underflow obtained from the screen

On computation of equations 5.5.1, 5.5.2 and 5.5.3, we get

$$\frac{P}{F} = \frac{f-u}{p-u} \quad \text{and} \quad \frac{U}{F} = \frac{p-f}{p-u}$$

The recovery of oversize material into the screen overflow is referred as Screen Efficiency (or Screen Effectiveness), η_p, based on the oversize material

$$\eta_p = \frac{\text{Oversize particles obtained in overflow}}{\text{Oversize particles present in feed ore}} = \frac{Pp}{Ff} = \frac{p(f-u)}{f(p-u)} \tag{5.5.4}$$

The recovery of undersize material into the screen underflow is referred as Screen Efficiency (or Screen Effectiveness), η_u, based on the undersize material

$$\eta_u = \frac{\text{Undersize particles obtained in underflow}}{\text{Undersize particles present in feed ore}} = \frac{U(1-u)}{F(1-f)} = \frac{(1-u)(p-f)}{(1-f)(p-u)} \tag{5.5.5}$$

A combined overall efficiency, or overall effectiveness, η, is then obtained by multiplying the above two equations together

$$\eta = \frac{p(f-u)}{f(p-u)} \times \frac{(1-u)(p-f)}{(1-f)(p-u)} = \frac{p(f-u)(1-u)(p-f)}{f(p-u)^2(1-f)} \tag{5.5.6}$$

If there are no broken or deformed apertures and screen is perfectly made, no single coarse particle will pass through the screen, i.e., $u=0$. Then the formula for fines recovery, η_u, and the formula for overall efficiency, η, both reduce to

$$\eta = \frac{p-f}{p(1-f)} \tag{5.5.7}$$

This formula is widely used and implies that recovery of the coarse material in the overflow is 100%. Examples 5.5.2 to 5.5.5 are illustrated with screen efficiency calculations in different cases.

Example 5.5.2: *Fine coal containing 75% by weight of −1.4 mm material is screened. If the weight percent of +1.4 mm material in oversize and undersize products are 60% and 5% respectively, estimate the effectiveness of the screen.*

Solution:

Given

Fraction of +1.4 mm material in the feed $= f = 1 - 0.75 = 0.25$
Fraction of +1.4 mm material in the overflow product $= p = 0.60$
Fraction of +1.4 mm material in the underflow product $= u = 0.05$

In this example, the objective is to remove the fines.

∴ Effectiveness of the screen in removing fines (undersize material)

$$\eta_u = \frac{(1-u)(p-f)}{(1-f)(p-u)} = \frac{(1-0.05)(0.60-0.25)}{(1-0.25)(0.60-0.05)} = 0.806 \Rightarrow 80.6\%$$

While removing fines, some quantity of coarse is obtained with fines. If this coarse material is also taken into account, overall effectiveness is to be calculated for effectiveness of the screen. i.e. effectiveness of the screen in separating both coarse and fines

$$\eta = \frac{p(f-u)(1-u)(p-f)}{f(p-u)^2(1-f)}$$
$$= \frac{0.4(0.2-0.1)(1-0.1)(0.4-0.2)}{0.2(0.4-0.1)^2(1-0.2)} = 0.5 \Rightarrow 50\%$$

Example 5.5.3: *Calculate screen efficiency when quartz mixture is screened through a 1.5 mm screen to obtain +1.5 mm fraction. The size analyses of feed, overflow and underflow is obtained by sampling and sieve analysis. The results are given in Table 5.5.3.1. If the feed rate to the screen is 100 tons/hr, calculate the tonnage of fines remained in overflow.*

Table 5.5.3.1 Size analyses of feed, overflow and underflow for example 5.5.3.

	Weight percent retained this size		
Screen size mm	Feed	Overflow	Underflow
3.3	3.5	7.0	–
2.3	13.5	36.0	–
1.5	33.0	37.0	15.0
1.0	22.7	13.0	43.0
0.8	16.0	4.0	25.0
0.6	5.4	3.0	8.0
0.4	2.1	–	3.0
0.2	1.8	–	2.0
−0.2	2.0	–	4.0

Solution:

From the values of Table 5.5.3.1

Fraction of +1.5 mm material in the feed $= f = 3.5 + 13.5 + 33.0 = 50\% \Rightarrow 0.5$
Fraction of +1.5 mm material in the overflow product $= p = 7.0 + 36.0 + 37.0$
$\qquad\qquad\qquad\qquad\qquad\qquad\qquad\qquad\qquad = 80\% \qquad \Rightarrow 0.8$
Fraction of +1.5 mm material in the underflow product $= u = 15\% \qquad \Rightarrow 0.15$
Flow rate of the feed material $\qquad\qquad\qquad\qquad\qquad = F = 100\,\text{tons/hr}$

In this example, the objective is to obtain coarse material

$\therefore$ Efficiency of the screen in separating coarse (oversize material)

$$\eta_p = \frac{p(f-u)}{f(p-u)} = \frac{0.8(0.5-0.15)}{0.5(0.8-0.15)} = 0.862 \quad \Rightarrow \quad 86.2\%$$

Total material balance $\qquad\qquad F = P + U$

+1.5 mm material balance $\qquad\quad Ff = Pp + Uu$

On substitution of the given values the equations becomes

$$100 = P + U$$

$$100 \times 0.5 = P \times 0.8 + U \times 0.15$$

Solving the above two equations $\Rightarrow \qquad P = 53.8\ \text{tons/hr}$

Tonnage of fines remained in overflow $\quad = P\,(1-p)$
$$= 53.8\,(1-0.8) = 10.76\ \text{tons/hr}$$

Example 5.5.4:   *A crusher product of a coarse ore consists 50% of $-1'' + 3/4''$ size. It is screened in a double deck screen using $1''$ and $3/4''$ screens. Sampling and size analysis of three fractions results 10% and 90% of $-1'' + 3/4''$ in top and middle fractions. Bottom fraction is pure $-3/4''$ size. Calculate the ratio of middle fraction to feed. Also calculate the efficiency of double deck screen in separating $-1'' + 3/4''$ material.*

Solution:

Given

Fraction of $-1'' + 3/4''$ in the feed $\qquad\ = f = 0.5$
Fraction of $-1'' + 3/4''$ in top fraction $\quad = u = 0.1$
Fraction of $-1'' + 3/4''$ in middle fraction $= p = 0.9$

As only middle fraction is required, the other two fractions together will be a second fraction. Fraction of $-1'' + 3/4''$ in other two fractions together is to be calculated.

Fraction of $-1'' + 3/4''$ in other fractions $= u = \dfrac{0.1 + 0.0}{2} = 0.05$

Ratio of middle fraction to feed $= \dfrac{P}{F} = \dfrac{f-u}{p-u} = \dfrac{0.5-0.05}{0.9-0.05} = 0.53$

Efficiency of the double deck screen in separating the $-1'' + 3/4''$ material

$$\eta_p = \frac{p(f-u)}{f(p-u)} = \frac{0.9(0.5-0.05)}{0.5(0.9-0.05)} = 0.953 \quad \Rightarrow \quad 95.3\%$$

Example 5.5.5: *Analyses of vibrating screen's products is given in Table 5.5.5.1.*

Table 5.5.5.1 Analyses of vibrating screen's products for example 5.5.5.

	Vibrating screen products		
Analysis	+1/4″ Wt in kg	−1/4″ + 1/8″ Wt in kg	−1/8″ Wt in kg
+1/4″	6.0	–	–
+1/4″ + 1/8″	0.75	5.5	0.1
−1/8″	0.25	0.5	6.9

Calculate:
(a) Recovery of the −1/4″ + 1/8″ material
(b) Percentage removal of +1/4″ and −1/8″ material
(c) Efficiency of screen in separating the fraction −1/4″ + 1/8″

Solution:

(a) Since recovery of the −1/4″ + 1/8″ material in middle fraction is to be calculated, the other two screen product fractions +1/4″ and −1/8″ are to be treated as rejects to be removed.

From the given data of Table 5.5.5.1

$$
\begin{aligned}
\text{Weight of } +1/4'' \text{ fraction} &= 7.0\,\text{kg}; \\
\text{Weight of } -1/8'' \text{ fraction} &= 7.0\,\text{kg}; \\
\text{Weight of } -1/4'' + 1/8'' \text{ fraction} &= 6.0\,\text{kg}; \\
\text{Weight of feed} = 7 + 7 + 6 &= 20.0\,\text{kg};
\end{aligned}
$$

$$F = 20\,\text{kg}; \qquad P = 6\,\text{kg}; \qquad U = 7 + 7 = 14\,\text{kg};$$

f, p and u are fractions of $-1/4'' + 1/8''$ material in F, P and U respectively

$$f = \frac{0.75 + 5.5 + 0.1}{20} = 0.3175; \quad p = \frac{5.5}{6} = 0.917; \quad u = \frac{0.75 + 0.1}{14} = 0.0607;$$

$$\text{Recovery of } -1/4'' + 1/8'' \text{ material} = \frac{Pp}{Ff} = \frac{6 \times 0.917}{20 \times 0.3175} = 0.8665 \Rightarrow 86.65\%$$

(b) Percent removal of +1/4″ and −1/8″ material means recovery of these two size materials in other two screen fractions.

$$\frac{U(1-u)}{F(1-f)} \times 100 = \frac{14(1 - 0.0607)}{20(1 - 0.3175)} \times 100 = 96.34\%$$

(c)   Efficiency of separating the $-1/4'' + 1/8''$ fraction is nothing but recovery of $-1/4'' + 1/8''$ fraction. i.e. 86.65%.

However, $-1/4'' + 1/8''$ material also occurs in other two fractions i.e 0.75 kg in $-1/4''$ fraction and 0.1 kg in $+1/8''$ fraction. If the loss of $-1/4 + 1/8''$ material in other two fractions are also taken into account, screen efficiency is the overall efficiency in separating $-1/4'' + 1/8''$ material and $-1/4''$ and $+1/8''$ material.

$$\eta = \frac{Pp}{Ff} \times \frac{U(1-u)}{F(1-f)} \times 100 = \frac{6 \times 0.917}{20 \times 0.3175} \times \frac{14(1 - 0.0607)}{20(1 - 0.3175)} \times 100 = 83.47\%$$

If f, p, and u are expressed in terms of the fractions of undersize material in feed, overflow and underflow respectively, by forming the mass balance equations, three expressions for efficiency of the screen can be obtained as

$$\eta_p = \frac{P(1-p)}{F(1-f)} = \frac{(1-p)(f-u)}{(1-f)(p-u)} \tag{5.5.8}$$

$$\eta_u = \frac{Uu}{Ff} = \frac{u(p-f)}{f(p-u)} \tag{5.5.9}$$

$$\eta = \frac{u(u-f)(1-p)(f-p)}{f(u-p)^2(1-f)} \tag{5.5.10}$$

In examples of 5.5.2, 5.5.3 and 5.5.4, if these formulae are used by taking fraction of undersize for f, p and u, the same result will be obtained.

The formulae 5.5.4 to 5.5.10 do not give an absolute value of the efficiency, as no allowance is made for the difficulty of the separation. A feed composed mainly of particles of a size near to that of the screen aperture – **near size material (near mesh material)** – presents a more difficult separation than a feed composed mainly of very coarse and very fine particles with a screen aperture intermediate between them. In such cases, it is proposed to define the efficiency as the ratio of the near size material taken out by the screen to the near size material present in the feed.

5.6 TROMP CURVE

Since the feed to the screen contains particles of different sizes, the separation efficiency may be different for different particles. Hence there is a need to take into account the amount of misplaced material that can occur or the difficulty of separation of some of the particles. In 1937, Tromp introduced a graphical method of assessing separation efficiency which is universally used and is alternatively referred as **Performance Curve, Partition Curve, Efficiency Curve,** or **Tromp Curve** [6]. The Tromp curve was originally used to evaluate the efficiency of gravity concentration of minerals. It was further found that they may also be used to describe the other processes including screening and classification where the feed and final products can be separated into narrow size fractions. Tromp curve constructed by a large number of narrow fractions give the maximum information about the performance of a mineral processing device.

Table 5.6.1 Size analyses of overflow and underflow streams of a screen.

Size in microns	Weight%	
	O/F	U/F
+20000	–	–
−20000 + 16000	37.5	0.5
−16000 + 8000	32.0	1.0
−8000 + 4000	13.0	10.6
−4000 + 2000	7.4	12.1
−2000 + 1000	3.6	15.0
−1000 + 500	2.5	18.0
−500 + 250	2.0	20.0
−250 + 125	1.5	19.8
−125	0.5	3.0
	100.0	100.0

The amount of misplaced material to an output stream is referred to as the **partition coefficient** (also called the **distribution factor** or **probability factor**). The partition coefficient is then defined as

$$\text{Partition Coefficient} = \frac{\text{Amount of material of particular size in a stream}}{\text{Amount of material of same size in the feed}}$$

It may be expressed as a fraction or a percentage. The partition coefficient is essentially the recovery of a given size material to a stream. The partition coefficient thus obtained is plotted against the mean separating size of the fraction to generate the performance curve. The mean values plotted may either be the arithmetic mean or the geometric mean. Plotting of partition curve can best be explained with an example.

A screen, producing 62.5% overflow, has the size analyses of overflow and underflow streams as shown in Table 5.6.1.

The following are the details of calculations (with reference to the Table 5.6.1.1 that follows) necessary to draw Tromp Curve.

Column D & E are the size analyses of overflow and underflow in relation to the feed material. Column D is to be calculated by multiplying the values of Column B by 0.625 as this is the fraction of feed reporting to overflow. Similarly Column E is to be calculated by multiplying the values of Column C by 0.375 as this is the fraction of feed reporting to underflow. Column F is the size analysis of the feed material and is calculated by adding the corresponding values of Column D and Column E. Column G is the arithmetic mean of the corresponding sieve size ranges. Column H, partition coefficient, is to be calculated by dividing each weight% in Column D by the corresponding weight% in Column F.

Plot the graph between mean size (Column G) and the partition coefficient (Column H) on semi-log graph sheet to accommodate wide range of size. This is the Tromp Curve of the screen and is shown in Fig. 5.6.1.1 for this example.

Table 5.6.1.1 Calculated values to draw Tromp curve.

A	B	C	D	E	F	G	H
	Weight%		Wt% of feed		Calculated		Partition
Size in microns	O/F	U/F	O/F B×0.625	U/F C×0.375	Feed D+E	Mean size	Coefficient D/F
+20000	–	–	–	–	–	–	–
−20000+16000	37.5	0.5	23.44	0.19	23.63	18000	0.99
−16000+8000	32.0	1.0	20.00	0.37	20.37	12000	0.98
−8000+4000	13.0	10.6	8.13	3.97	12.10	6000	0.67
−4000+2000	7.4	12.1	4.63	4.54	9.17	3000	0.50
−2000+1000	3.6	15.0	2.25	5.62	7.87	1500	0.28
−1000+500	2.5	18.0	1.56	6.75	8.31	750	0.19
−500+250	2.0	20.0	1.25	7.50	8.75	375	0.14
−250+125	1.5	19.8	0.94	7.43	8.37	187	0.11
−125	0.5	3.0	0.31	1.13	1.44		0.21
			62.50	37.50			

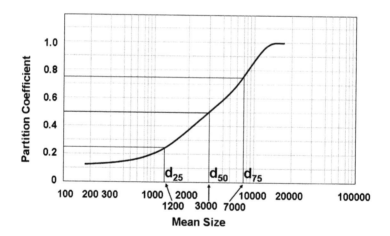

Figure 5.6.1.1 Tromp curve.

From this Tromp curve, d_{50} is determined corresponding to the partition coefficient of 0.5. d_{50} is a cut point or separation size, and is defined as that the particles of that size have an equal probability of passing through or retained by the aperture. d_{50} in this example is 3000 microns. d_{50} is unlikely to correspond with the size of the screen aperture, the cut-point usually being less than the aperture size. The Tromp curve effectively models the screen, and can be used for simulation and design purposes.

The Tromp curve is a convenient way of showing the sharpness of separation. However, a numerical figure is better for describing the deviation from ideal behavior. These numerical figures are based on the error between the actual curve and the line of perfect separation, and are termed the **probable error, error area** or **ecart probability**.

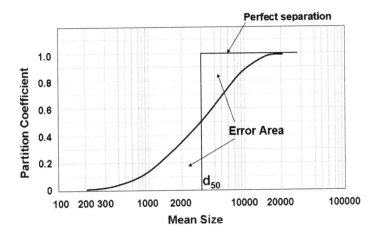

Figure 5.6.1.2 Tromp curve showing perfect separation and error area.

Let us consider a screen with 2 mm aperture. In a perfect separation, any particle of less than 2 mm size should pass through the screen and hence the amount remaining on the screen (with the oversize product) at the completion of the process should be zero. Any particle of greater than 2 mm size should remain on the screen and hence the amount of this material in the oversize product should be 100%. That is, the partition coefficient for −2 mm material in the oversize product will be zero and the partition coefficient for +2 mm material in the oversize product will be 1.0 to 100%. The performance curve then will have the shape of the solid line as shown in Fig. 5.6.1.2. That is, there will be a sharp jump from 0 to 1.0 (or 100%) at the separation size. This separation size is referred as d_{50}.

In an actual screening, deviation from the perfect separation is quantified by determining the area between the performance curve and the ideal curve. This area is termed as error area. If several performance curves are plotted on the same axes, then this area provides a means of comparing the sharpness of separation.

Another method of characterizing the performance curve is to determine 50% of the difference between the separation size at a partition coefficient of 0.75 (or 75%) and 0.25 (or 25%). This is referred as **Ecart Probability** (E_p), **probable error** or **probable deviation**

$$E_p = \frac{d_{75} - d_{25}}{2} \tag{5.6.11}$$

If the performance curve is a straight line between the d_{75} and d_{25} points, then the probable error is a measure of the slope of this curve and is proportional to the reciprocal of the slope. As the slope of the performance curve approaches the vertical, the probable error approaches zero. The smaller the probable error, the greater is the sharpness of separation. Another term used for sharpness of separation, **Imperfection**, I, is expressed as

$$I = \frac{d_{75} - d_{25}}{2d_{50}} \tag{5.6.12}$$

For the example considered here, the various values are:

$$d_{50} = 3000\,\mu\text{m};\ d_{75} = 7000\,\mu\text{m};\ d_{25} = 1200\,\mu\text{m}$$

$$\text{Ecart Probability} = E_p = \frac{d_{75} - d_{25}}{2} = \frac{7000 - 1200}{2} = 2900$$

$$\text{Imperfection} = I = \frac{d_{75} - d_{25}}{2d_{50}} = \frac{7000 - 1200}{2 \times 3000} = 0.97$$

Example 5.6.1: *A sample of a material is being screened by using vibrating screen. The size analyses of the feed, overflow and underflow streams are as given in Table 5.6.1.2. If the overflow is 70.9% of the feed, draw the partition curve and determine Ecart probability and Imperfection.*

Table 5.6.1.2 Size analyses of Feed, overflow and underflow streams of a screen.

Size mm	Weight % retained		
	Feed	Overflow	Underflow
+50.8	6.3	7.4	0.0
−50.8 + 25.4	10.1	15.4	0.0
−25.4 + 20.0	20.6	27.5	0.6
−20.0 + 12.7	20.3	28.3	5.8
−12.7 + 6.35	24.2	17.5	37.8
−6.35 + 4.72	4.6	1.0	13.9
−4.72	13.9	2.9	41.9

Solution:

Required values are calculated and presented in Table 5.6.1.3.

Table 5.6.1.3 Calculated values for example 5.6.1.

A	B	C	D	E	F	G	H	J
		Weight%		Wt% of feed				
						Calculated	Mean	Partition
				O/F	U/F	Feed	size	Coefficient
Size in mm	Feed	O/F	U/F	C×0.709	D×0.291	E+F	mm	E/G
+50.8	6.3	7.4	0.0	5.2	0.0	5.2	–	1.0
−50.8 + 25.4	10.1	15.4	0.0	10.9	0.0	10.9	38.1	1.0
−25.4 + 20.0	20.6	27.5	0.6	19.5	0.2	19.7	22.7	0.99
−20.0 + 12.7	20.3	28.3	5.8	20.1	1.7	21.8	16.4	0.92
−12.7 + 6.35	24.2	17.5	37.8	12.4	11.0	23.4	9.5	0.53
−6.35 + 4.72	4.6	1.0	13.9	0.7	4.0	4.7	5.5	0.15
−4.72	13.9	2.9	41.9	2.1	12.2	14.3	–	0.15
	100.0	100.0	100.0	70.9	29.1	100.0		

Size analysis of calculated feed is in good agreement with the measured feed. With values of Table 5.6.1.3, Partition curve is drawn and shown in Figure 5.6.1.3.

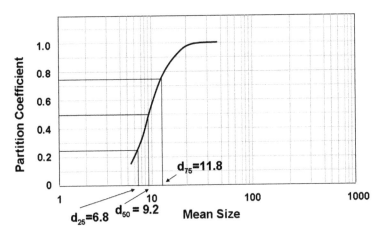

Figure 5.6.1.3 Partition curve for the example 5.6.1.

From the Fig. 5.6.1.3, $d_{50} = 9.2$ mm, $d_{75} = 11.8$ mm, $d_{25} = 6.8$ mm.

$$\text{Ecart Probability} = E_p = \frac{d_{75} - d_{25}}{2} = \frac{11.8 - 6.8}{2} = 2.5$$

$$\text{Imperfection} = I = \frac{d_{75} - d_{25}}{2d_{50}} = \frac{11.8 - 6.8}{2 \times 9.2} = 0.272$$

5.7 PROBLEMS FOR PRACTICE

5.7.1:   *An ore containing 45% of −0.5 mm material is screened on 0.5 mm screen at the rate of 700 tons per day. If 280 tons per day of undersize is obtained and it contains no oversize, determine the efficiency of the screen.*

[88.9%]

5.7.2:   *Road metal product obtained from a crusher consists of 25% of −1/4″ size is to be used for preparation of concrete. In order to get +1/4″ fraction required for the preparation of concrete, it is screened by using 1/4″ screen. The overflow and underflow products of the screen contain 80% of +1/4″ size and 95% of −1/4″ size respectively. If the feed to the screen is 55 tons, what is the +1/4″ size metal lost with underflow of the screen? Estimate the effectiveness of the screen.*

[0.18 tons; 99.56%]

5.7.3: *−10 mesh material is fed to a double deck vibrating screen for separation. The desired product is −40 + 60 mesh material. A 40 mesh and a 60 mesh screens are therefore used. The feed is introduced on the 40 mesh screen. From the sieve analyses of the feed and the three products given in table 5.7.3.1, calculate the effectiveness of the double deck screen in separating −40 + 60 mesh material.*

Table 5.7.3.1 Sieve analyses of feed and three products for problem 5.7.3.

| Mesh | Mass fraction | | | |
	Feed	Oversize from 40 mesh screen	Oversize from 60 mesh screen	Undersize from 60 mesh screen
−10 + 20	0.097	0.197	0.026	0.0005
−20 + 30	0.186	0.389	0.039	0.0009
−30 + 40	0.258	0.337	0.322	0.0036
−40 + 60	0.281	0.066	0.526	0.3490
−60 + 85	0.091	0.005	0.061	0.2990
−85 + 100	0.087	0.006	0.026	0.3470

[23.08%]

Chapter 6

Density

Density is important next to the size of the particle. Density of the particle can be mass density or weight density. **Mass density** is the mass of the particle per unit volume whereas **Weight density** is the weight of the particle per unit volume. **Specific gravity** is defined as the ratio of the density of the particle to the density of water at 4°C. In C.G.S system of units, numerical value of density and specific gravity is same. Hence, it is normal practice to indicate specific gravity for density.

In all beneficiation operations, particularly in gravity concentration operations, density has an important role together with the size and the shape of the particles. As the ore contains different minerals, the density of an ore varies depending on the minerals it contains. Before the ore is to be beneficiated, the density of the ore and the density of the different minerals present in the ore are to be determined.

A bulk solid (bulk material) is combination of particles and space. For a bulk material, the average particle density can be determined by dividing the weight of the material (solids) by the true volume occupied by the particles (not including the voids). This can be determined by using density bottle.

The stepwise procedure for determination of density of an ore is as follows:

1. Wash, dry and weigh the density bottle with stopper. Let this weight be w_1.
2. Thoroughly dry the ore sample.
3. Keep 5–10 grams of ore sample in the bottle and reweigh. Let this weight be w_2.
4. Now fill the bottle with a liquid of known density. The liquid used should not react with the ore.
5. Insert the stopper, allow the liquid to fall out of the bottle, wipe of excess liquid and weigh the bottle. Let this weight be w_3.
6. Remove ore and liquid from the bottle and fill the bottle with liquid alone and repeat step 5. Let this weight be w_4.

$w_2 - w_1$ is the weight of the ore sample
$w_4 - w_1$ is the weight of the liquid occupying whole volume of the bottle
$w_3 - w_2$ is the weight of the liquid having the volume equal to the volume of density bottle less volume of ore sample taken
$(w_4 - w_1) - (w_3 - w_2)$ is the weight of the liquid of volume equal to that of the ore sample

If ρ_l is the density of the liquid

$$\text{Density of the ore sample} = \frac{w_2 - w_1}{(w_4 - w_1) - (w_3 - w_2)} \times \rho_l \qquad (6.1)$$

The bulk material is a combination of particles and space, the fraction of the total volume not occupied by the particles is referred as the 'voidage' or 'void fraction'. Sometimes the term 'porosity' is applied to bulk material to mean the same as 'voidage'.

The **bulk density** is the overall density of a material kept in large quantities, which can be defined as the mass or weight of the material divided by its total volume (particles and voids) and depends upon the true density of the material and the pore space between the particles. It is a measure of the storage capacity.

Three kinds of bulk density that apply to materials handling calculations are:

1   **Aerated density**

When the sample of the bulk material is carefully poured into a measuring cylinder to measure its volume, then the computed density is called 'Aerated', 'loose', or 'poured' bulk density (ρ_a).

2   **Packed density**

If the sample is packed by dropping the cylinder vertically a number of times from a height of one or two centimeters on to a table, then the computed density is called 'packed' or 'tapped' bulk density (ρ_c).

3   **Dynamic** or **Working density**

The dynamic or working density (ρ_d) is a function of Aerated and Packed densities. Its value is intermediate to that of aerated and packed densities. The dynamic or working density (ρ_d) is expressed as

$$\rho_d = (\rho_c - \rho_a)C + \rho_a \qquad (6.2)$$

where C is the Compressibility expressed as

$$C = \frac{(\rho_c - \rho_a)}{\rho_c} \qquad (6.3)$$

The lower the compressibility, the material is more free flowing. The dividing line between free flowing (granular) and non-free flowing (powder) is about 0.20–0.21 compressibility. A higher value indicates a powder that is non-free flowing and will be likely to bridge in a hopper. The compressibility of a material often helps to indicate uniformity in size and shape of the material, its deformability, surface area, cohesion and moisture content.

For the five materials given in the Table 6.1, compressibility and working density are calculated and based on compressibility values, five materials are arranged in the increasing order of flowability.

Table 6.1 Aerated and packed densities of five materials.

Material code	A	B	C	D	E
Aerated density, kg/m³	1390	1800	1430	1440	1390
Packed density, kg/m³	1730	2050	1610	1650	1690

Compressibility of material $A = C_A$

$$= \frac{(\rho_c - \rho_a)}{\rho_c} = \frac{(1730 - 1390)}{1730} = 0.1965$$

Working density $= \rho_d = (\rho_c - \rho_a)C_A + \rho_a$

$$= (1730 - 1390)0.1965 + 1390 = 1456.8 \, \text{kg/m}^3$$

Compressibility of material $B = C_B$

$$= \frac{(\rho_c - \rho_a)}{\rho_c} = \frac{(2050 - 1800)}{2050} = 0.122$$

Working density $= \rho_d = (\rho_c - \rho_a)C_B + \rho_a$

$$= (2050 - 1800)0.122 + 1800 = 1830.5 \, \text{kg/m}^3$$

Compressibility of material $C = C_C$

$$= \frac{(\rho_c - \rho_a)}{\rho_c} = \frac{(1610 - 1430)}{1610} = 0.1118$$

Working density $= \rho_d = (\rho_c - \rho_a)C_C + \rho_a$

$$= (1610 - 1430)0.1118 + 1430 = 1450.1 \, \text{kg/m}^3$$

Compressibility of material $D = C_D$

$$= \frac{(\rho_c - \rho_a)}{\rho_c} = \frac{(1650 - 1440)}{1650} = 0.1273$$

Working density $= \rho_d = (\rho_c - \rho_a)C_D + \rho_a$

$$= (1650 - 1440)0.1273 + 1430 = 1456.7 \, \text{kg/m}^3$$

Compressibility of material $E = C_E$

$$= \frac{(\rho_c - \rho_a)}{\rho_c} = \frac{(1690 - 1390)}{1690} = 0.1775$$

Working density $= \rho_d = (\rho_c - \rho_a)C_E + \rho_a$

$$= (1690 - 1390)0.1775 + 1390 = 1443.3 \, \text{kg/m}^3$$

Increasing order of flowability: $A \, E \, D \, B \, C$

6.1 SOLIDS AND PULP

Solid particles of less than 2 mm size are named as **Sands**, **Slimes** and **Colloids**. The following are the approximate size ranges:

Particles	Size ranges
Sands	between 2 mm and 74 microns
Slimes	between 74 and 0.1 microns
Colloids	between 0.1 and 0.001 micron

Most of the mineral beneficiation operations are wet. Water is added to the ore particles to aid beneficiation. The mixture of water and solid particles is known as **Pulp**.

Other terms commonly used are:

Suspension: When the solid particles are held up in the water, the pulp is called suspension. In other words, in suspension, the solid particles are well dispersed throughout.

Slurry: A mixture of fine solids (called slimes) and water

Sludge: Thick pulp i.e., pulp with less quantity of water

Pulp or slurry density is most easily measured in terms of weight of the slurry per unit volume (gm/cm^3 or kg/m^3). A sample of slurry taken in container of known volume is weighed to give slurry density directly. Marcy Scale available in the market gives direct reading for the density of the slurry and % solids in the slurry.

The composition of a slurry is often represented as the fraction (or percent) of solids by weight. It is determined by sampling the slurry, weighing, drying and reweighing.

$$C_w = \text{fraction of solids by weight} = \frac{\text{Weight of the particles}}{\text{Weight of the slurry}}$$

$$C_v = \text{fraction of solids by volume} = \frac{\text{Volume of the particles}}{\text{Volume of the slurry}}$$

Knowing the densities of the slurry (ρ_{sl}), water (ρ_w) and dry solids (ρ_p), the fraction of solids (C_w) by weight can be calculated. A material balance equation in terms of the volume can be written as

Volume of the solids + Volume of the water = Volume of the slurry

$$\frac{C_w}{\rho_p} + \frac{1 - C_w}{\rho_w} = \frac{1}{\rho_{sl}} \tag{6.4}$$

$$\Rightarrow \qquad C_w = \frac{\rho_p(\rho_{sl} - 1)}{\rho_{sl}(\rho_p - 1)} \qquad [\because \rho_w = 1\,\text{gm/cm}^3] \tag{6.5}$$

$$\Rightarrow \qquad \rho_{sl} = \frac{\rho_p}{\rho_p + C_w(1 - \rho_p)} \tag{6.6}$$

Similarly, a material balance equation in terms of the weight can be written as

Weight of the solids + Weight of the water = Weight of the slurry

$$C_v \rho_p + (1 - C_v)\rho_w = \rho_{sl} \tag{6.7}$$

$$\Rightarrow \qquad C_v = \frac{(\rho_{sl} - 1)}{\rho_p - 1} \quad [\because \rho_w = 1\,\text{gm/cm}^3] \tag{6.8}$$

$$\Rightarrow \qquad \rho_{sl} = 1 + C_v(\rho_p - 1) \tag{6.9}$$

Dilution ratio is the ratio of the weight of the water to the weight of the solids in the slurry.

$$\text{Dilution ratio} = DR = \frac{1 - C_w}{C_w} \tag{6.10}$$

Substituting the value of C_w of equation 6.5

$$\text{Dilution ratio} = DR = \frac{\rho_p - \rho_{sl}}{\rho_p(\rho_{sl} - 1)} \tag{6.11}$$

Dilution ratio is particularly important as the product of dilution ratio and weight of the solids in the slurry is equal to the weight of the water in the slurry.

The examples 6.1.1 to 6.1.11 illustrate the applications of density calculations in different situations.

Example 6.1.1: *When 2 liter slurry sample of hematite of specific gravity of 4.8 is filtered, 250 gm of wet filter cake is obtained. Subsequently the filter cake is perfectly dried. The weight of dried hematite is 180 gm.*

Calculate (a) *% solids by volume in the slurry*
(b) *% solids by weight in the slurry*
(c) *Liquid solid ratio by volume & by weight*
(d) *Density of the slurry*
(e) *% moisture in the filter cake*
(f) *Bulk density of the filter cake*

Solution:

This problem is worked out by using density definition alone

Weight of the wet filter cake	$= 250\,\text{gm}$
Volume of the slurry sample	$= 2\,\text{litre} = 2000\,\text{cm}^3$
Weight of dried hematite	$= 180\,\text{gm}$
Density of hematite	$= 4.8\,\text{gm/cm}^3$
Volume of hematite	$= 180/4.8 = 37.5\,\text{cm}^3$
Volume of water in the slurry	$= 2000 - 37.5 = 1962.5\,\text{cm}^3$
% hematite by volume	$= 37.5 \times 100/2000 = 1.875\%$
Weight of water	$= 1962.5\,\text{gm}$

Weight of Hematite	$= 180\,\text{gm}$
% hematite by weight	$= 180 \times 100/2142.5 = 8.4\%$
Liquid solid ratio by volume	$= 1962.5/37.5 = 52.3$
Liquid solid ratio by weight	$= 1962.5/180 = 10.9$
Density of the slurry	$= 2142.5/2000 = 1.07125$
Weight of moisture in filter cake	$= 250 - 180 = 70\,\text{gm}$
% moisture by weight in filter cake	$= 70 \times 100/250 = 28\%$
Volume of filter cake	$= 37.5 + 70 = 107.5\,\text{cm}^3$
Bulk density of the filter cake	$= 250/107.5 = 2.33$

Example 6.1.2: *Calculate the pulp density if the % quartz by weight is 40 in the pulp and the specific gravity of particles is 2.65.*

Solution:

Given

% quartz by weight in pulp $= 40\%$ $\Rightarrow$ $C_w = 0.4$
Density of the particle $= \rho_p = 2.65\,\text{gm/cm}^3$

Substituting these values in equation 6.4

$$\frac{C_w}{\rho_p} + \frac{1 - C_w}{\rho_w} = \frac{1}{\rho_{sl}} \quad \Rightarrow \quad \frac{0.4}{2.65} + \frac{1 - 0.4}{1.0} = \frac{1}{\rho_{sl}} \quad \Rightarrow \quad \rho_{sl} = 1.33\,\text{gm/cm}^3$$

Pulp density $= \rho_{sl} = 1.33\,\text{gm/cm}^3$

Alternately by using equation 6.6

$$\text{Pulp density} = \rho_{sl} = \frac{\rho_p}{\rho_p + C_w(1 - \rho_p)} = \frac{2.65}{2.65 + 0.40(1 - 2.65)} = 1.33\,\text{gm/cm}^3$$

Example 6.1.3: *Determine % solids by weight in the pulp if the sp.gr. of solids is 4.95 and the pulp density is 1.9 gm/cm³. Also calculate % water by volume in the pulp.*

Solution:

Given

Density of solids $= \rho_p = 4.95\,\text{gm/cm}^3$
Density of the pulp $= \rho_{sl} = 1.90\,\text{gm/cm}^3$

Substituting ρ_p and ρ_{sl} values in equation 6.4

$$\frac{C_w}{\rho_p} + \frac{1 - C_w}{\rho_w} = \frac{1}{\rho_{sl}} \quad \Rightarrow \quad \frac{C_w}{4.95} + \frac{1 - C_w}{1.0} = \frac{1}{1.9} \quad \Rightarrow \quad C_w = 0.5936 \quad \Rightarrow \quad 59.36\%$$

Alternately, $C_w = \dfrac{\rho_p(\rho_{sl} - 1)}{\rho_{sl}(\rho_p - 1)} = \dfrac{4.95(1.9 - 1)}{1.9(4.95 - 1)} = 0.5936 \quad \Rightarrow \quad 59.36\%$

% Solids by weight in the pulp $= 59.36\%$

Substituting the values of density of solids and pulp in equation 6.7

$C_v\rho_p + (1 - C_v)\rho_w = \rho_{sl} \quad \Rightarrow \quad C_v \times 4.95 + (1 - C_v) \times 1.0 = 1.9$

$\Rightarrow \quad C_v = 0.2278 \qquad \Rightarrow \quad 22.78\%$

Alternately, $C_v = \dfrac{(\rho_{sl} - 1)}{\rho_p - 1} = \dfrac{(1.9 - 1)}{4.95 - 1} = 0.2278 \quad \Rightarrow \quad 22.78\%$

% Solids by volume in the pulp $\quad = 22.78\%$
% water by volume $= 100.00 - 22.78 = 77.22\%$

Example 6.1.4: *How much quantity of galena of density of 7.5 gm/cm³ must be added to water to obtain 1 litre of pulp at a pulp density of 2.0 gm/cm³?*

Solution:

Given

Density of galena $\quad = \rho_p \quad = 7.5$ gm/cm³
Density of the pulp $= \rho_{sl} \quad = 2.0$ gm/cm³
Volume of the pulp $= 1$ litre $= 1000$ cm³

Substituting these values in equation 6.4

$\dfrac{C_w}{\rho_p} + \dfrac{1 - C_w}{\rho_w} = \dfrac{1}{\rho_{sl}} \quad \Rightarrow \quad \dfrac{C_w}{7.5} + \dfrac{1 - C_w}{1} = \dfrac{1}{2} \quad \Rightarrow \quad C_w = 0.5769$

Fraction of galena by weight in the pulp $\quad = 0.5769$

Weight of the pulp $= 1000 \times 2.0 = 2000$ gm $\quad = 2.0$ kg

Weight of galena to be added $= 2.0 \times 0.5769 = 1.154$ kg

Alternately

Substituting the values of density of galena and the pulp in equation 6.7

$C_v\rho_g + (1 - C_v)\rho_w = \rho_{sl} \quad \Rightarrow \quad C_v \times 7.5 + (1 - C_v) = 2 \quad \Rightarrow \quad C_v = 0.1538$

Fraction of galena by volume in the pulp $= 0.1538$

Weight of galena to be added for 1 cm³ of pulp $= 0.1538 \times 7.5 = 1.154$ gm

Weight of galena to be added for 1 litre of pulp $= 1.154 \times 1000 = 1154$ gm
$= 1.154$ kg

Example 6.1.5: *What is the density of a slurry containing 25% by weight SiO_2 (density = 2.65 gm/cm³), 15% by weight Magnetite (density = 5.18 gm/cm³), and 15% by weight Galena (density = 7.5 gm/cm³)?*

Solution:

Given

Density of SiO_2 $= \rho_1 = 2.65$ gm/cm³
Density of magnetite $= \rho_2 = 5.18$ gm/cm³
Density of galena $= \rho_3 = 7.50$ gm/cm³
Density of water $= \rho_4 = 1.00$ gm/cm³
Fraction of SiO_2 $= C_1 = 0.25$
Fraction of magnetite $= C_2 = 0.15$
Fraction of galena $= C_3 = 0.15$
Fraction of water $= C_4 = 1.0 - 0.25 - 0.15 - 0.15 = 0.45$

Equation 6.4 can be written as follows for three solids and water mixture

$$\frac{C_1}{\rho_1} + \frac{C_2}{\rho_2} + \frac{C_3}{\rho_3} + \frac{C_4}{\rho_4} = \frac{1}{\rho_{sl}}$$

Substituting all the values in the above equation

$$\Rightarrow \quad \frac{0.25}{2.65} + \frac{0.15}{5.18} + \frac{0.15}{7.5} + \frac{0.45}{1.0} = \frac{1}{\rho_{sl}}$$

$$\Rightarrow \quad\quad\quad\quad \rho_{sl} = 1.69 \text{ gm/cm}^3$$

Density of the slurry $= \rho_{sl} = 1.69$ gm/cm³

Example 6.1.6: *An ore pulp fed to a concentrator consists of valuable particles, gangue particles and water at the ratio of 2:3:5 by volume. If the specific gravity of valuable particles and gangue particles are 7.5 and 2.5 respectively, calculate the pulp density and percent valuables by weight.*

Solution:

Given

Density of valuable particles $= \rho_1 = 7.50$ gm/cc
Density of gangue particles $= \rho_2 = 2.50$ gm/cc

As the volume ratio is 2:3:5,
total volume can be taken as $= 2 + 3 + 5 = 10$ cm³

Equation 6.7 can be written as

$$C_1\rho_1 + C_2\rho_2 + C_3\rho_3 = 10\rho_{sl}$$

Substituting the values in the above equation

$$\Rightarrow \quad 2 \times 7.50 + 3 \times 2.50 + 5 \times 1.00 = 10\rho_{sl} \quad \Rightarrow \quad \rho_{sl} = 2.75 \text{ gm/cc}$$

$$\text{Pulp density} = \rho_{sl} = 2.75 \text{ gm/cc}$$

$$\% \text{ valuables in ore pulp} = \frac{2 \times 7.50}{2 \times 7.50 + 3 \times 2.50 + 5 \times 1.00} \times 100 = 54.55\%$$

Example 6.1.7: *Magnetite slurry is flowing at the rate of 2 litres/sec in a benefi- ciation plant. The density of magnetite and slurry are 5000 kg/m³ and 1500 kg/m³ respectively. Calculate the tonnage of the magnetite treated in the plant in an hour.*

Solution:

Given

$$\text{Flow rate of magnetite} \quad = 2 \text{ litres/sec} \ = 2000 \text{ cm}^3/\text{sec}$$
$$\text{Density of magnetite} = \rho_p \ = 5000 \text{ kg/m}^3 = 5.0 \text{ gm/cm}^3$$
$$\text{Density of the slurry} = \rho_{sl} = 1500 \text{ kg/m}^3 = 1.5 \text{ gm/cm}^3$$

Substituting these values in equation 5.4

$$\frac{C_w}{\rho_p} + \frac{1 - C_w}{\rho_w} = \frac{1}{\rho_{sl}} \quad \Rightarrow \quad \frac{C_w}{5.0} + \frac{1 - C_w}{1.0} = \frac{1}{1.5} \quad \Rightarrow \quad C_w = 0.417$$

Fraction of magnetite in the pulp $= C_w = 0.417$

Weight of magnetite per cc of slurry $= 1.5 \times 0.417 = 0.6255 \text{ gm/cm}^3$

Weight flow rate of magnetite in the slurry

$$= 0.6255 \times 2000 = 1251 \text{ gm/sec} = 1.251 \text{ kg/sec} = 4.5036 \text{ tons/hr}$$

Example 6.1.8: *Determine the density of the pulp having dilution ratio of 3 and solids density of 2650 kg/m³.*

Solution:

Given

$$\text{Density of the solids} = 2650 \text{ kg/m}^3 = 2.65 \text{ gm/cm}^3$$
$$\text{Dilution ratio} = 3 = \frac{1 - C_w}{C_w} \quad \Rightarrow \quad C_w = 0.25$$

Fraction of solids in the pulp $= C_w = 0.25$

Substituting density of solids and fraction of solids in equation 6.4

$$\frac{C_w}{\rho_p} + \frac{1 - C_w}{\rho_w} = \frac{1}{\rho_{sl}} \quad \Rightarrow \quad \frac{0.25}{2.65} + \frac{1 - 0.25}{1.0} = \frac{1}{\rho_{sl}} \quad \Rightarrow \quad \rho_{sl} = 1.18 \text{ gm/cm}^3$$

$$\text{Density of the pulp} = \rho_{sl} = 1.18 \text{ gm/cm}^3$$

Example 6.1.9: *In a laboratory flotation test following data was obtained: Pulp consists of 30% galena and 40% quartz by weight Specific gravities of galena and quartz are 7.5 and 2.65 Calculate pulp density and pulp dilution.*

Solution:

Given

Density of the galena $= \rho_1 = 7.50\,\text{gm/cm}^3$
Density of the quartz $= \rho_2 = 2.65\,\text{gm/cm}^3$
Fraction of galena $\quad = C_1 = 0.30$
Fraction of quartz $\quad = C_2 = 0.40$
Fraction of water $\quad = C_3 = 1.0 - 0.30 - 0.40 = 0.30$

Equation 6.4 can be written as $\qquad \dfrac{C_1}{\rho_1} + \dfrac{C_2}{\rho_2} + \dfrac{C_3}{\rho_3} = \dfrac{1}{\rho_{sl}}$

Substituting the all the values in the above equation

$$\Rightarrow \quad \frac{0.30}{7.50} + \frac{0.40}{2.65} + \frac{0.30}{1.00} = \frac{1}{\rho_{sl}} \quad \Rightarrow \quad \rho_{sl} = 2.04\,\text{gm/cm}^3$$

Pulp density $= \rho_{sl} = 2.04\,\text{gm/cm}^3$

Fraction of solids in the pulp $= C_w = C_1 + C_2 = 0.30 + 0.40 = 0.70$

$\therefore$ Pulp dilution $=$ Dilution ratio $= \dfrac{1 - C_w}{C_w} = \dfrac{1 - 0.70}{0.70} = 0.43$

Example 6.1.10: *A sample from a slurry stream containing solids of specific gravity 2.65 is collected in a one litre can. The can has been filled in 6 seconds and the slurry so collected weighs 1.3 kg. Calculate the weight flow rate of solids within the slurry in kg/hr.*

Solution:

Given

Density of the solids $\qquad = \rho_p \qquad = 2.65\,\text{gm/cm}^3$
Weight of the slurry collected $= 1.3\,\text{kg} = 1300\,\text{gm}$
Volume of the slurry collected $= 1\,\text{litre} = 1000\,\text{cm}^3$

Density of the slurry $= \rho_{sl} = 1300\,\text{gm}/1000\,\text{cm}^3 = 1.3\,\text{gm/cm}^3$

Substituting the density values in equation 6.4

$$\frac{C_w}{\rho_p} + \frac{1 - C_w}{\rho_w} = \frac{1}{\rho_{sl}} \quad \Rightarrow \quad \frac{C_w}{2.65} + \frac{1 - C_w}{1.0} = \frac{1}{1.3} \quad \Rightarrow \quad C_w = 0.371$$

% solids by weight in the slurry $= 0.371 \quad \Rightarrow \quad 37.1\%$

Slurry flow rate $= 1$ litre per 6 seconds

$$= \frac{1000}{6} \, cm^3/sec = \frac{1000}{6} \times 1.3 = 216.7 \, gm/sec$$

Weight flowrate of solids $= 216.7 \times 0.371 = 80.4 \, gm/sec = 289.4 \, kg/hr$

Example 6.1.11: *Two slurry streams, one has a flowrate of 5.3 m³/hr and the other has a flow rate of 4.2 m³/hr, are discharged to a sump where from it is pumped to a beneficiation plant for treatment. On sampling and analyses of two streams, it is found that percent solids by weight in two streams are 45 and 50 respectively and the specific gravity of the solids is 3.2. Calculate the tonnage of dry solids pumped per hour and the percent solids by weight of the total slurry pumped from the sump.*

Solution:

Given

Volumetric flow rate of slurry stream 1 $= 5.3 \, m^3/hr$
Volumetric flow rate of slurry stream 2 $= 4.2 \, m^3/hr$
Fraction of solids in stream 1 by weight $= 0.45$
Fraction of solids in stream 2 by weight $= 0.50$
Density of the solids $= \rho_p = 3.2 \, gm/cm^3$

For stream 1 $\qquad \dfrac{C_w}{\rho_p} + \dfrac{1 - C_w}{\rho_w} = \dfrac{1}{\rho_{sl}} \Rightarrow \dfrac{0.45}{3.2} + \dfrac{1 - 0.45}{1.0} = \dfrac{1}{\rho_{sl}}$

$$\Rightarrow \rho_{sl} = 1.45 \, gm/cm^3 = 1450 \, kg/m^3$$

Weight flowrate of stream 1 $= 5.3 \times 1450 = 7685 \, kg/hr$

Weight flowrate of solids in stream 1 $= 7685 \times 0.45 = 3458.3 \, kg/hr$

For stream 2 $\qquad \dfrac{C_w}{\rho_p} + \dfrac{1 - C_w}{\rho_w} = \dfrac{1}{\rho_{sl}} \Rightarrow \dfrac{0.50}{3.2} + \dfrac{1 - 0.50}{1.0} = \dfrac{1}{\rho_{sl}}$

$$\Rightarrow \rho_{sl} = 1.52 \, gm/cc = 1520 \, kg/m^3$$

Weight flowrate of stream 2 $= 4.2 \times 1520 = 6384 \, kg/hr$

Weight flowrate of solids in stream 2 $= 6384 \times 0.50 = 3192 \, kg/hr$

Tonnage of dry solids pumped from the sump
$\qquad = $ Tonnage of dry solids in both the streams
$\qquad = 3458.3 + 3192 = 6650.3 \, kg/hr$

Weight flowrate of two streams $= 7685 + 6384 = 14069$ kg/hr

% solids in the slurry pumped from the sump $= \dfrac{6650.3}{14069} \times 100 = 47.3\%$

6.2   RETENTION TIME

From the grinding stage onwards, most mineral beneficiation operations are carried out on slurry streams. The slurry is transported through the circuit via pumps and pipelines. The water acts as a transportation medium. The volume of the slurry flowing through the circuit will affect the residence time in unit operations. Volumetric flowrate can be measured by diverting the stream of the slurry into a suitable container for a measured period of time. The ratio of the volume of the slurry collected to the time taken to collect the slurry gives the flowrate of the slurry. This volumetric flowrate is important in calculating retention time of the slurry in any operation. For instance, if $180 \, \text{m}^3/\text{hr}$ of the slurry is fed to a mixing tank of volume of $15 \, \text{m}^3$, then on an average, the retention time of particles in the tank will be

$$\text{Retention time} = \frac{\text{Tank Volume}}{\text{Flow rate}} = \frac{15}{180} = \frac{1}{12} \, \text{hr} = 5 \, \text{minutes}$$

That means, any part of the slurry takes 5 minutes from the time it enters the tank to the time it leaves the tank. The example 6.2.1 illustrates the calculation of volume of the mixing tank required for a retention time of 5 minutes.

Example 6.2.1:   *A Mineral processing plant treats 250 tons of solids per hour. The feed pulp containing 40% solids by weight is sent to a tank for mixing for 5 minutes before it is pumped to the next operation. Calculate the volume of the mixing tank required. Density of solids is $2650 \, kg/m^3$.*

Solution:

Given

Density of solids $\quad = 2650 \, \text{kg/m}^3$
Percent solids in slurry $= 40\%$
Tonnage of solids $\quad = 250 \, \text{tons/hr}$
Time of mixing $\quad = 5 \, \text{minutes}$

$$\text{Volumetric flowrate of solids} = \frac{250 \times 1000}{2650} = 94.34 \, \text{m}^3/\text{hr}$$

Weight flowrate of water $=$ Weight flowrate of solids $\times$ dilution ratio

$$= 250 \times \frac{1 - 0.40}{0.40} = 375 \, \text{tons/hr}$$

$$\text{Volumetric flowrate of water} = \frac{375 \times 1000}{1000} = 375 \, \text{m}^3/\text{hr}$$

$$\text{Volumetric flowrate of slurry} = 375 + 94.34 = 469.34 \, \text{m}^3/\text{hr}$$

$$\text{Retention time} = 5 \text{ minutes}$$

$$\text{Volume of the mixing tank} = 469.34 \times \frac{5}{60} = 39.1 \, \text{m}^3$$

6.3 MISCIBLE LIQUIDS

When it is required to prepare a liquid of definite density by mixing two miscible liquids of known densities as in case of float and sink analysis, the equation is written as

$$C_v \rho_1 + (1 - C_v)\rho_2 = \rho_{12} \tag{6.12}$$

where
ρ_1 = density of liquid 1
ρ_2 = density of liquid 2
ρ_{12} = density of the resultant liquid after mixing two liquids
C_v = fraction of liquid 1 by volume

$$\Rightarrow \qquad C_v = \frac{\rho_{12} - \rho_2}{\rho_1 - \rho_2} \tag{6.13}$$

If C_w is the fraction of liquid 1 by weight, the equation is

$$\frac{C_w}{\rho_1} + \frac{1 - C_w}{\rho_2} = \frac{1}{\rho_{12}} \tag{6.14}$$

$$\Rightarrow \qquad C_w = \frac{\rho_1(\rho_{12} - \rho_2)}{\rho_{12}(\rho_1 - \rho_2)} \tag{6.15}$$

Example 6.3.1 shows the calculation of quantity of two miscible liquids required to be added to yield a solution of desired density.

Example 6.3.1: *Calculate the volume of benzene and carbon tetrachloride required to prepare 100 cc of solution of heavy liquid of specific gravity 1.2. Specific gravities of benzene and carbon tetrachloride are 0.8 and 1.6 respectively.*

Solution:

Given

Density of benzene $= \rho_1 = 0.8 \, \text{gm/cm}^3$
Density of Carbon tetrachloride $= \rho_2 = 1.6 \, \text{gm/cm}^3$
Density of resulting solution $= \rho_{12} = 1.2 \, \text{gm/cm}^3$
Volume of the solution required $= 100 \, \text{cc}$

Substituting density values in equation 6.12

$$C_v \rho_1 + (1 - C_v)\rho_2 = \rho_{12}$$

$$\Rightarrow \quad C_v \times 0.8 + (1 - C_v) \times 1.6 = 1.2$$

$$\Rightarrow \quad C_v = 0.50 \quad \Rightarrow \quad 50\%$$

i.e. quantity of the benzene required is 50% by volume.

As the total quantity of the solution is 100 cc, 50 cc of benzene and 50 cc of carbon tetrachloride are required to prepare 100 cc of solution.

6.4 PROBLEMS FOR PRACTICE

6.4.1: *For the following values of Aerated and Packed densities of six materials, calculate Compressibility and Working density and arrange the materials in the decreasing order of flowability.*

Material code	A	B	C	D	E	F
Aerated density, gm/cc	1.45	1.68	1.75	1.35	1.82	1.52
Packed density, gm/cc	1.75	1.92	1.98	1.61	2.09	1.71

[F C B E D A]

6.4.2: *450 gm of solids are taken and prepared a pulp of 500 cc by adding water to it. If the weight of 500 cc pulp is 800 gm, what is the specific gravity of solids? Determine the density of solids in kg/m³. Calculate percent solids in the pulp by weight and by volume.*

[3.0, 3000 kg/m³, 56.25%, 30%]

6.4.3: *Determine % ferrosilicon by weight to be added to prepare the suspension of specific gravity of 1.80 if the specific gravity of the ferrosilicon is 7.5.*

[51.3%]

6.4.4: *It is required to prepare a pulp of 30% solids by volume with 4.2 litres of water. Determine the weight of solids to be added in kilograms if the specific gravity of solids is 2.65.*

[4.77 kg]

6.4.5: *A quartz slurry flowing at the rate of 720 kg/hr is sampled and found its density as 1400 kg/m³. If the specific gravity of the quartz is 2.65, calculate the weight flowrate of quartz within the slurry.*

[330.5 kg/hr]

6.4.6: A pump is fed by a slurry having a flowrate of 5 m³/hr containing 40% solids by weight. Specific gravity of solids in the slurry is 3.0. Calculate the tonnage of dry solids pumped per hour.

[2.728 tons/hr]

6.4.7: Calculate the pulp density and dilution ratio if the pulp consists of 20% magnetite and 30% quartz by weight. Specific gravities of magnetite and quartz are 5.0 and 2.65.

[1.53 gm/cm³, 1.0]

6.4.8: Two liquids of specific gravities 1.26 and 1.6 are to be mixed to obtain 300 cc solution of specific gravity 1.4. Calculate the quantities of two liquids required by volume and by weight.

[176.4 cc, 123.6 cc, 222.3 gm, 197.7 gm]

6.4.9: Calculate the quantity of bromoform and benzene required to prepare 100 cc of solutions of heavy liquids with specific gravities of 1.4, 1.6 and 1.8. The specific gravities of bromoform and benzene are 2.85 and 0.8 respectively.

[29.3 cc, 70.7 cc, 39 cc, 61 cc, 48.8 cc, 51.2 cc; 83.5 gm, 56.5 gm, 111.2 gm, 48.8 gm, 139.1 gm, 40.9 gm]

Chapter 7

Liberation

Liberation is the first and the most important step in beneficiation of minerals and coal. The second step, separation, is impracticable if the first step, liberation, is not accomplished successfully.

Liberation can be defined as freeing or detachment of dissimilar mineral grains. The operation employed to liberate the dissimilar mineral grains is Size reduction or Comminution.

Free particles: If the particles of ore consist of a single mineral, they are termed as Free particles.

Locked particles: If the particles of ore consist of two or more minerals, they are termed as locked particles. If the locked particles contain valuable mineral at considerable quantity, they are termed as middling particles.

Grain size: It is the size of a mineral as it occurs in the ore.

Particle size: It is the size of any particle whether free or locked particle.

Grain and Grain size pertain to uncrushed ore and Particle and Particle size pertain to crushed or ground ore.

Liberation size: It is the size of a mineral particle at which that mineral is completely liberated. It is the size of a free particle of required (valuable) mineral.

Various mineral grains, present in the ore, exist in physical combination with each other. To detach the valuable mineral grains from all other gangue mineral grains, it is essential to reduce the size of the ore particles.

If one mineral species in an ore is to be separated physically from all other species in the ore, all grains of the desired species must be physically detached from all remaining species in the ore. In an ore containing mineral A, B, & C (Fig. 7.1) if all grains

Figure 7.1 A particle of an Ore containing A, B, & C minerals.

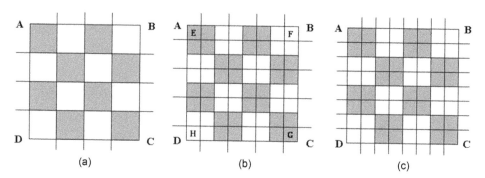

Figure 7.2 Liberation methods.

of mineral A (considered as valuable mineral) is to be separated from the ore, all grains of A must be detached from other minerals B & C (gangue minerals). When such detachment is complete, mineral A is said to be liberated. However liberation of mineral A does not require liberation of other minerals B & C in the ore.

Let us consider a particle in two dimensions (Fig. 7.2). Let ABCD be 8 × 8 cm size particle consists of valuable mineral particles (shaded portion) and gangue mineral particles (white portion) of 2 × 2 cm size each in equal proportion. This particle is reduced by size reduction to a 2 × 2 cm size. If the fracture takes place along the boundary lines as shown in Fig. 7.2(a), it yields 16 particles of 2 × 2 cm size each. All 16 particles are free particles out of which 8 particles are of valuable mineral and 8 particles are of gangue mineral. Here all the valuable mineral particles are liberated. This type of liberation is called liberation by detachment.

In this case, **Grain size, Particle size** and **Liberation size** are equal.

Suppose if the fracture takes place across the grains as shown in Fig. 7.2(b), it yields 25 particles out of which 4 particles E, F, G, H are of 1 × 1 cm size, 12 particles are of 2 × 1 cm size and remaining 9 particles are of 2 × 2 cm size. Here

> E and G are free particles of valuable mineral
> F and H are free particles of gangue mineral
> All others are locked particles

Hence valuable mineral has been liberated partially to a lesser extent.

If the resulting particles are again crushed and fracture takes place along the boundary lines as shown in Fig. 7.2(c), it yields 64 particles, all are free particles of 1 × 1 cm size out of which 32 are valuable mineral particles and 32 are gangue mineral particles. Here all valuable mineral particles are liberated. This happens when the particles are reduced to 1 × 1 cm size which is less than grain size. 1 × 1 cm is the particle size and also the liberation size.

If the fracture during second time also does not take place along the boundary lines, it yields still locked particles which need further reduction in size. This reduction is continued till all valuable mineral particles occur as free particles. Then the particle size, hence liberation size, is less than 1 × 1 cm. This happens because the fracture

Figure 7.3 Typical comminution product.

takes place across the mineral grain. This type of liberation is called liberation by size reduction. In this case, **Liberation size** (also **Particle size**) is less than the **Grain size**.

It is important to note that the particles are reduced by size reduction in both the cases but it matters whether the fracture takes place along the boundaries or across the grains. Locked particles will be produced when the fracture takes place across the grains.

Fig. 7.3 shows an example of typical comminution product wherein black refers valuable mineral and white refers gangue mineral. Only 4 are liberated free valuable mineral particles, 8 are free gangue mineral particles, 6 are locked particles containing valuable and gangue minerals. These locked particles are known as **middling particles.**

Degree of liberation   quantitatively referred to as **percent liberation,** of a certain mineral, is defined as the percent of that mineral liberated and occurring as free particles in relation to the total amount of the same mineral present in the ore.

For example, let the amount of valuable mineral present in one ton of ore is 300 kg. When it is comminuted to certain size, the amount of valuable mineral exists as free particles is 294 kg. The remaining 6 kg of valuable mineral exists as locked particles with gangue minerals.

$$\text{Then the percent liberation} = \frac{294}{300} \times 100 = 98\%$$

Beneficiation is carried out at this size to separate 98% valuable minerals so that 2% of the valuable minerals are lost. To separate 100% valuable minerals, the ore is to be further reduced in size to get 100% liberation which consumes additional power. If the cost of this additional power is more than the cost of 2% valuable minerals, more than 98% liberation is not at all desirable.

Chrome ore of different size fractions are analysed for liberation of chromite particles by optical microscopic examination. Data showing liberation of chromite at various size fractions is presented in Table 7.1. The percentage chromite liberation is calculated using the relationship

$$\text{Chromite liberation} \% = \frac{\text{Free chromite} \%}{\text{Free chromite} \% + \text{Locked chromite} \%} \times 100$$

The data reveals that the chrome ore required grinding up to −80 # for adequate liberation of the chromite particles.

Table 7.1 Liberation of the chrome ore at different particle size fractions [7].

Particle size (mesh #)	Free chromite particles %	Locked particles %	Free gangue %	Chromite liberation %
−8 + 10	15.38	58.91	25.70	20.70
−10 + 25	30.4	34.16	35.43	47.09
−25 + 50	39.2	22.38	38.41	63.66
−50 + 60	49.0	10.23	40.71	82.72
−60 + 80	50.1	9.0	40.8	84.77
−80	52.2	6.4	41.3	89.08

Table 7.2 Liberation of the manganese ore at different particle size fractions [8].

Particle size (mesh #)	Free particles	Locked particles	Degree of liberation %
−22 + 44	12	40	23.08
−44 + 60	32	51	38.55
−60 + 100	34	30	53.12
−100 + 150	38	10	79.17
−150 + 200	58	8	87.88
−200	68	8	89.47

Analysis of comminution product of Manganese ore is carried out on individual sieve fraction by mounting the mineral grains on square glass plate and observed under reflected light microscope to know the degree of liberation. Data showing liberation of manganese mineral at various size fractions is presented in Table 7.2.

The data reveals that the manganese ore under investigation required grinding up to −200 # for adequate liberation of the manganese mineral particles.

Beneficiation method to be used depends on liberation size of the ore which in turn depends on type of the ore. Ore types can be conveniently classified as follows:

Massive ores: In these ores, reasonable amount of crushing makes the valuables liberated.
                  Example: Coal, bedded iron ores

Intergrown Ores: In these ores, valuables can be freed only partially by crushing and require certain amount of grinding to complete the liberation.
                  Example: Chrome ore

Disseminated ore: In these ores, valuables are sparely distributed through a waste rock matrix and require fine grinding to liberate the valuables.
                  Example: Gold ore

To liberate the valuables, the ore particles are to be reduced in size by the application of the forces. When the forces are applied on the ore particle, fracture takes place depending upon the method of application of the forces.

Chapter 8

Comminution

The operation of applying a force on the particle to break it is called size reduction. **Comminution** is a general term for size reduction that may be applied without regard to the actual breakage mechanism involved.

In any industrial comminution operation, the breakage of any individual particles is occurring simultaneously with that of many other particles. The breakage product of any particle is intimately mixed with those of other particles. Thus an industrial comminution operation can be analyzed only in terms of a distribution of feed particles. However, each individual particle breaks as a result of the stresses applied to it and it alone.

8.1 OBJECTIVES OF COMMINUTION

The following are some of the objectives of comminution:

1. Reduction of large lumps into small pieces
2. Production of solids of desired size range
3. Liberation of valuable minerals from gangue minerals
4. Preparation of feed material for different beneficiation operations
5. Convenience in handling and transportation

The energy consumed for the comminution operation is high when compared to other operations such as screening, beneficiation, dewatering, conveying etc. in mineral and mining industries. Hence much attention is to be paid to minimize the production of fines (finer than required) which will consume additional power for reducing to fines.

8.2 LAWS OF COMMINUTION

Laws of comminution are concerned with the relationship between energy input and the sizes of feed and product particles. Three laws of comminution for energy requirements have been put forward by Rittinger, Kick and Bond respectively. None of these three laws is applicable over a wide range of sizes. The Rittinger and Kick laws while tenable in some cases, were never of much use as practical tools. The third law proposed by F.C. Bond in 1952 is based on a detailed compilation and study of numerous laboratory and plant comminution data and provides the technician with a reasonably accurate measure of power requirements.

Rittinger's Law: Rittinger [9] postulated that the energy required for size reduction of a solid is directly proportional to the new surface area created. Mathematically, this can be expressed as

$$\frac{E}{m} = W = K_r \left[\frac{1}{d_P} - \frac{1}{d_F} \right] \tag{8.1.1}$$

where

E = Energy or Power, kW
m = Feed rate to machine, tons/hr
W = Energy or power, kWhr/ton
d_P = Size of the product particles, microns
d_F = Size of the feed particles, microns
K_r = Rittinger's constant.

The reciprocal of K_r is called the **Rittinger's number**. Rittinger's law is best applicable to coarse and intermediate size reduction.

Kick's Law: Kick [10] proposed that the energy required to produce analogous changes of configuration of geometrically similar bodies of equal technologic state varies as the volumes or weights of these bodies. If a particle is only deformed so that only its configuration or surface area is altered, then the above statement remains to be true. However, in comminution where the number of particles increases due to breakage of large particles and new surfaces are created, the above statement ceases to be applicable.

The modified Kick's law that can be applied to crushing states that the work required for crushing a given quantity of material is constant for a given reduction ratio irrespective of original size. Mathematically, this law can be expressed as

$$\frac{E}{m} = W = K_k \log \frac{d_F}{d_P} \tag{8.1.2}$$

where K_k is Kick's constant.

According to this law, comminution energy depends only on the reduction ratio and is independent of the original size of the feed. It is obviously ridiculous to say that the energy required for reducing a 100 mm particle to 50 mm size will be the same as that for reducing a 1 mm particle to 0.5 mm. In fact, higher amount of energy is required for reducing fine particles to still finer size than for breaking down large pieces of rock. Kick's law is reasonably accurate in the crushing range above about 1 cm in diameter.

Bond's Law: Bond [11] states that the total work useful in breakage that has been applied to a given weight of homogeneous broken material is inversely proportional to the square root of the average size of the product particles, directly proportional to the length of crack tips formed and directly proportional to the square root of new surface created. Mathematically, Bond's law is expressed as

$$\frac{E}{m} = W = K_b \left[\frac{1}{\sqrt{d_P}} - \frac{1}{\sqrt{d_F}} \right] \tag{8.1.3}$$

where K_b is Bond's constant.

To apply Bond's law, Bond's constant has to be evaluated. Bond's constant is evaluated by defining what is called work index, W_i. It is defined as the gross energy in kWhr/short ton of feed necessary to reduce a very large feed to such a size that 80% of product particles passes 100 microns screen.

Based on this definition, it can be written that

If $d_F = \infty$, and $d_P = 100$ microns, $\dfrac{E}{m} = W_i$ kWhr/short ton

On substitution in 8.1.3,

$$\frac{E}{m} = W_i = K_b\left[\frac{1}{\sqrt{100}} - \frac{1}{\sqrt{\infty}}\right] \Rightarrow W_i = \frac{K_b}{10} \Rightarrow K_b = 10\,W_i$$

Thus, if 80% of feed particles passes d_F micron screen and 80% of product passes d_P micron screen, then

$$\frac{E}{m} = W = 10\,W_i\left[\frac{1}{\sqrt{d_P}} - \frac{1}{\sqrt{d_F}}\right] \tag{8.1.4}$$

The work index includes the friction in the crusher and the power W is gross power. An experimental procedure for determination of Work Index of a given material as suggested by Bond has been given in Annexure. Bond's law is applicable reasonably in the range of conventional rod mill and ball mill grinding.

It is a fact that most of the energy supplied to a comminution machine is absorbed by the machine itself to move various parts of the machine, and only a small fraction of the total energy is available for breaking the material. For example, in a ball mill, less than 1% of the total energy input is used for actual size reduction, the bulk of the energy is utilized in running the mill and in the production of heat.

8.3 TYPES OF COMMINUTION OPERATIONS

The run-of-mine ore is quite coarse and cannot be reduced to fine size in one stage. It may require three or more stages. Each stage requires separate equipment. The comminution operations are divided into two broad groups as **Crushing** wherein large lumps are reduced to fragments or smaller particles and **Grinding** wherein relatively coarse particles are reduced to the ultimate fineness.

The machines used for crushing and grinding are entirely different. It is to be noted that the energy required for comminution of unit mass of smaller particles is more than the energy required for unit mass of coarser particles. However, the energy required to reduce coarser particle is more than that of smaller particle. Hence the machines used for crushing (crushers) must be massive and rugged and the machines used for grinding (mills) must be capable of dispersing energy over a large area. In crushers, the breakage forces are applied either by compression or impact whereas in grinding mills shear forces are predominantly applied. The comminution laws are of vital importance in calculating the power consumption in comminution operations. Examples 8.3.1 to 8.3.6 illustrates the calculations of power consumption.

Example 8.3.1: *A crusher fed at the rate of 1 ton/hr is crushing 12 mm cubes to a product size of 80% retained on 3 mm screen, 10% retained on 2 mm screen, and 10% retained on 1 mm screen. If the crusher requires 4 HP for this operation, what power is required to crush the same material at the same rate from a feed size of 8 mm to product size of 1 mm. Estimate the power using (a) Rittinger's law and (b) Kick's law.*

Solution:

Given

Feed size during first operation	$= d_{F1} = 12$ mm;
Feed size during second operation	$= d_{F2} = 8$ mm;
Product size during second operation	$= d_{P2} = 1$ mm;
Energy required during first operation	$= E_1 = 4$ HP;
Energy required during second operation	$= E_2 = ?$
Mass flow rate during both operations	$= m_1 = m_2 = 1$ ton/hr

Average size of the product during first operation $= d_{P1} = \dfrac{100}{\sum \dfrac{w}{d}}$

$$= \frac{100}{\dfrac{80}{3} + \dfrac{10}{2} + \dfrac{10}{1}} = 2.4 \text{ mm}$$

(a) As per Rittinger's law $\dfrac{E}{m} = K_r \left(\dfrac{1}{d_P} - \dfrac{1}{d_F} \right)$

Substituting the values in this equation, gives

$$\frac{4}{1} = K_r \left(\frac{1}{2.4} - \frac{1}{12} \right) \tag{I}$$

$$\frac{E_2}{1} = K_r \left(\frac{1}{1} - \frac{1}{8} \right) \tag{II}$$

Eq. (II)/Eq. (I) gives $E_2 = 10.5$ HP

Power required to crush from a feed size of 8 mm to 1 mm product $= 10.5$ HP

(b) As per Kick's Law $\dfrac{E}{m} = K_k \log \dfrac{d_F}{d_P}$

Substituting the values in this equation, gives

$$\frac{4}{1} = K_k \log \frac{12}{2.4} \tag{III}$$

$$\frac{E_2}{1} = K_k \log \frac{8}{1} \tag{IV}$$

Eq. (IV)/Eq. (III) gives $E_2 = 5.168$ HP

Power required to crush from a feed size of 8 mm to 1 mm product $= 5.168$ HP

Example 8.3.2: *A rock of average particle size of 25 mm is crushed to a product of average particle size of 6 mm at a rate of 10 tons/hr. At this rate the mill takes 18 kW power. It requires 0.5 kW to run the mill empty. What will be the power consumption if the same feed was crushed to a particle diameter to 10 mm? Assume that Rittinger's Law is valid.*

Solution:

Given

Feed size for both operations	$= d_F$	$= 25\,mm$
Product size during first operation	$= d_{P1}$	$= 6\,mm$
Product size during second operation	$= d_{P2}$	$= 10\,mm$

Energy required during first operation $= E_1$		$= 18\,kW$
Energy required for running the mill empty		$= 0.5\,kW$
Flow rate during both operations	$= m_1$	$= m_2 = 10\,ton/hr$

Energy utilized for crushing $= 18.0 - 0.5 = 17.5\,kW$

As per Rittinger's law $\quad \dfrac{E}{m} = K_r \left(\dfrac{1}{d_P} - \dfrac{1}{d_F} \right)$

Substituting values in this equation, gives

$$\frac{17.5}{10} = K_r \left(\frac{1}{6} - \frac{1}{25} \right) \tag{I}$$

$$\frac{E}{10} = K_r \left(\frac{1}{10} - \frac{1}{25} \right) \tag{II}$$

Eq. (II)/Eq. (I) gives $E = 8.29\,kW$
Power consumption to crush the given material to 10 mm size
$$= 8.29 + 0.5 = 8.79\,kW$$

Example 8.3.3: *A rock of 5 cm size is fed to a gyratory crusher. The differential screen analysis of the product is given in Table 8.3.3.1 under column (3). The power requirement for crushing is 432 kW/ton. The crusher requires 12 kW of power on no load. By reducing the clearance between the crushing head and the cone, the differential screen analysis of the product becomes as shown in column (4) of Table 8.3.3.1. Calculate the power requirement for the second operation using Kick's Law.*

Table 8.3.3.1 Screen analysis of gyratory crusher product.

(1) Mesh No.	(2) Size of opening mm	(3) Weight % Retained	(4) Weight % Retained
4	4.70	–	–
6	3.33	3.1	–
8	2.36	10.3	3.3
10	1.65	20.0	8.2
14	1.17	18.6	11.2
20	0.83	15.2	12.3
28	0.59	12.0	13.0
35	0.42	9.5	19.5
48	0.30	6.5	13.5
65	0.21	4.3	8.5
100	0.15	0.5	6.2
150	0.10	–	4.0
−150	–	–	0.3

Solution:

Given

Size of the feed particle	$= d_F = 5\,\text{cm}$	
Power required on no load	$= 12\,\text{kW}$	
Power consumption for crushing	$= 432\,\text{kW/ton}$	

Let d_i be the average size of each fraction and w_i be the weight percent of each fraction. Various values necessary are calculated and given in Table 8.3.3.2.

Table 8.3.3.2 Calculated values for example 8.3.3.

Mesh No.	Average size (d_i) mm	for (3) w_i	for (3) w_i/d_i	for (4) w_i	for (4) w_i/d_i
+4	+4.70	–	–	–	–
−4 + 6	(4.70 + 3.33)/2 = 4.015	3.1	0.772	–	–
−6 + 8	(3.33 + 2.36)/2 = 2.845	10.3	3.620	3.3	1.160
−8 + 10	(2.36 + 1.65)/2 = 2.005	20.0	9.975	8.2	4.090
−10 + 14	(1.65 + 1.17)/2 = 1.410	18.6	13.192	11.2	7.943
−14 + 20	(1.17 + 0.83)/2 = 1.000	15.2	15.200	12.3	12.300
−20 + 28	(0.83 + 0.59)/2 = 0.710	12.0	16.901	13.0	18.310
−28 + 35	(0.59 + 0.42)/2 = 0.505	9.5	18.812	19.5	38.614
−35 + 48	(0.42 + 0.30)/2 = 0.360	6.5	18.056	13.5	37.500
−48 + 65	(0.30 + 0.21)/2 = 0.255	4.3	16.863	8.5	33.333
−65 + 100	(0.21 + 0.15)/2 = 0.180	0.5	2.778	6.2	34.444
−100 + 150	(0.15 + 0.10)/2 = 0.125	–	–	4.0	32.000
−150	(0.10 + 0.00)/2 = 0.050	–	–	0.3	6.000
		100.0	116.169	100.0	225.694

Size of the product particle $= d_{P1} = \dfrac{100}{\sum \dfrac{w_i}{d_i}} = \dfrac{100}{116.169} = 0.8608\,\text{mm} = 0.08608\,\text{cm}$

Applying Kick's law $\quad W = K_k \ln \dfrac{d_F}{d_P} = K_k \ln \dfrac{5}{0.08608} = K_k\,4.062$

Power utilized for crushing = Power consumption for crushing

$$- \text{Power to run empty crusher}$$

$$= 432 - 12 = 420\,\text{kW/ton}$$

Kick's law gives $\quad 420 = 4.062 K_k$

$$\therefore \quad K_k = \dfrac{420}{4.062} = 103.4\,\text{kW/ton}$$

Size of the product particle when the clearance is reduced (second operation)

$$= d_{P2} = \dfrac{100}{\sum \dfrac{w_i}{d_i}} = \dfrac{100}{225.694} = 0.443\,\text{mm} = 0.0443\,\text{cm}$$

Again applying Kicks law $\quad W = K_k \ln \dfrac{d_F}{d_P} = 103.4 \times \ln \dfrac{5}{0.0443} = 488.7\,\text{kW/ton}$

$\therefore$ Power utilized for crushing during second operation $= 488.7\,\text{kW}$

Power required for second operation = Power utilized for crushing

$$+ \text{Power required for empty crusher}$$

$$= 488.7 + 12 = 500.7\,\text{kW/ton}$$

Example 8.3.4: *The following results were obtained in a laboratory experiment designed to find the work index of an ore:*

1000 grams of copper zinc ore were ground in a laboratory mill for 12 minutes. The power input to the empty mill (contained ball charge only) was 226.9 watts. The power input to the mill when grinding the sample was 283.2 watts. Feed to the mill was 80% minus 1530 microns. Product from the mill was 80% minus 79 microns.

Find (a) Net kWhr per ton of ore required to grind ore from 1530 microns to 79 microns.

(b) kWhr per ton of ore to grind from infinite size to 100 microns, i.e. Work Index.

Solution:

Given

Weight of copper zinc ore grounded	$= 1000\,\text{gm} = 1\,\text{kg}$
Time of grinding	$= 12$ minutes
Power input to the empty mill with ball charge	$= 226.9$ watts
Power input for grinding the ore	$= 283.2$ watts
Feed size to the mill	$= d_F = 1530$ microns
Product from the mill	$= d_P = 79$ microns

(a)   Net wattage input $= 283.2 - 226.9 = 56.3$ watts

$$\text{Net power used} = 56.3 \times 12 \text{ watt minutes} = \frac{56.3 \times 12}{1000 \times 60}$$

$$= 0.01126 \, \text{kWhr}$$

0.01126 kWhr power is used to grind 1000 gm (1 kg)

$\therefore$ Net power required to grind one ton of ore $= 0.01126 \times 1000$

$$= 11.26 \, \text{kWhr}$$

(b)   Work Index $= W_i = \dfrac{W}{10\left(\dfrac{1}{\sqrt{d_P}} - \dfrac{1}{\sqrt{d_F}}\right)}$

$$= \frac{11.26}{10\left(\dfrac{1}{\sqrt{79}} - \dfrac{1}{\sqrt{1530}}\right)}$$

$$= 12.95 \, \text{kWhr/ton}$$

Example 8.3.5: *Calculate the power required to crush the phosphate rock from a feed size of 80% passing 4″ to a product size of 80% passing 1/8″ at a rate of 100 tons/hr. The work index of phosphate rock is 2.74. Also calculate the power required to crush the product further to 80% passing 1000 microns.*

Solution:

Given

80% passing size of the product $= d_P = 1/8'' = 3175$ microns;
80% passing size of the feed $\quad = d_F = 4'' = 101600$ microns;
Work Index of the limestone $\quad = W_i = 2.74 \, \text{kWhr}$

As per the Bond's Law $W = 10 W_i \left(\dfrac{1}{\sqrt{d_P}} - \dfrac{1}{\sqrt{d_F}}\right)$

$$= 10 \times 2.74 \left(\frac{1}{\sqrt{3175}} - \frac{1}{\sqrt{101600}}\right) = 0.4003 \, \text{kWhr/ton}$$

Power per 100 tons/hr $= 0.4003 \times 100 = 40.03 \, \text{kW}$

For the second operation

$$80\% \text{ passing size of the product} = d_P = 1000 \text{ microns}$$
$$80\% \text{ passing size of the feed} \quad = d_F = 1/8'' = 3175 \text{ microns};$$

$$W = 10\,W_i\left(\frac{1}{\sqrt{d_P}} - \frac{1}{\sqrt{d_F}}\right)$$

$$= 10 \times 2.74\left(\frac{1}{\sqrt{1000}} - \frac{1}{\sqrt{3175}}\right) = 0.3803 \text{ kWhr/ton}$$

Power per 100 tons/hr $= 0.3803 \times 100 = 38.03 \text{ kW}$

$\therefore$ Total power required to crush phosphate rock from a feed size of $4''$ to a product size of 1000 microns $= (40.03 + 38.03) \text{ kW} = 78.06 \text{ kW}$

Example 8.3.6: *A lead zinc ore is being ground from a feed size of 80% passing 3 mesh to a product size of 80% passing 100 mesh for flotation treatment and power consumption is 7.2 kWhr per ton of ore ground. Because of the changing nature of the ore with the increased depth of mining, the metallurgical recovery can only be maintained by finer grinding. Test work has indicated that by crushing the feed to 80% passing 4 mesh and grinding in a ball mill to 80% passing 325 mesh, recoveries will be satisfactory. Calculate the power required to grind one ton of ore under the new conditions, according to bond's law.*

Mesh sizes for 3, 4, 100 and 325 are 6730, 4760, 149 and 37 microns respectively.

Solution:

Given
$$80\% \text{ passing size of the feed} \quad = d_P = 6730 \text{ microns}$$
$$80\% \text{ passing size of the product} = d_F = 149 \text{ microns}$$
$$\text{Power consumption} \quad\quad\quad = W = 7.2 \text{ kWhr/ton}$$
For second operation
$$80\% \text{ passing size of the feed} \quad = d_P = 4760 \text{ microns}$$
$$80\% \text{ passing size of the product} = d_F = 37 \text{ microns}$$

As per the Bond's Law $\quad W = K_b\left(\dfrac{1}{\sqrt{d_P}} - \dfrac{1}{\sqrt{d_F}}\right)$

For first operation $\quad 7.2 = K_b\left(\dfrac{1}{\sqrt{149}} - \dfrac{1}{\sqrt{6730}}\right)$ $\qquad$ (I)

For second operation $\quad W = K_b\left(\dfrac{1}{\sqrt{37}} - \dfrac{1}{\sqrt{4760}}\right)$ $\qquad$ (II)

Eq. (II)/Eq. (I) gives $\quad W = 15.49 \text{ kWhr/ton}$

Power required under new conditions of size reduction $= 15.49 \text{ kWhr/ton}$

8.4 PROBLEMS FOR PRACTICE

8.4.1: *A certain crusher accepts a feed of rock having a volume-surface mean diameter of 0.75" and discharges a product of volume-surface mean diameter of 0.2". The power required to crush 12 tons/hr is 9.3 HP. What should be the power consumption if the capacity is reduced to 10 tons/hr and the volume-surface mean diameter of the product to 0.15"? The mechanical efficiency remains unchanged. Use Rittinger's law.*

[8.4 kW]

8.4.2: *A crusher was used to crush a material with a feed size of −5.08 cm +3.81 cm and the power required was 3.73 kWhr/ton. The screen analysis of the product is given in Table 8.4.2.1.*

Table 8.4.2.1 Screen analysis of crusher product.

	through	on	on	on	on	on	on
Size of aperture, cm	0.63	0.38	0.203	0.076	0.051	0.025	0.013
% product	Nil	26	18	23	8	17	8

What would be the power required to crush one ton per hour of the same material from a feed −4.44 cm + 3.81 cm to a product of average size 0.051 cm? Use Rittinger's Law.

[5.78 kWhr/ton]

8.4.3: *What is the power required to crush 100 tons/hr of limestone if 80% of the feed passes 2" screen and 80% of the product passes 1/8" screen? The work index for limestone may be taken as 12.74 kWhr.*

[169.6 kW]

8.4.4: *If a ball mill requires 3 kWhr per ton of feed to reduce a feed of 1600 micron size to a product of 400 microns size, calculate the work index.*

[12 kWhr/ton]

8.4.5: *A comminution process consumed 460 HP/ton power to crush a rock from 7.5 cm to 0.75 cm. If the power consumed was 697 HP/ton to reduce the feed of the same size, calculate the product size in the event.*

[4101.2 microns]

Chapter 9

Crushing

The crushing is to reduce large lumps to fragments, nearly to $1/2$ inch size in many cases. The crushing action in all crushing machines (crushing) results from forces applied to the particles by some moving part working against a stationary or some other moving part. It is the first stage of size reduction. It has to crush run-of-mine ore contains large size particles. It requires greater force to apply on particles. Hence the crushers are very rugged and massive. Crushing is generally a dry operation. No crusher is capable of producing a very fine product from an extremely coarse feed and is usually performed in two or three stages depending upon the extent of size reduction needed.

The extent of size reduction achieved by any crushing operation is described by the **reduction ratio**. The reduction ratio is broadly defined as the ratio of the feed size to the product size in any crushing operation. It is very useful in determining what a crusher can do, or is doing, in the way of size reduction. It can also be used as a partial indicator of the stresses the crusher will be subjected to during operation, an element in determining the crusher capacity and as an indicator of crusher efficiency. However, there is no one method of calculation which will provide a useful figure for all of these considerations, so there are various types of reduction ratios in use. Few of these ratios are defined as follows:

$$\text{Limiting reduction ratio} = \frac{\text{Size passing all feed}}{\text{Size passing all product}}$$

$$\text{Average reduction ratio} = \frac{\text{Average size of the feed particles}}{\text{Average size of the product particles}}$$

$$80\% \text{ passing reduction ratio} = \frac{80\% \text{ passing size of the feed}}{80\% \text{ passing size of the product}}$$

Reduction ratio is a convenient measure for comparing the performance of different crushers.

As the crushing is performed in stages, crushing may be divided into primary, secondary, and tertiary stages based on the particle size. The crushers can be classified into the following five groups according to the size of the product they produce.

1 **Primary Crushers:**   Jaw crusher, Gyratory crusher
2 **Secondary Crushers:** Reduction gyratory, Cone crusher, Rolls crusher

3   **Tertiary Crushers:**   Short-head cone crusher
4   **Fine Crushers:**   Impact crushers
5   **Special Crushers:**   Bradford Breaker, Toothed Roll crusher,
                             Gravity stamps

Lumps of run-of-mine ore usually of 1 m size is reduced to 100–200 mm size in heavy duty primary crushers. The usual size of feed to secondary crushers is 600 mm and the product is usually 10–100 mm size. In tertiary crushers, particles of 250 mm size are reduced to 3–25 mm size.

Fine crushers reduce the coarse particles to fine even to 200 mesh in some cases. Special crushers are designed for specific ores, for example, rotary breaker and toothed rolls crusher for coal and gravity stamps for gold ore milling.

9.1   TYPES OF CRUSHERS

Jaw, gyratory, cone, and rolls crushers are different types of nipping machines. Types of crushers, their characteristics, range of reduction ratios and their principal applications are shown in Table 9.1.1.

In nipping machines, the distance between two crushing surfaces at the feed opening is called **gape** and at the discharge opening it is called **set**. The distance between two crushing surfaces when they are close to each other is called **closed set** and when they are away from each other it is called **open set**. In a roll crusher set is constant.

The roll crusher is one of the nipping machines. The angle between the two crushing surfaces is called the **angle of nip** (Figure 9.1.1). For the selection of size of roll crusher for the reduction of different sizes of feed, an expression for the angle of nip can be obtained as follows considering a spherical particle. Let D be the diameter of the rolls, d be the diameter of the spherical feed particle, n be the angle of nip, and s be the distance between two rolls.

Neglecting the gravity force, the two forces acting on the particle when it is fed between two rolls are the tangential force, T, and the normal force, N. If the resultant of these two forces, R, is directed downward, the particle will be nipped and get crushed. If the resultant force R is directed upward, the particle will be thrown out from the rolls and there will be no crushing. The limiting condition is therefore when resultant force R is just horizontal. Under this limiting condition, the angle between the tangents is called angle of nip, n, since this is the maximum angle beyond which the particle will not be nipped. Under these limiting condition, the vertical components of N and T will balance with each other. Thus,

$$N \sin \frac{n}{2} = T \cos \frac{n}{2} \qquad \Rightarrow \qquad \tan \frac{n}{2} = \frac{T}{N}$$

Since $\dfrac{T}{N}$ is the coefficient of friction, μ, then $\tan \dfrac{n}{2} = \mu$ (9.1.1)

Hence in a roll crusher, the necessary and sufficient condition for the particle to be nipped is obviously

$$\tan \frac{n}{2} \leq \mu \qquad\qquad (9.1.2)$$

Table 9.1.1 Types of Crushers.

Crusher	Characteristics	Reduction Ratio	Applications
Jaw Crusher Blake (double toggle)	Swing jaw pivoted at top; Flywheel evens the power draft; Standard crusher; Intermittent machine; Yields relatively coarse product	4–9	Primary Suitable for hard, tough abrasive, and sticky feeds; dry operation
Blake (single toggle)	Swing jaw pivoted at top; Flywheel evens the power draft; Intermittent machine	4–9	Primary; restricted to laboratory use; dry operation
Dodge	Swing jaw pivoted at bottom; Yields closed size product; Prone to choking	4–9	Primary; restricted to laboratory use; Dry operation
Gyratory	Consists two vertical, truncated, conical shells; Apex of outer pointing down; Apex of inner pointing up; Outer shell is stationary; Inner shell gyrates or rotates; Suspended spindle; Yields coarse product; Continuous machine; High capacity	3–10	Primary; More suitable for slabby feeds; Dry operation
Reduction Gyratory	Has bell shaped heads and concaves; continuous	6–8	Secondary; Dry operation
Cone	Modified gyratory; Spindle is supported in a curved, universal bearing at the bottom	6–8	Secondary; dry operation
Standard	Has stepped liners to allow coarse feed		
Short-head	Heads has steeper angle to prevent choking; fine feeds		Tertiary; Dry operation
Rolls	Two horizontal cylinders or rolls revolve towards each other; continuous	2–4	Best for secondary; dry or wet operation
Toothed Rolls	Rolls with toothed surface	–	Best for soft rock like coal
Hammer mill	Consists of swing hammers mounted on a rotating shaft	large	Fine crusher; Used for non-abrasive ores dry operation
Bradford breaker	Cylinder with perforations like tommel	–	Used for coarse breaking of coal

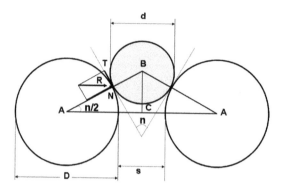

Figure 9.1.1 Angle of nip of Roll Crusher.

Considering the triangle ABC in Figure 9.1.1,

$$\cos\frac{n}{2} = \frac{\dfrac{D}{2}+\dfrac{s}{2}}{\dfrac{D}{2}+\dfrac{d}{2}} \quad \Rightarrow \quad \cos\frac{n}{2} = \frac{D+s}{D+d} \tag{9.1.3}$$

The theoretical capacity of a roll crusher in kg/hr is given by

$$C = 60\pi\,DNWs\rho \tag{9.1.4}$$

where N is the number of revolutions of the rolls per minute, D is the diameter of the rolls, W is the width of the roll face, s is the set, all in meters and ρ is the density of the particle in kg/m^3. In practice, allowing for voids between the particles, loss of speed in gripping the feed, etc., the actual capacity is usually about 25% of the theoretical capacity. Smooth-surfaced roll crushers are generally used for fine crushing whereas corrugated or toothed roll crushers are used for coarse crushing.

Calculations by using above formulae are illustrated in Examples 9.1.1 to 9.1.7.

Example 9.1.1: *Determine the minimum diameter of the rolls in a roll crusher to reduce 3.81 cm pieces of rock to 1.27 cm if the coefficient of friction between the rock and steel is 0.4.*

Solution:

Given

    Size of the feed particle   = 3.81 cm
    Size of the product particle = 1.27 cm
    Coefficient of friction between rock and steel = 0.4

Coefficient of friction $= f = \tan n/2 = 0.4$

$\Rightarrow \quad n/2 = 21°48'$ $\therefore \quad \cos n/2 = \cos 21°48' = 0.9285$

But $\cos\dfrac{n}{2} = \dfrac{D+s}{D+d}$ $\Rightarrow \quad 0.9285 = \dfrac{D+1.27}{D+3.81}$

                        $\Rightarrow \quad D = 31.72$ cm

Diameter of the roll to reduce 38.1 mm pieces of rock to 12.7 mm = 31.72 cm

Example 9.1.2: *In a smoothed rolls crusher, crushing rolls of 100 cm diameter each are set at 12.5 cm apart. If the angle of nip is 31°, calculate the maximum size of the feed to the roll crusher.*

Solution:

Given

Diameter of the roll $= D = 100$ cm
Set of the rolls $\quad = s \ = 12.5$ cm
Angle of nip $\qquad = n \ = 31°$

$$\Rightarrow \qquad \frac{n}{2} = \frac{31}{2} = 15.5°$$

$$\cos\frac{n}{2} = \cos 15.5° = 0.9636$$

But $\cos\dfrac{n}{2} = \dfrac{D+s}{D+d} \qquad \Rightarrow \quad 0.9636 = \dfrac{100+12.5}{100+d}$

$$\Rightarrow \quad d = 16.75 \text{ cm}$$

∴ Maximum size of the feed particle $= 16.75$ cm

Example 9.1.3: *Calculate the angle of nip in case of a roll crusher of the following specifications.*

Diameter of the roll $= 7.5\ cm$
Feed size $\qquad\qquad = 0.6\ cm$
Product size $\qquad\quad = 0.05\ cm$

Solution:

Given

Diameter of the roll $\qquad\qquad = D = 7.5$ cm
Diameter of the feed particle $= d \ = 0.6$ cm
Set of the rolls $\qquad\qquad\quad = s \ = 0.05$ cm

$$\cos\frac{n}{2} = \frac{D+s}{D+d} = \frac{7.5+0.05}{7.5+0.6} = 0.9321$$

$$\Rightarrow \qquad \frac{n}{2} = 21°14' \quad \Rightarrow \quad n = 42°28'$$

Angle of nip in roll crusher $= 42°28'$

Example 9.1.4: *When a double roll crusher having rolls of 140 cm diameter and 50 cm face width crushes the particles of size 60 cm, determine the product size obtained. The coefficient of friction is 0.28. Compare the result when the coefficient of friction is 0.32.*

Solution:

Given

Size of the feed particle $= 60$ cm
Diameter of the rolls $\quad = 140$ cm
Coefficient of friction $\ = 0.28$

$$\text{Coefficient of friction} = f = \tan\frac{n}{2} = 0.28 \quad \Rightarrow \frac{n}{2} = 15°39'$$

$$\therefore \quad \cos\frac{n}{2} = \cos 15°39' = 0.9634$$

$$\text{But } \cos\frac{n}{2} = \frac{D+s}{D+d} \qquad \Rightarrow \quad 0.9634 = \frac{140+s}{140+60}$$

$$\Rightarrow \quad s = 52.68\,\text{cm}$$

$$\text{Product size of the particle} = 52.68\,\text{cm}$$

If the coefficient of friction is 0.32, $\tan\dfrac{n}{2} = 0.32$

$$\Rightarrow \quad \frac{n}{2} = 17°45' \qquad \therefore \quad \cos\frac{n}{2} = \cos 17°45' = 0.953$$

$$\text{But } \cos\frac{n}{2} = \frac{D+s}{D+d} \qquad \Rightarrow \quad 0.953 = \frac{140+s}{140+60}$$

$$\Rightarrow \quad s = 50.6\,\text{cm}$$

$$\text{Product size of the particle} = 50.6\,\text{cm}$$

From the above result, it is evident that with the increase of coefficient of friction, smaller and smaller product sizes can be obtained using the same feed size and the same crusher.

Example 9.1.5: *A crushing rolls crushes a rock of specific gravity 2.5 from 3.81 cm with a reduction ratio of 3. The angle of nip is 39° and the rolls operate at 60 rpm to give a capacity of 40 tons/hr. Calculate the coefficient of friction between steel and the rock, product particles size, the diameter and width of the rolls.*

Solution:

Given

Size of the feed particle $= d$	$= 3.81\,\text{cm}$	
Reduction ratio	$= RR = 3$	
Angle of nip	$= n$	$= 39°$
Speed of the rolls	$= N$	$= 60\,\text{rpm}$
Mill capacity	$= C$	$= 40\,\text{tons/hr} = 40000\,\text{kg/hr}$
Density of the rock	$= \rho$	$= 2.5 \times 1000 = 2500\,\text{kg/m}^3$

$$\text{Coefficient of friction} = f = \tan\frac{n}{2} = \tan\frac{39}{2} = 0.3541$$

$$\text{Size of the product particle} = \frac{d}{RR} = \frac{3.81}{3} = 1.27\,\text{cm}$$

$$\frac{n}{2} = \frac{39}{2} = 19.5 \quad \Rightarrow \quad \cos\frac{n}{2} = 0.9426$$

$$\text{But } \cos\frac{n}{2} = \frac{D+s}{D+d} \quad \Rightarrow \quad 0.9426 = \frac{D+1.27}{D+3.81} \quad \Rightarrow \quad D = 40.44\,\text{cm}$$

$\therefore$ Diameter of the roll crusher $= 40.44\,cm$

Theoretical capacity of the crushing rolls $= C = 60\pi DNWs\rho$

$$\Rightarrow W = \frac{C}{60\pi DNs\rho}$$

$$\Rightarrow W = \frac{40000}{60\pi \times 40.44/100 \times 60 \times 1.27/100 \times 2500} = 0.2754\,m = 27.54\,cm$$

$\therefore$ Width of the roll face $= 27.54\,cm$

Example 9.1.6:   *A certain set of crushing rolls of 150 cm diameter and 50 cm width of the face are set in such a way that the crushing surfaces are 1.25 cm apart at the narrowest point. If the crushing rolls are crushing a rock of specific gravity 2.35 with angle of nip of 30° and run at 100 rpm speed,*

   a)   What is the allowable size of the feed particles or rock?
   b)   What is the actual capacity in tons/hr if the actual capacity is 15% of the theoretical capacity?

   After long use, the tires on the rolls of mill have become roughened so that the angle of nip is changed to 32° 30'. Then what will be the maximum permissible size of feed and the capacity of the rolls?

Solution:
   Given

Set of a roll crusher	$= s$	$= 1.25\,cm$
Diameter of the rolls	$= D$	$= 150\,cm$
Width of the roll face	$= W$	$= 50\,cm$
Speed of the roll	$= N$	$= 100\,rpm$
Density of limestone	$= \rho$	$= 2.35\,gm/cm^3 = 2.35\,tons/m^3$
Angle of nip	$= n$	$= 30°$

$$\cos\frac{n}{2} = \cos\frac{30}{2} = \cos 15 = 0.9659$$

But $\cos\dfrac{n}{2} = \dfrac{D+s}{D+d}$ $\Rightarrow$ $0.9659 = \dfrac{150+1.25}{150+d}$

$$\Rightarrow \quad d = 6.59\,cm$$

Maximum size of the feed particle $= 6.59\,cm$

Theoretical capacity of crushing roll if it runs at maximum rpm $=$

$$C = 60\pi DNWs\rho = 60 \times \frac{22}{7} \times 1.5 \times 100 \times 0.5 \times 0.0125 \times 2.35$$

$$= 415.45\,tons/hr$$

Actual capacity of crushing roll $= 415.45 \times 0.15 = 62.32$ tons/hr

Angle of nip after long use $= 32°30'$

$$\cos \frac{n}{2} = \cos \frac{32°30'}{2} = \cos 16°15' = 0.9605$$

$$\text{But } \cos \frac{n}{2} = \frac{D+s}{D+d} \quad \Rightarrow \quad 0.9605 = \frac{150 + 1.25}{150 + d}$$

$$\Rightarrow \quad d = 7.47 \text{ cm}$$

Since the capacity is independent of feed size, it remains unchanged. So the capacity of the crusher $= 62.32$ tons/hr

Example 9.1.7: *The breadth of the rolls of a roll crusher crushing 1.5" diameter to 0.5" diameter spheres is 1 ft and the peripheral speed is 1000 ft/min. If the actual capacity is 20% of theoretical capacity, what is the actual volumetric capacity of the crusher? Calculate the roll speed in rpm. The diameter of the roll is 2 ft.*

Solution:

Given

Diameter of the feed particle	$= d = 1.5'' = 3.81$ cm
Diameter of the product particle	
$\quad = $ gap between the rolls	$= s = 0.5'' = 1.27$ cm
Diameter of the rolls	$= 2$ ft $= 60.96$ cm
Breadth of the rolls	$= 1$ ft $= 30.48$ cm
Peripheral speed	$= \pi DN = 1000$ ft/minute
	$= 30480$ cm/minute

$$\text{Volumetric capacity} = \frac{C}{\rho} = 60\pi DNWs$$

$$= 60 \times 30480 \times 30.48 \times 1.27$$

$$= 70.79 \times 10^6 \text{ cm}^3/\text{minute} = 70.79 \text{ m}^3/\text{minute}$$

Actual capacity $= 20\%$ of theoretical capacity

$$= 70.79 \times 0.2 = 14.158 \text{ m}^3/\text{minute}$$

$$\text{Roll speed} = \frac{\text{Peripheral speed}}{\pi D} = \frac{30480}{\pi \times 60.96} = 159.1 \text{ rpm}$$

9.2 CRUSHING OPERATION

Slow compression is the characteristic action of the nipping crushers. Crushers are usually operated dry. When the material fed to the crusher is at a slow rate, the individual particles are crushed freely. The crushed product is quickly removed from the crushing zone. This type of crushing, known as **free crushing**, avoids the production of excessive fines by limiting the number of contacts.

When the material fed to the crusher is at a high rate, the crusher is choked and prevents the complete discharge of crushed product. This results the crushing between ore particles as well as between ore particle and crushing surface. This type of operation is called **choked crushing** and it increases the amount of fines produced. Choked crushing is preferred in some cases as it reduces the reduction stages. It is obvious that the reduction ratio of choked crushing operation is more when compared to free crushing operation.

Examples 9.2.1 and 9.2.2 illustrates the calculation of reduction ratios when jaw crusher crushing chromite ore and copper ore.

Example 9.2.1: *Chromite ore of $-2^{1}/_{2}'' + 2''$ size has been crushed in a laboratory jaw crusher having length of the receiving opening of 7'' and set of 1''. The size analysis of the product is given in Table 9.2.1.1:*

Table 9.2.1.1 Size analysis of the jaw crusher product.

Size	Weight of each fraction, gm
$-2'' + 1^{3}/_{4}''$	155
$-1^{3}/_{4}'' + 1^{1}/_{2}''$	170
$-1^{1}/_{2}'' + 1^{1}/_{4}''$	275
$-1^{1}/_{4}'' + 1''$	485
$-1'' + {}^{3}/_{4}''$	1105
$-{}^{3}/_{4}'' + {}^{1}/_{2}''$	285
$-{}^{1}/_{2}'' + 10\,mm$	145
$-10 + 6\,mm$	155
$-6 + 3\,mm$	105
$-3 + 1\,mm$	50
$-1\,mm$	70

Calculate: (a) *Average size of the product*   (b) *Limiting reduction ratio*
           (c) *Mean reduction ratio*   (d) *80% reduction ratio*

Solution:

Given

Size passing all feed $= 2.5''$

Average size of the feed $= \dfrac{2.5 + 2.0}{2} = 2.25''$

80% passing size of the feed $= 2.5''$

To determine average size and 80% passing size of the product, the necessary values are calculated from the Table 9.2.1.1 and presented in Table 9.2.1.2.

Table 9.2.1.2 Calculated values for Table 9.2.1.1.

Size	Average size (d), mm	Weight of each fraction, gm	Weight% w	Cum. wt% passing	w/d
$-2'' + 1\frac{3}{4}''$	47.625	155	05.17	100.00	0.10856
$-1\frac{3}{4}'' + 1\frac{1}{2}''$	41.275	170	05.67	94.83	0.13737
$-1\frac{1}{2}'' + 1\frac{1}{4}''$	34.925	275	09.16	89.16	0.26228
$-1\frac{1}{4}'' + 1''$	28.575	485	16.17	80.00	0.56588
$-1'' + \frac{3}{4}''$	22.225	1105	36.83	63.83	1.65714
$-\frac{3}{4}'' + \frac{1}{2}''$	15.875	285	09.50	27.00	0.59843
$-\frac{1}{2}'' + 10\,mm$	11.350	145	04.83	17.50	0.42555
$-10 + 6\,mm$	08.000	155	05.17	12.67	0.64625
$-6 + 3\,mm$	04.500	105	03.50	07.50	0.77778
$-3 + 1\,mm$	02.000	50	01.67	04.00	0.83500
$-1\,mm$	00.500	70	02.33	02.33	4.66000
		3000	100.00		10.67424

$$\text{Average size of the product} = \frac{w}{\sum \dfrac{w}{d}} = \frac{100}{10.67424} = 9.37\,\text{mm}$$

80% passing size of the product $= 1.25''$ (from Table 9.2.1.2)

$$\text{Limiting reduction ratio} = \frac{\text{Size passing all feed}}{\text{Size passing all product}} = \frac{2.5}{2} = 1.25$$

$$\text{Mean reduction ratio} = \frac{\text{Average size of the feed}}{\text{Average size of the product}} = \frac{2.25 \times 25.4}{9.37} = 6.1$$

$$80\% \text{ reduction ratio} = \frac{80\% \text{ passing size of the feed}}{80\% \text{ passing size of the product}} = \frac{2.5}{1.25} = 2$$

Example 9.2.2: *Data from a crushing test conducted on a sample of copper ore when it is crushed in a jaw crusher is given in Table 9.2.2.1.*

Table 9.2.2.1 Crushing test data.

Feed		Product	
Screen size	Weight of material, gm	Screen size	Weight of material, gm
$+2.0''$	3556		
$-2.0'' + 1.5''$	4214		
$-1.5'' + 1.0''$	14355	$+1/2''$	563
$-1.0'' + 3/8''$	9592	$-1/2'' + 3/8''$	3322
$-3/8'' + 4\,mesh$	3370	$-3/8'' + 4\,mesh$	9456
$-4 + 10\,mesh$	1857	$-4 + 10\,mesh$	2272
-10	1906	-10	3812

The aperture sizes of 4 mesh and 10 mesh are 4760 microns and 1680 microns respectively. Considering 80% passing size of the feed and the product, calculate the reduction ratio of the Jaw crusher.

Solution:

Screen sizes and corresponding differential weight percentages and cumulative weight percentages are calculated for both the feed and the product and presented in Table 9.2.2.2:

Table 9.2.2.2 Calculated values for Table 9.2.2.1.

Feed				Product			
Screen size	Weight gm	wt%	Cum. wt% passing	Screen size	Weight gm	wt%	Cum. wt% passing
+2.0″	3556	9.15	100.00	+1/2″	563	2.90	100.00
−2.0″	4214	10.85	90.85	−1/2″	3322	17.10	97.10
−1.5″	14355	36.95	80.00	−3/8″	9456	48.68	80.00
−1.0″	9592	24.69	43.05	−4 mesh	2272	11.70	31.32
−3/8″	3370	8.67	18.36	−10 mesh	3812	19.62	19.62
−4 mesh	1857	4.78	9.69				
−10 mesh	1906	4.91	4.91				
	38850	100.00			19425	100.00	

From the above table

80% passing size of the feed   = 1.5″
80% passing size of the product = 3/8″

$$\therefore \text{Reduction ratio} = \frac{80\% \text{ passing size of the feed}}{80\% \text{ passing size of the product}}$$

$$= \frac{1.5}{3/8} = 4$$

9.3   OPEN AND CLOSED CIRCUIT CRUSHING OPERATIONS

Usually each stage of size reduction is followed by a screen which forms a circuit. Crushing may be conducted in open or closed circuit as shown in Figure 9.3.1.

In an open circuit crushing operation, the feed material is reduced by one crusher. The product of this crusher is screened and only oversize material is crushed by another crusher of small size as the **throughput** (the quantity of material crushed in a given time) is less. The crushed product from the second crusher and undersize material from the screen together form the final product. In case of closed circuit crushing operation, the oversize material from the screen is re-fed to the same crusher. The undersize material from the screen is the required final product.

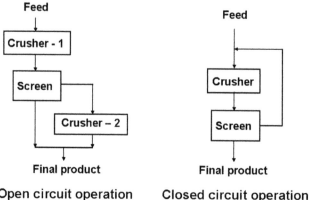

Open circuit operation   Closed circuit operation

Figure 9.3.1 Open circuit and closed circuit crushing operations.

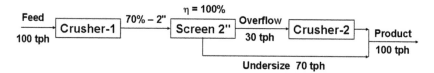

Figure 9.3.1.1 Open circuit crushing details when screen efficiency is 100%.

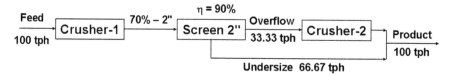

Figure 9.3.1.2 Open circuit crushing details when screen efficiency is 90%.

9.3.1 Open circuit crushing operation

Let us consider a crusher set at 4″ is crushing an ore at the rate of 100 tons/hr to a product size consists of 70% of –2″. The product of the crusher is sent to a screen of 2″ square opening. The oversize from the screen is fed to another crusher to give a product of all pass 2″. The circuit details assuming 100% and 90% screen efficiencies are shown in the Figures 9.3.1.1 & 9.3.1.2.

It is important to note that the overflow of the screen when screen efficiency is 90% also consists of undersize material. When screen efficiency is 90%, 100 tph of material is crushed in primary crusher and 33.33 tph of material is crushed in secondary crusher. This can be arrived by considering screen efficiency as follows:

Efficiency of the screen based on oversize

$$\eta = \frac{\text{Weight of +2'' material in crusher product}}{\text{Weight of +2'' material obtained from the screen (overflow)}}$$

Therefore, weight of +2'' material obtained from the screen (overflow) and crushed by secondary crusher

$$= \frac{\text{Weight of +2'' material in crusher product}}{\eta} = \frac{30}{0.9} = 33.33 \text{ tph}$$

Similar calculations are illustrated in example 9.3.1.

Example 9.3.1: *Crushed product of a primary crusher having 57% of +1'' is sent to 1'' screen. The overflow of the screen is again crushed in a secondary crusher to −1'' size and sent along with underflow of the screen as final product. Draw the crushing circuit with all details. Calculate the rate of the material crushed in secondary crusher if the effectiveness of the screen is 80% based on oversize material and the feed rate to a primary crusher is 100 tons/hr.*

Solution:

Crushing circuit is drawn as shown in Figure 9.3.1.3.

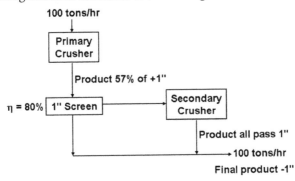

Figure 9.3.1.3 Crushing circuit for example 9.3.1.

Weight of +1'' material in feed to screen = 57% of 100 tons/hr = 57 tons/hr

$$\text{Screen efficiency} = \eta = 0.80 = \frac{\text{Weight of oversize material present in feed}}{\text{Weight of overflow obtained from the screen}}$$

$$\therefore \text{Weight of overflow obtained from the screen} = \frac{57}{0.80} = 71.25 \text{ tons/hr}$$

Material crushed in secondary crusher = 71.25 tons/hr
Underflow from the screen = 100 − 71.25 = 28.75 tons/hr

9.3.2 Closed circuit crushing operation

There are two types of closed circuit crushing operations called **regular** and **reverse**. In regular type, the new feed is fed to the crusher, crusher product is fed to the screen and the overflow from the screen is fed back to the same crusher. The quantity of the overflow material fed back to the crusher is called **circulating load**. The underflow from the screen is the required final product and is equal to the new feed. This regular type is usually employed when the feed contains less percentage of undersize material. Initially the quantity of final product produced is less than the quantity of the feed material. As the operation proceeds further the quantity of the final product gradually increases and will be equal to the quantity of the feed material after some time. After attaining this equilibrium condition, the quantity of the final product is always equal to the quantity of the feed material and the circulating load is constant. The circulating load, expressed as a percentage of the quantity of feed material is called **percent circulating load**.

This is explained as follows by taking an example. Consider a crusher reduces 75% of feed material to –2″ size and assume 100% screen efficiency. For an initial feed of 100 tons, crusher product contains 75 tons of undersize material and 25 tons of oversize material. The oversize of 25 tons is returned to the same crusher. Further assume that the crusher reduces 75% of this 25 tons to –2″ on the second time. Hence 25% of 25 tons (6.25 tons) is returned to the crusher as oversize. On the next pass, 25% of 6.25 tons (1.56 tons) is returned to the crusher as oversize and so on. It reaches a stage where oversize fraction returned to the crusher is constant afterwards. This is called circulating load. The crushing operation when the circulating load is constant can be considered as equilibrium stage. Details of calculations are shown in the Table 9.3.2.

The objectives of employing closed circuit operation are to minimize the production of fines and to reduce the energy consumption by avoiding size reduction of already reduced particles to the required size.

Closed circuit crushing operation when screen efficiency is 100% can be represented as shown in Figure 9.3.2.1.

Table 9.3.2 Closed circuit crushing calculations.

Stage No.	tph new feed	tph Oversize from previous stage	tph Total feed to this stage	Products from this stage	
				tph Oversize	tph Undersize
1	100	–	100.0	25.0	75.0
2	100	25.0	125.0	31.25	93.75
3	100	31.25	131.25	32.81	98.44
4	100	32.81	132.81	33.20	99.61
5	100	33.20	133.20	33.30	99.90
6	100	33.30	133.30	33.325	99.975
7	100	33.325	133.325	33.331	99.994
8	100	33.331	133.331	33.333	99.998
9	100	33.333	133.333	33.3332	99.9998
∝	100	33.3333	133.3333	33.3333	100.0

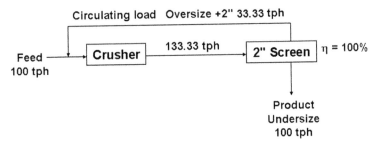

Figure 9.3.2.1 Closed circuit crushing details when screen efficiency is 100%.

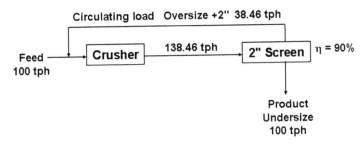

Figure 9.3.2.2 Closed circuit crushing details when screen efficiency is 90%.

Let C be the circulating load. The amount of material crushed in a crusher at equilibrium stage is $(100 + C)$. As already assumed, the crusher product contains 25% oversize. This oversize is retained on the screen as 100% screen efficiency is assumed.

$$\therefore \quad C = (100 + C) \times 0.25$$
$$\Rightarrow \quad C = 33.33 \, \text{tph}$$

Thus, to crush 100 tph of ore by closed circuit crushing operation, it is necessary to use a crusher of capacity 133.33 tph. The circulating load is 33.33 tph.

When the screen efficiency is assumed as 90%, then

$$C = \frac{(100 + C) \times 0.25}{0.90}$$
$$\Rightarrow \quad C = 38.46 \, \text{tph}$$

In this case, a crusher of 138.46 tph capacity is necessary to crush 100 tph of ore. The circulating load is 38.46 tph. Closed circuit crushing operation when screen efficiency is 90% can be represented as shown in Figure 9.3.2.2.

Two types of common closed circuit crushing operations are considered for the calculations in the Examples 9.3.2 and 9.3.3.

Example 9.3.2: *The details of crushing plant employing gyratory crusher are as follows:*

2″ square screen is in closed circuit with crusher
New feed to crusher = 100 T.P.H
Crusher product contains 54% < 2″ and 46% > 2″ fed to screen
Screen efficiency = 85% (based on oversize)

Draw the flow diagram and find circulating load.
Also find circulating load if the screen efficiency is based on undersize.

Solution:

Flow diagram is shown in Figure 9.3.2.3.

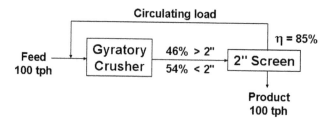

Figure 9.3.2.3 Flow diagram for example 9.3.2.

Let C be the circulating load.
Load on the gyratory crusher $= C + 100$ tons/hr

Circulating load if the screen efficiency is based on oversize

$+2″$ material in crusher product $= 0.46(C + 100)$ tons/hr

$$\text{Screen efficiency} = \eta = 0.85 = \frac{+2″ \text{ material present in feed to the screen}}{+2″ \text{ fraction obtained from the screen}}$$
$$= \frac{0.46(C + 100)}{C}$$

$\Rightarrow \qquad\qquad C = 117.95$ tons/hr
Circulating load if the screen efficiency is based on oversize $= 117.95$ tons/hr

Circulating load if the screen efficiency is based on undersize

$-2″$ material in crusher product $= 0.54(C + 100)$ tons/hr

$$\text{Screen efficiency} = \eta = \frac{-2″ \text{ fraction obtained from the screen}}{-2″ \text{ material present in the feed to the screen}}$$

$\Rightarrow \qquad\qquad 0.85 = \dfrac{100}{0.54(C + 100)}$

$\Rightarrow \qquad\qquad C = 117.86$ tons/hr

Circulating load if the screen efficiency is based on undersize $= 117.86$ tons/hr

Alternately

$$\eta = \frac{-2'' \text{ fraction obtained from the screen}}{-2'' \text{ material present in the feed to the screen}}$$

$-2''$ material present in the feed to the screen

$$= \frac{-2'' \text{ fraction obtained from the screen}}{\eta} = \frac{100}{0.85}$$

If F is the feed to the screen, $-2''$ material present in the feed to the screen

$$= F \times 0.54 = \frac{100}{0.85}$$

$\Rightarrow$
$$F = \frac{100}{0.85 \times 0.54} = 217.86 \text{ tons/hr}$$

Circulating load $= 217.86 - 100 = 117.86$ tons/hr

Circulating load if the screen efficiency is based on undersize $= 117.86$ tons/hr

Example 9.3.3: *In the crushing circuit shown in Figure 9.3.3, calculate the total load on $1/2''$ screen. Screen efficiencies are based on oversize material.*

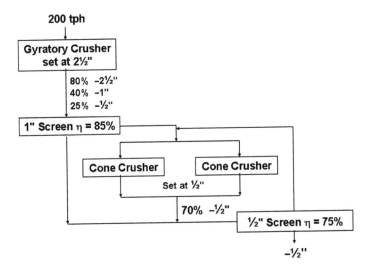

Figure 9.3.3 Crushing circuit for example 9.3.3.

Solution:

$+1''$ material in the feed to $1''$ screen $= 0.60 \times 200 = 120$ tons/hr

Overflow from $1''$ screen $= \dfrac{120}{0.85} = 141.2$ tons/hr

Underflow from $1''$ screen $= 200 - 141.2 = 58.8$ tons/hr

$-\tfrac{1}{2}''$ material in the feed to $1''$ screen $= 0.25 \times 200 = 50$ tons/hr

$\therefore +\tfrac{1}{2}''$ material in underflow from $1''$ screen $= 58.8 - 50 = 8.8$ tons/hr

Let C be the circulating load i.e., overflow from $\tfrac{1}{2}''$ screen

Total load on cone crushers $= 141.2 + C$

$+\tfrac{1}{2}''$ material in cone crusher product $= 0.30(C + 141.2)$

Load of $+\tfrac{1}{2}''$ to $\tfrac{1}{2}''$ screen $= 0.30(C + 141.2) + 8.8$

$$\text{Screen efficiency} = \eta = 0.75 = \frac{\text{Load of} + 1/2'' \text{ to } 1/2'' \text{ screen}}{\text{circulating load}}$$

$$= \frac{0.30(C + 141.2) + 8.8}{C}$$

$$\Rightarrow \qquad\qquad\qquad C = 113.7 \text{ tons/hr}$$

Total load to $\tfrac{1}{2}''$ screen $= 141.2 + 113.7 + 58.8 = 313.7$ tons/hr

In the reverse type closed circuit crushing operation the new feed is fed to the screen, overflow from the screen is fed to crusher and crusher product is re-fed to the screen. The screen overflow which is fed to crusher and fed back to the screen is the circulating load. The screen underflow is the final product and is equal to the new feed. The reverse type circuit is usually employed in the case where the feed contains a relatively high percentage of undersize material which needs to be removed prior to its introduction into the crusher.

Figure 9.3.3.1 shows the typical example of reverse type of closed circuit crushing operation when screen efficiency is assumed as 100%.

It is to be noted that in this type of crushing circuit not only the crushing of undersize material of the feed material (25% in this case) is avoided but also the lower capacity crusher (75 tph) is required to install.

Example 9.3.4 illustrates the calculations when the screen efficiency is 85% and the crusher in closed circuit crushes the material not to all pass required size ($-\tfrac{3}{4}''$).

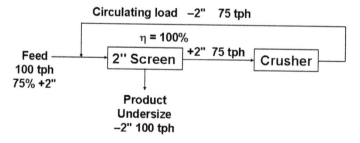

Figure 9.3.3.1 Closed circuit crushing operation.

Example 9.3.4: *The circuit shown in Figure 9.3.4 is adopted to crush an ore.*

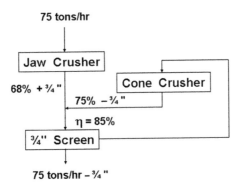

Figure 9.3.4 Crushing circuit for example 9.3.4.

a) At steady state condition, what is the total load goes through the cone crusher?
b) What is the percent circulating load?
c) Of the original 75 tph of jaw crusher product, how many tons by-pass the cone crusher?
d) At steady state conditions, what is the total load which goes over the screen?
Efficiency of the screen is based on oversize material.

Solution:

(a) Let C be the circulating load

$+3/4''$ material in feed to the screen $= 0.25C + 0.68 \times 75 = 0.25C + 51$

$$\text{Screen efficiency} = \eta = 0.85 = \frac{+3/4'' \text{ material in feed to the screen}}{+3/4'' \text{ obtained from the screen}}$$

$$= \frac{0.25C + 51}{C}$$

$\Rightarrow \qquad\qquad C = 85 \text{ tons/hr}$

∴ Load goes through cone crusher $= 85$ tons/hr

(b) % circulating load $= \dfrac{85}{75} \times 100 = 113.3\%$

(c) Oversize in jaw crusher product $= 0.68 \times 75 = 51$ tons/hr

Screen overflow out of 75 tons/hr $= \dfrac{51}{0.85} = 60$ tons/hr

Out of 75 tons/hr 60 tons/hr goes to cone crusher
∴ Material by passed the cone crusher $= 75 - 60 = 15$ tons/hr

(d) Total load to the screen $= 75 + 85 = 160$ tons/hr

9.4 PROBLEMS FOR PRACTICE

9.4.1:   *A material of 1.5″ diameter spherical particles is intended to crush in a roll crusher. Assuming coefficient of friction between the rock and the steel to be 0.29 and reduction ratio to be 3:1, calculate the diameter of rolls under free crushing conditions.*

[24.2″]

9.4.2:   *A crushing rolls of 24″ diameter is set at 0.5″ apart. If the angle of nip is 31°, calculate the maximum size of the feed particle to the rolls.*

[1.4″]

9.4.3:   *Find out the maximum size of the limestone particles fed to a crushing rolls run at 100 rpm, having rolls of 150 cm diameter and 50 cm width of the roll set at 1 cm. The angle of nip is 30° and the specific gravity of limestone is 2.66. Also calculate the theoretical capacity of the crushing rolls.*

[6.33 cm, 376.2 tons/hr]

9.4.4:   *Rolls of 100 cm diameter and 38 cm face width of a crushing rolls are set so that crushing surfaces are 1.25 cm apart. They are to crush a rock of 2.35 specific gravity and the angle of nip of 30°. As per manufacturer recommendation, they may be run at 50 to 100 rpm. What is the permissible size of the feed to the crusher. If the actual capacity of the crusher is 12% of the theoretical capacity, calculate the maximum actual capacity of the crusher in tons/hr.*

[4.82 cm, 25.26 tons/hr]

9.4.5:   *Data obtained from a crushing test conducted on a sample of Iron ore in laboratory is given in Table 9.4.5.1.*

Table 9.4.5.1   Crushing Test Data for problem 9.4.5.

Feed		Product	
Screen size	Weight of the material retained, gm	Screen size	Weight of the material retained, gm
5″	1112		
4″	1668		
3″	3614	1½″	696
2″	2641	1″	1624
1″	2363	½″	4060
½″	1807	¼″	3132
−½″	695	−¼″	2088

Considering 80% passing size of the feed and the product, calculate reduction ratio.

[4]

9.4.6: *A 3″ square screen is in closed circuit with a gyratory crusher which is fed at a rate of 200 tons/hr. If the crusher product consists 70% of −3″ particles and the screen efficiency is 90% based on undersize material, draw the circuit diagram in detail and calculate the circulating load. Also calculate the circulating load if the screen efficiency is 85% based on oversize.*

[117.46 tons/hr, 109 tons/hr]

9.4.7: *Manganese ore at the rate of 300 tons/hr is crushed in a gyratory crusher set at 4″ to a size 58% of +1½″ and fed to a 1½″ screen of 75% efficiency based on oversize material. The overflow of the screen is crushed in another gyratory crusher set at 2″ to a size 30% of +1½″ and re-fed to the screen along with the product of first gyratory crusher. If the finished product is 300 tons/hr of −1½″ size, draw the flow diagram and calculate the total load on the screen.*

[686.67 tons/hr]

Chapter 10

Grinding

Grinding is the last stage of the comminution process. The particles are reduced from a maximum upper feed size range of approximately 9,000 to 10,000 microns (3/8 inch), to some upper limiting product size ranging between 35 mesh and 200 mesh (420 microns and 74 microns). Grinding machines most frequently used are **tumbling mills** (also called **grinding mills**). A tumbling mill reduces particle size by applying impact and attrition stresses to the materials to be ground. It is designed to strike the particles with sharp blows of short duration and to produce a rubbing action under as high a unit pressure as possible. Accordingly, grinding mill consists of a horizontal rotating steel shell supported by end bearings on which hallow trunnions revolve. Loose crushing bodies, known as **grinding medium**, are placed inside the shell. Either steel balls/rods or pebbles are used as grinding medium. They are free to move inside the rotating shell making the particles to break by repetitive blows and by rolling and sliding one over the other. Attrition or shearing forces which result from the application of forces by rolling and sliding bodies tend to produce more fine particles than impact forces applied on particles by repetitive blows. The interior of tumbling mill is lined by replaceable liners primarily to protect the mill body for wear and damage.

Grinding mills are classified as Ball mills, Rod mills, Tube mills, Pebble mills and Autogenous mills based on type of grinding medium, shell length to diameter ratio and method of discharge. In rod mills, steel rods usually about 6 inch shorter than the length of the grinding chamber are used as grinding medium and the shell length to diameter ratio is between 1.5 and 2.5. Spherical steel balls are used as grinding medium in ball mills and tube mills. Shell length to diameter ratio in ball mills is 1.0 to 1.5 and in tube mills it is 3 to 5. Tube mills may have multiple compartments and thus represent a series of ball mills.

Pebbles of hard rock or other nonmetallic material is used as grinding medium in pebble mills. Autogenous mills use coarse ore particles as grinding medium. Autogenous grinding is defined as the action of a material grinding upon itself, as occurs when pieces of ore of different sizes are rotated together in a tumbling mill. The same action takes place in pebble mill grinding when sized pieces of the ore are used as grinding medium. Autogenous mills are of very large diameter range upto 36 ft, with lengths as little as one-third, or even one-fourth of the mill diameter. Semi-autogenous (SAG) mills use a combination of the ore and a reduced charge of balls as a grinding medium to overcome the difficulties encountered in autogenous grinding.

Usually the material is fed at one end of the mill and discharged at the other end. The feed enters through a hallow trunnion at the centre of the feed end. In dry grinding mill, the feed is by vibrating feeder. Three types of feeders are in use to feed the material to the wet grinding mills. In a **spout feeder**, the material is fed by gravity through the spout which consists of a cylindrical chute projecting directly in to the trunnion liner. In **drum feeder**, the entire mill feed enters the drum and an internal spiral carries it and fed to the mill. Grinding balls are conveniently added through this feeder during operation. In case of **scoop feeder**, material is fed to the drum and the scoop picks it up and fed to the mill.

The discharge of the ground product from the mill is primarily of two types viz., overflow discharge and grate discharge. In an overflow discharge mill, the ground product pass through a hallow trunnion at the discharge end of the mill. In a grate discharge mill, the ground product pass through the slots of a retaining grid or grate fixed at the discharge end of the mill. Another method for discharging ground material especially in rod mills is by peripheral discharge, either end peripheral discharge or center peripheral discharge. This type of product discharge is used in both wet and dry grinding operations. However, it is more common in dry grinding applications.

10.1 GRINDING ACTION

When the grinding mill is rotated, the mixture of grinding medium, raw material, and water (in case of wet grinding) is lifted, the magnitude depends on the rotational speed of the mill.

Due to the rotation and friction of the mill shell, the balls are lifted along the rising side of the mill until a position of dynamic equilibrium is reached and then drop down to the toe of the mill charge. Grinding of ore particles takes place due to simple rolling of one ball over the other (cascading) and by the free fall of balls (cataracting). Cascading leads to fine grinding whereas cataracting leads to coarse grinding. Figure 10.1.1 illustrates the motion of the charge in ball mill.

As the speed of the mill increases, the balls are lifted to more height and a stage is reached where the balls are carried around the shell and never allowed to fall. That

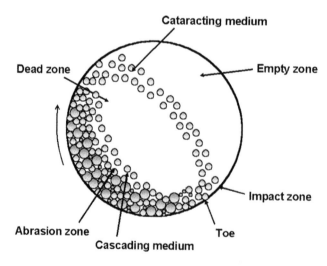

Figure 10.1.1 Motion of the charge in ball mill.

means centrifuging occurs. The balls will rotate as if they are part of the shell. There will be no relative movement of the balls. The speed at which centrifuging occurs is known as **critical speed**.

Let m = mass of the ball, kg
R = Radius of the mill shell, metre
N = speed of the mill, rpm
v = linear velocity of the ball, m/sec = $\dfrac{2\pi RN}{60}$

Let us consider a ball which is lifted up the shell. When the ball moving around inside the mill, it follows a circular path till it reaches certain height and then it changes its circular path and follows a parabolic path while dropping on to the toe (Fig. 10.1.2).

When the ball is at the point of changing its path, centrifugal and centripetal forces acting on it will just balance. The centrifugal force acting on the ball is $\dfrac{mv^2}{R}$. The component of weight of the body opposite to the direction of centrifugal force is the centripetal force i.e. $mg \cos \alpha$, where α is the angle made by the line of action of centripetal force with the vertical. P is a point where the ball changes its circular path to parabolic path. At this point, both the forces are equal (Fig. 10.1.3).

Therefore, $\dfrac{mv^2}{R} = mg \cos \alpha$

$$\Rightarrow \quad \frac{\left(\dfrac{2\pi RN}{60}\right)^2}{R} = g \cos \alpha \qquad \Rightarrow \quad N^2 = \frac{(60)^2 g \cos \alpha}{4\pi^2 R}$$

If D is the diameter of the mill, d is the diameter of the ball, the radius of the circular path of outer most ball is $\dfrac{(D-d)}{2}$. On substitution

$$N^2 = \frac{(60)^2 g \cos \alpha}{2\pi^2 (D-d)} \qquad\qquad (10.1)$$

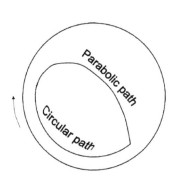

Figure 10.1.2 Path of a ball.

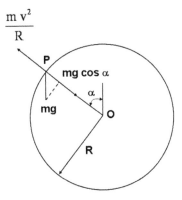

Figure 10.1.3 Forces on a ball.

Critical speed of the mill occurs when $\alpha = 0$. i.e. the ball will rotate along with the mill. At this point $\cos \alpha = 1$. Then $N = N_c$.

$$\therefore \quad N_c^2 = \frac{(60)^2 g}{2\pi^2 (D - d)} \tag{10.2}$$

If D and d are taken in metres, it is reduced to

$$\text{Critical speed} = N_c = \frac{42.3}{\sqrt{(D - d)}} \text{ revolutions/minute} \tag{10.3}$$

If D and d are taken in feet, the same expression is reduced to

$$\text{Critical speed} = N_c = \frac{76.65}{\sqrt{(D - d)}} \text{ revolutions/minute} \tag{10.4}$$

Ball mills are operated at 60–80% of critical speed. Rod mills are operated at 50–65% of critical speed. Autogenous mills are operated at much higher speeds at 80–85% of critical speed. In examples 10.1.1 to 10.1.4, calculations concerned to critical speed are illustrated.

Example 10.1.1: *What is the critical speed of a ball mill of 48″ I.D. charged with 3″ balls?*

Solution:

Given

Diameter of the ball mill $= 48″ = 48/12 = 4\,\text{ft}$
Diameter of the balls $\quad = 3″ \quad = 3/12 \quad = 0.25\,\text{ft}$

$$\text{Critical speed of the ball mill} = N_c = \frac{76.65}{\sqrt{(D - d)}}$$

$$= \frac{76.65}{\sqrt{(4 - 0.25)}} = 39.58\,\text{rpm}$$

Example 10.1.2: *A lead beneficiation plant installed a ball mill of 2 metres diameter. If it wishes to use 10 cm balls as a grinding media and operate at 70% of the critical speed, what would be the operating speed of the ball mill.*

Solution:

Given

Diameter of the ball mill $= 2\,\text{m}$
Diameter of the balls $\quad = 0.1\,\text{m}$

$$\text{Critical speed of the ball mill} = N_c = \frac{42.3}{\sqrt{(D - d)}}$$

$$= \frac{42.3}{\sqrt{(2 - 0.1)}} = 30.69\,\text{rpm}$$

Operating speed $= 70\%$ of critical speed $= 0.7 \times 30.69$
$$= 21.48\,\text{rpm}$$

Example 10.1.3: *What rotational speed, in revolutions per minute, would you recommend for a ball mill of 1200 mm in diameter charged with 75 mm balls?*

Solution:

Given

Diameter of the ball mill = 1200 mm = 1.2 m
Diameter of the balls   = 75 mm   = 0.075 m

Critical speed of the ball mill = $N_c = \dfrac{42.3}{\sqrt{(D-d)}}$

$$= \dfrac{42.3}{\sqrt{(1.2-0.075)}} = 39.9 \text{ rpm}$$

Operating speed of the ball mill is 60 to 80% of the critical speed

Operating speed = 60 to 80% of 39.9 rpm = 24 to 32 rpm

Example 10.1.4: *In a ball mill of 2000 mm diameter, 100 mm diameter steel balls are being used for grinding at a speed of 15 rpm. At what speed will the mill have to run if the 100 mm balls are replaced with 50 mm balls, all the other conditions remaining same?*

Solution:

Given

Diameter of the ball mill = 2000 mm = 2 m
Diameter of the balls   = 100 mm = 0.1 m

Critical speed of the ball mill = $N_c = \dfrac{42.3}{\sqrt{(D-d)}}$

$$= \dfrac{42.3}{\sqrt{(2-0.1)}} = 30.69 \text{ rpm}$$

As the mill is running at 15 rpm,

the percent of critical speed the mill is operated $= \dfrac{15}{30.69} \times 100 = 48.9\%$

Critical speed of the ball mill if the balls are of 50 mm diameter

$$= N_c = \dfrac{42.3}{\sqrt{(D-d)}}$$

$$= \dfrac{42.3}{\sqrt{(2-0.05)}} = 30.26 \text{ rpm}$$

Operating speed of the mill $= 30.26 \times \dfrac{48.9}{100} = 14.79 \text{ rpm}$

10.2 WET AND DRY GRINDING

The grinding may be wet or dry depending on the subsequent process and the nature of the product. Wet grinding is generally used in mineral processing plants as subsequent operations for most of the ores are carried out wet. Wet grinding is usually carried out with 60%–75% solids by weight. The chief advantages of wet grinding are increased capacity (as much as 15%) for a given size of equipment and less power consumption per ton of the product. Less power consumption is due to the penetration of water into the cracks of the particles which reduces the bond strength at the crack tip. The Bond Index of the material in wet grinding is 75% of the Bond Index in dry grinding. Mill can run upto 75–80% of the critical speed in the case of wet grinding whereas in dry grinding maximum limit is 65–70% only. Dry grinding is used whenever physical or chemical changes in the material occur if water is added. It causes less wear on the liners and grinding media. Dry grinding mills are often employed to produce an extremely fine product. This arises from the high settling rate of solids suspended in air as compared with solids suspended in water.

10.3 GRINDING CIRCUITS

Mesh of grind (m.o.g) is the term used to designate the size of the grounded product in terms of the percentage of the material passing a given mesh. **Optimum mesh of grind** defines the mesh of grind at which a maximum profit is made on sales, when both the working costs and the effect of grinding on the recovery of values have been considered. In grinding, there are always some particles which may repeatedly be reduced to fine size whereas some other particles may not be reduced. The primary objective of a grinding mill is to reduce all particles to the stated size i.e., mesh of grind. When a grinding mill is fed with a material, it should be at a rate calculated to produce the correct product in one pass in which case it is known as **Open circuit grinding** (Fig. 10.3.1.1).

There is no control on product size distribution in the open circuit grinding. If the residence time of the material in the mill is reduced, it results underground product with low degree of liberation. If the residence time is increased, it results over-ground product. Much energy is wasted in the process by grinding the material to a size below the size required. If the material is removed from the mill as soon as it is grounded to the required size, over-grinding can be eliminated or minimized. This is done by employing a classifier. This operation is known as **Closed circuit grinding** operation (Fig. 10.3.1.2).

Separation takes place in the classifier based on the principle of settling of solids in fluids. Classifier separates the feed material into two fractions viz., fine or undersize material and coarse or oversize material wherein oversize material is returned to the same grinding mill and undersize material is the finished product. The tonnage or weight flow rate of the overflow or undersize solids of a classifier is always equal to the tonnage or weight flow rate of the solids fed to the grinding mill in a closed circuit operation under steady state conditions.

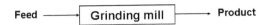

Figure 10.3.1.1   Open circuit grinding.

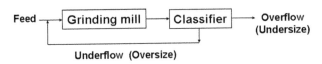

Figure 10.3.1.2   Closed circuit grinding.

A material balance equation can be written over the classifier as follows:

$$\text{Slurry balance} \qquad F = P + U \qquad (10.3.1)$$

$$\text{Solids balance} \qquad Ff = Pp + Uu \qquad (10.3.2)$$

where F, P and U are the pulp or slurry flow rates of the feed, overflow (or undersize) and underflow (or oversize) of the classifier respectively and f, p and u are the fraction or percent solids in the feed, overflow and underflow of the classifier.

$$\text{Solving 10.3.1 \& 10.3.2} \qquad \frac{U}{P} = \frac{f - p}{u - f} \qquad (10.3.3)$$

The tonnage or weight flow rate of the solids fed back or returned to the grinding mill is called **circulating load** and its weight expressed as a percentage of the tonnage or weight flow rate of new feed solids fed to the grinding mill is called **% circulating load**.

% circulating load

$$= \frac{\text{Tonnage or weight flow rate of underflow solids from classifier}}{\text{Tonnage or weight flow rate of new feed solids fed to grinding mill}} \times 100$$

$$= \frac{\text{Tonnage or weight flow rate of underflow solids from classifier}}{\text{Tonnage or weight flow rate of overflow solids from classifier}} \times 100$$

$$= \frac{Uu}{Pp} \times 100 = \frac{u(f - p)}{p(u - f)} \times 100 \qquad (10.3.4)$$

If F, P and U are the weight flow rates of solids in feed, overflow and underflow, and DR_F, DR_P and DR_U are the dilution ratios of feed, overflow and underflow, then

$$\text{Solids balance} \qquad F = P + U \qquad (10.3.5)$$

$$\text{Water balance} \qquad F \times DR_F = P \times DR_P + U \times DR_U \qquad (10.3.6)$$

$$\text{\% circulating load} = \frac{U}{NF} \times 100 = \frac{U}{P} \times 100 \qquad (10.3.7)$$

where NF = Weight flow rate of new feed solids fed to the mill $= P$
     = Weight flow rate of overflow solids discharged from the classifier

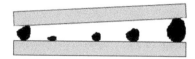

Figure 10.3.1.3 Grinding action of rods.

Percent circulating load is called **circulating ratio** if expressed as ratio.

Solving equations 10.3.5 and 10.3.6,

$$\text{Circulation ratio} = \frac{U}{P} = \frac{DR_P - DR_F}{DR_F - DR_U} \qquad (10.3.8)$$

The grinding mills are generally operated at circulating loads of 200–500% in order to have the grinding correctly to the required size.

As the particle size decreases, the energy required to fracture each particle decreases but the energy per unit mass rises more rapidly. Hence in closed circuit grinding, considerable energy is saved by avoiding over grinding. Grinding in the mining industry is almost always in closed circuit. In case of rod mill, the rods are kept apart by the coarsest particles. The grinding action results from line contact of the rods on the ore particles and is exerted preferentially on the coarsest particles. Smaller and fine particles do not ground till the coarsest particle is reduced in size (Fig. 10.3.1.3). Thus the rod mill produces a more closely sized product with little oversize or slimes. Hence the rod mills may be considered as coarse grinding machines.

As discussed in closed circuit crushing, the closed circuit grinding is of **regular** type when new feed is fed to grinding mill, grounded product is classified in a classifier and classifier underflow is re-fed to grinding mill. Regular type closed circuit grinding is considered in examples 10.3.1 and 10.3.2 and illustrated performing calculations based on percent solids and size analysis of the streams.

Example 10.3.1:   *100 dry tons per hour of new crude ore is fed to a ball mill which is in closed circuit with a classifier. On sampling and analyses, the percent solids by weight in the feed to the classifier, in the classifier overflow (fines) and underflow (coarse) are found as 50, 25 and 84 respectively. Calculate the circulating load tonnage.*

Solution:

Given

Solids feed rate to the ball mill $= NF = 100$ tons/hr
Solids in feed to the classifier $= f = 50\% = 0.5$
Solids in overflow from classifier $= p = 25\% = 0.25$
Solids in underflow from classifier $= u = 84\% = 0.84$

Closed circuit grinding diagram is shown in Fig. 10.3.1.4.

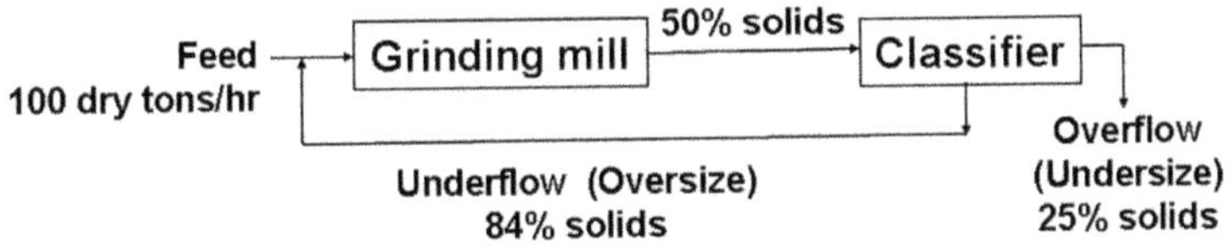

Figure 10.3.1.4 Closed circuit grinding for example 10.3.1.

In closed circuit grinding operation,
New feed solids (Feed solids to the grinding mill)
$\quad$ = Product solids (Overflow solids from the classifier) = 100 dry tons/hr

Weight flow rate of overflow (water + solids) product $= P = \dfrac{100}{0.25}$

$$= 400 \text{ tons/hr}$$

Let F = Total weight flow rate of feed (water + solids) to the classifier
$\quad U$ = Total weight flow rate of underflow (water + solids) from the classifier

Total material (solids + water) balance $\quad F = P + U \quad \Rightarrow \quad F = 400 + U$

Solids balance $\quad Ff = Pp + Uu \quad \Rightarrow \quad F(0.5) = 400(0.25) + U(0.84)$

Solving above two equations $\quad \Rightarrow \quad U = 294.12 \text{ tons/hr}$

$\quad$ Solids in underflow $= 0.84 \times 294.12 = 247.1 \text{ dry tons/hr}$

$$\% \text{ circulating load} = \frac{\text{Cirulating load}}{\text{New feed}} \times 100 = \frac{247.1}{100} \times 100 = 247.1\%$$

Alternately
Let $\quad F$ = Weight flow rate of feed solids to the classifier
$\quad\quad P$ = Weight flow rate of overflow solids from the classifier
$\quad\quad U$ = Weight flow rate of underflow solids from the classifier

And DR_F, DR_P and DR_U are the dilution ratios of feed, overflow and underflow.

$$\text{Dilution ratio of feed} \quad = DR_F = \frac{1 - 0.5}{0.5} = 1.0$$

$$\text{Dilution ratio of overflow} \quad = DR_P = \frac{1 - 0.25}{0.25} = 3.0$$

$$\text{Dilution ratio of underflow} = DR_U = \frac{1 - 0.84}{0.84} = 0.1905$$

Solids balance $F = P + U \quad \Rightarrow \quad F = 100 + U$

Water balance $F \times DR_F = P \times DR_P + U \times DR_U \Rightarrow F(1.0) = 100(3.0) + U(0.1905)$

Solving above two equations $\quad\quad U = 247.1 \text{ tons/hr}$

$$\% \text{ circulating load} = \frac{\text{Cirulating load}}{\text{New feed}} \times 100 = \frac{247.1}{100} \times 100 = 247.1\%$$

Example 10.3.2: *A mill in closed circuit with a classifier receives 300 dry tons of crude ore per day. The sieve analyses of three samples from mill discharge, classifier overflow and underflow are shown in Table 10.3.2.*

Table 10.3.2   Screen analyses of three samples for example 10.3.2.

Mesh	Mill discharge wt%	Classifier overflow wt%	Classifier underflow wt%
+48	42.30	1.20	55.70
48/150	30.50	26.20	31.90
150/200	6.10	12.40	4.05
−200	21.10	60.20	8.35

Calculate the tonnage of the circulating load.

Solution:

Let F, P and U be the feed, overflow and underflow solids of classifier

$$\text{Solid balance} \quad F = P + U$$

New feed to the mill = overflow from classifier = P = 300 dry tons/day

$$\therefore \quad F = P + U = 300 + U$$

Let f, p and u be % solids of any size in feed, overflow and underflow of classifier

$$\therefore \quad Ff = Pp + Uu$$

Balance of +48 mesh material gives

$$F(42.30) = P(1.20) + U(55.70)$$
$$\Rightarrow \quad (300 + U)(42.30) = 300(1.20) + U(55.70)$$
$$\Rightarrow \quad U = 920 \text{ dry tons/day}$$
$$\therefore \quad \text{Circulating load} = 920 \text{ dry tons/day}$$

The same result can also be obtained by balancing any size material.

In **reverse** type closed circuit grinding operation, the new feed enters the classifier, underflow of the classifier is fed to the grinding mill, the mill discharges to the classifier and the classifier overflow is the finished product of the grinding circuit. Calculation of reverse type of circuit is illustrated in example 10.3.3.

Example 10.3.3: *An ore grounded in rod mill contains 15% of −90 microns material. It is further grounded in a rake classifier ball mill circuit as shown in Figure 10.3.3 at the rate of 50 tons/hr. If the percent*

–90 microns material in classifier underflow, overflow and ball mill discharge are 20, 75 and 50 respectively, calculate the percent circulating load.

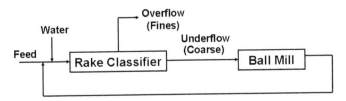

Figure 10.3.3 Circuit diagram for example 10.3.3.

Solution:

In closed circuit grinding operation feed solids to the grinding circuit will be equal to the overflow solids from the rake classifier at steady state. Therefore

Tonnage of classifier overflow solids $= 50$ tons/hr

Let the tonnage of classifier underflow solids be S tons/hr.

Balance of -90 microns material over the rake classifier

$$50 \times 15 + S \times 50 = 50 \times 75 + S \times 20$$

$$\Rightarrow \qquad S = 100 \text{ tons/hr}$$

Classifier underflow solids $= 100$ tons/hr

$$\text{Percent circulating load in the ball mill circuit} = \frac{100}{50} \times 100 = 200\%$$

Grinding circuits in which all of the plant feed is ground are called **primary grinding circuits**. The principal ones where ball mills are used are single stage ball mill, rod mill-ball mill, and autogenous or semiautogenous mill-ball mill systems.

Single stage ball mill circuit (Fig. 10.3.1.2) is the simplest primary grinding circuit where crushed ore is ground in closed circuit in a single stage ball mill to the size required for mineral liberation. Single stage ball mills are more sensitive to changes in top size and particle size distribution of the mill feed than rod mills.

Different primary grinding circuits with rod mill-ball mill are used depending on characteristics of the ore and the product size requirements. For these circuits (Fig. 10.3.3.1) the raw ore needs crushing to feed to the rod mill.

In Fig. 10.3.3.2, an autogenous or a semiautogenous primary mill is also used to perform most of the crushing.

In some processing plants, the products from the beneficiation operation are required to grind for further liberation of the mineral to be recovered, to obtain higher

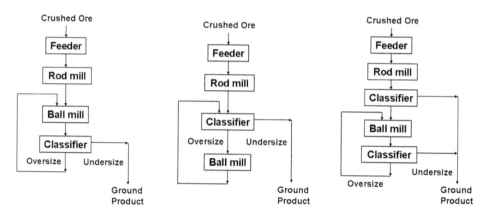

Figure 10.3.3.1   Rod mill–Ball mill primary grinding circuits.

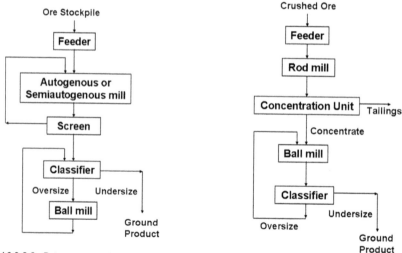

Figure 10.3.3.2 Primary grinding circuit with Autogenous mill.

Figure 10.3.3.3 Regrinding circuit.

concentrate grade, to reduce tailing losses and/or to obtain liberation of other miner-als such as pyrite in lead-zinc circuit and magnetite in copper circuit. In these cases, circuits are called **regrinding circuits** (Fig. 10.3.3.3). Another example of regrinding circuit is grinding the iron ore concentrate to the required particle size specifications as practiced in pelletization.

Regular type grinding circuit with ball mill followed by classifier is considered in example 10.3.4 and reverse type circuit with classifier followed by ball mill is considered in examples 10.3.5 and 10.3.6 for illustration.

Example 10.3.4: *An integrated circuit consisted of a crusher, grinding mill and a classifier. The underflow from the classifier was returned to the mill for re-grinding. The classifier feed, underflow and overflow are sampled and found that they contain solids of 45%, 80% and 20% by weight respectively. Calculate the %circulating load.*

Solution:

Given

% solids by weight in classifier feed $= f = 45\%$
% solids by weight in classifier underflow $= u = 80\%$
% solids by weight in classifier overflow $= p = 20\%$

$$\% \text{ circulating load} = \frac{u(f-p)}{p(u-f)} \times 100 = \frac{80(45-20)}{20(80-45)} \times 100 = 285.7\%$$

Alternately

$$\text{Dilution ratio of feed} \qquad = DR_F = \frac{100-45}{45} = 1.2222$$

$$\text{Dilution ratio of underflow} = DR_U = \frac{100-80}{80} = 0.25$$

$$\text{Dilution ratio of overflow} \quad = DR_P = \frac{100-20}{20} = 4.00$$

$$\% \text{ Circulating load} = \frac{DR_P - DR_F}{DR_F - DR_U} \times 100 = \frac{4.0 - 1.2222}{1.22 - 0.25} \times 100 = 286.4\%$$

Example 10.3.5: *For the grinding circuit shown in Figure 10.3.5, the material of specific gravity 3.2 is fed at the rate of 250 tph.*

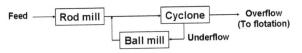

Figure 10.3.5 Grinding circuit for example 10.3.5.

If the ratio of solids underflow of cyclone to the fresh feed to the feed to the rod mill is 4 and the % solids in overflow and underflow of cyclone by weight are 36.5% and 80% respectively, Calculate

a) percent solids by weight in the feed to the cyclone
b) slurry tonnage of feed to the cyclone, overflow and underflow from the cyclone
c) % solids by volume in feed

Solution:

Given

Tonnage of feed to the circuit $= F = 250$ tph
Specific gravity of the solids $= \rho_p = 3.2$

Ratio of cyclone underflow to new feed $= 4$

% solids by weight in cyclone overflow $= p = 36.5\%$

% solids by weight in cyclone underflow $= u = 80\%$

Solids balance

Solids tonnage in cyclone underflow $= U = F \times 4 = 250 \times 4 = 1000\,\text{tph}$

Solids tonnage in cyclone overflow $\ = P = \text{Feed to rod mill} = F = 250\,\text{tph}$

Solids tonnage in feed to the cyclone $= F + 4F = 5F = 5 \times 250 = 1250\,\text{tph}$

Water balance

Dilution ratio of cyclone overflow $\ = DR_P = \dfrac{100 - p}{p} = \dfrac{100 - 36.5}{36.5} = 1.74$

Dilution ratio of cyclone underflow $= DR_u = \dfrac{100 - u}{u} = \dfrac{100 - 80}{80} = 0.25$

Tonnage of water in cyclone overflow $\ = P \times DR_P = 250 \times 1.74 \ = 435\,\text{tph}$

Tonnage of water in cyclone underflow $= U \times DR_U = 1000 \times 0.25 = 250\,\text{tph}$

Tonnage of water in cyclone feed $\qquad = 435 + 250 \qquad\qquad\qquad = 685\,\text{tph}$

Slurry balance

Tonnage of overflow slurry $\qquad = 250 + 435 \ = 685\,\text{tph}$

Tonnage of underflow slurry $\qquad = 1000 + 250 = 1250\,\text{tph}$

Tonnage of feed slurry to cyclone $= 1250 + 685 = 1935\,\text{tph}$

% solids by weight in feed to the cyclone $= \dfrac{1250}{1935} \times 100 = 64.6\%$

Volume of feed slurry $=$ Volume of solids $+$ volume of water

$$= \dfrac{1250}{3.2} + 685 = 390.6 + 685 = 1075.6\,\text{m}^3/\text{hr}$$

% solids by volume $\quad = \dfrac{390.6}{1075.6} \times 100 = 36.31\%$

Example 10.3.6: *The circuit shown in Fig 10.3.6 is fed with 25 tons/hr dry solids of density 3000 kg/m³. The feed to the cyclone contains 36% solids by weight. Sampling and analyses of three streams results 250 μm size in the rod mill discharge, ball mill discharge and cyclone feed is 27%, 5%, and 14% respectively. Determine the volumetric flow rate of feed (solid + water) to the cyclone.*

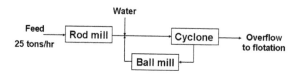

Figure 10.3.6 Circuit diagram for example 10.3.6.

Solution:

Given

Rate of dry solids feed to the Rod mill = 25 tph
Density of the solids = 3000 kg/m³   = 3 tons/m³
% solids in the feed to cyclone   = 36%
% −250 μm in rod mill discharge   = 27%
% −250 μm ball mill discharge   = 5%
% −250 μm in cyclone feed   = 14%

Solids balance

Cyclone feed (C) = Ball mill feed (B) + Rod mill feed (R)

$$C = B + 25 \qquad \text{(I)}$$

Balance of 250 μm size

$$C \times 14 = B \times 5 + 25 \times 27 \qquad \text{(II)}$$

Solving (I) & (II) equations   ⇒ $C = 61.1$ tons/hr

Volumetric flow rate of feed (dry) to cyclone = 61.1/3 = 20.37 m³/hr

Dilution ratio of feed to cyclone = $\dfrac{100 - 36}{36}$ = 1.78

Flow rate of water in feed to cyclone = 61.1 × 1.78 = 108.76 tph

Volumetric flow rate of water = 108.76 m³/hr

Volumetric flow rate of slurry fed to cyclone = 20.37 + 108.76
                         = 129.13 m³/hr

10.4 PROBLEMS FOR PRACTICE

10.4.1: Calculate the critical speed of a ball mill having 600 mm diameter with 25 mm size balls.

[55.51 rpm]

10.4.2: Determine the operating speed of a ball mill of 1800 mm diameter containing steel balls of 90 mm diameter. The mill operates at 70% of the critical speed.

[22.6 rpm]

10.4.3: Calculate the average circulating load of a mill, which is in closed circuit with a classifier and grinds 500 tons of dry ore per day. The screen analyses of the mill discharge, classifier overflow and underflow are given in Table 10.4.3.

Table 10.4.3 Screen analyses data for problem 10.4.3.

		Classifier	
Mesh	Mill discharge %	Overflow %	Underflow %
−65 + 150	15.2	19.6	13.8
−150 + 200	6.1	12.4	4.1

[1573.2 dry tons/day]

Chapter 11

Principles of settling

Separation of ore particles according to their sizes, called classification, and according to their density, called gravity concentration, are the two operations usually the ore is processed through to get the high grade concentrate. The basic principles of these operations are the principles of settling of particles in fluid medium. These basic principles of settling are discussed in this chapter.

The movement of a solid particle in a fluid depends on many parameters and the inter-relation among these parameters is pretty complex. A simplified analysis of the movement of the particle can be made by the following assumptions:

1. The shape and size of the particle are defined. For the simplest case, it is assumed that the particle is sphere of diameter 'd'.
2. The particle is non-porous and incompressible (i.e., its density remains constant). The particle is insoluble in the fluid and does not react chemically with it.
3. The density and viscosity of the fluid are constant.
4. The effect of surface characteristics or interfacial conditions between the solid and the fluid on the dynamics of the particle is neglected.
5. The particle is freely settling under gravity. Other particles are either absent or even if presents, do not interfere with the motion of the particle under consideration.
6. The fluid forms an infinite medium. In other words, the particle under consideration is at an infinite distance (very long distance) from the fluid boundaries and therefore the boundary effect or commonly called the wall effects on the dynamics of the particle can be neglected.
7. The fluid is a continuous medium, and the particle size is much larger than the mean free path of the fluid molecules. Thus the effect of slip between the particles and the fluid molecules can be neglected.

Consider a single homogeneous spherical particle of diameter 'd' and density 'ρ_p' falling under gravity in a viscous fluid of density 'ρ_f' and viscosity 'μ_f'. There are three forces acts on a particle:

1. **Gravity force,** $m_p g$, product of the mass of the particle (m_p) and acceleration due to gravity (g), acts downwards
2. **Buoyancy force,** $m_f g$, (by Archimedes' principle) product of the mass of the fluid displaced by the particle (m_f) and the acceleration due to gravity (g), which acts parallel and opposite to the gravity force

3   **Drag force, F_R,** (resistance to the motion), which acts on the surface of the particle and is parallel and opposite to the gravity force.

According to the Newton's second law of motion, the equation of motion of the particle is

$$m_p g - m_f g - F_R = m_p \frac{dV}{dt} \tag{11.1}$$

where V is the velocity of the particle and $\frac{dV}{dt}$ is the net or effective acceleration of the particle.

If the total downward force (gravity force) acting on the particle becomes just equal in magnitude and opposite in direction to the total upward force (the sum of the buoyancy force and drag force or resistance force) acting on the particle in a fluid, the net force acting on the particle is zero, the acceleration of the particle is also zero and the particle moves with a constant velocity.

This velocity is the maximum velocity attained by the particle. It is known as **maximum velocity** or **terminal velocity (V_m)**. When once the particle attains this velocity, it will fall with the same velocity thereafter.

When the acceleration is zero, the particle attains the terminal velocity.

Hence when $\frac{dV}{dt} = 0$ equation 11.1 becomes

$$F_R = g(m_p - m_f) \tag{11.2}$$

$$\Rightarrow \quad F_R = g\left[\frac{\pi}{6} d^3 \rho_p - \frac{\pi}{6} d^3 \rho_f \right] \text{ as the particle is assumed as sphere}$$

$$\Rightarrow \quad F_R = \frac{\pi}{6} g d^3 (\rho_p - \rho_f) \tag{11.3}$$

11.1   LAMINAR AND TURBULENT FLOWS

When a particle is falling in a fluid, the fluid flows around the particle. This flow may be either laminar (streamline flow round a particle) or turbulent. In laminar flow the fluid flows around the particle without forming any eddies or swirls. In turbulent flow the fluid breaks into eddies and swirls, around the particle. With increased rate of flow, eddies become larger and more complex and the flow becomes more turbulent. If the particle is small, the flow is usually viscous or laminar and the resistance to the fall of the particle mainly depends on the viscosity of the fluid. If the particle is large, the flow is usually turbulent and accompanied by the formation of eddies and vortices in the fluid. These eddies give large resistance to the fall of particle and the viscosity of the fluid becomes less important and can be neglected in determining the resistance under fully turbulent conditions.

Reynolds number was derived by Osborne Reynolds [12] and is given by

$$N_{Re} = \frac{dV \rho_f}{\mu_f} \tag{11.4}$$

The type of flow depends on the Reynolds number.

If $N_{Re} < 1.0$,   the flow is laminar
If $N_{Re} > 1000$,   the flow is turbulent
If $1.0 > N_{Re} < 1000$,   the flow is neither laminar nor turbulent

The laminar flow around the particle is possible when the particles are fine, whereas turbulent flow is possible with comparatively coarser particles. However, in most of the mineral beneficiation operations, intermediate range of particle sizes exists.

11.2   FLUID RESISTANCE

The nature of the resistance (or drag) depends on the velocity of descent. At low velocities, motion is smooth because the layer of fluid in contact with the body, moves with it, while the fluid, a short distance away, is motionless. Between these two positions is a zone of intense shear in the fluid all around the descending particle. Hence the resistance to the motion is due to the shear forces or viscosity of the fluid and is called **viscous resistance**. As the size of the particle increases, settling velocity increases. At high velocities, the main resistance is due to the displacement of fluid by the particle and is known as **turbulent resistance**. In this case, the viscous resistance is relatively small.
    Drag force increases with increase in the velocity of the particle and is often called the kinematic force. At the initial stages of settling, when the particle velocity is small, F_R is negligibly small. As a result, the magnitude of the effective acceleration is quite large. Thus the particle has a high acceleration and its velocity starts increasing rapidly. As the velocity of the particle increases, the kinematic force also increases and the rate of change of velocity diminishes. Ultimately, a stage is reached when the total downward force acting on the particle (gravity force) becomes just equal to the total upward force acting on it (the sum of the buoyancy force and kinematic force). Then the net force acting on the particle is zero, the particle moves with a constant velocity or zero acceleration.
    The kinematic force F_R can be expressed in general as

$$F_R = AKC_D \tag{11.5}$$

where A = characteristic area of the system
       K = characteristic kinetic energy per unit volume
       C_D = a dimensionless parameter

    The terms A and K are to be defined depending on the situation, the system and the process under consideration. For the free fall of a spherical particle in a fluid, Newton proposed that A is defined as projected area of the particle, measured in a plane perpendicular to the direction of motion of the particle.

$$\text{For the spherical particle} \quad A = \frac{\pi d^2}{4} \tag{11.6}$$

$$K = \frac{1}{2}\rho V^2 \tag{11.7}$$

where V is the relative velocity between the particle and the fluid. As it is assumed that the fluid is stationary, V is the free settling velocity of the particle. At the point of maximum velocity of the particle, it is V_m. C_D is the frictional drag or drag coefficient or coefficient of resistance.

$$\text{Therefore} \quad F_R = \frac{\pi d^2}{4} \frac{1}{2} \rho_f (V_m)^2 C_D \tag{11.8}$$

11.3 TERMINAL VELOCITY

Equation 11.3 can be written as

$$\frac{\pi d^2}{4} \frac{1}{2} \rho_f V_m^2 C_D = \frac{\pi}{6} g d^3 (\rho_p - \rho_f)$$

$$\Rightarrow \quad C_D = \frac{4}{3} \frac{gd}{V_m^2} \frac{(\rho_p - \rho_f)}{\rho_f} \tag{11.9}$$

$$\text{or} \quad V_m = \sqrt{\frac{4}{3} \frac{gd}{C_D} \frac{(\rho_p - \rho_f)}{\rho_f}} \tag{11.10}$$

This is a Newton's equation. This equation is applicable for the $N_{Re} > 1000$, and particle sizes of more than 2000 microns in general.

If the particle velocity is quite low, the resistance to the motion is due to the shear forces or viscosity of the fluid and is called **viscous resistance**. Stokes [13], an eminent English Physicist, deduced an expression for the drag force as $3\pi d\mu_f V$. Then equation 11.3 becomes, after replacing V by V_m, terminal velocity of the particle,

$$3\pi d\mu_f V_m = \frac{\pi}{6} g d^3 (\rho_p - \rho_f) \tag{11.11}$$

$$\text{On computation,} \quad V_m = \frac{d^2 g (\rho_p - \rho_f)}{18\mu_f} \tag{11.12}$$

This is a Stokes' equation. This equation describes settling of particles under laminar flow conditions where fluid viscosity provides the resistance to flow. This equation holds good upto 100 microns size of quartz particles and upto 74 microns size of galena particles.

Since the Stokes' equation has been developed for viscous resistance or laminar free settling of particles, its range of validity can be analysed by substituting this equation

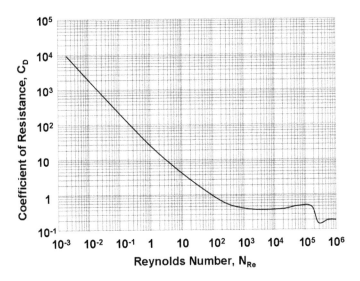

Figure 11.3.1 Relationship of coefficient of resistance to Reynolds number.

in equation 11.9

$$C_D = \frac{4}{3}\frac{gd}{V_m}\frac{(\rho_s - \rho_f)}{\rho_f}\frac{18\mu_f}{d^2 g(\rho_p - \rho_f)}$$

$$= \frac{24\mu_f}{dV_m\rho_f} = \frac{24}{dV_m\rho_f/\mu_f} = \frac{24}{N_{Re}}$$

$$C_D = \frac{24}{N_{Re}} \tag{11.13}$$

$N_{Re} = \dfrac{dV_m\rho_f}{\mu_f}$ is called particle Reynolds number

A chart of log C_D vs. log N_{Re} is a diagonal line of slope -1 for this Stokes relationship for spherical particles. Figure 11.3.1 shows a standard plot reproduced from 1937 report of the Committee on Sedimentation of the National Research Council [14].

From this plot, it can be seen that the plot is linear for N_{Re} less than 1.0 which can be marked as laminar settling region within which the Stokes' law is applicable. For large values of particle Reynolds number ($N_{Re} > 1000$), the plot is almost a horizontal line parallel to log N_{Re} axis which indicates that coefficient of resistance is constant and independent of the Reynolds number. This is called turbulent settling region.

For this condition, Sir Isaac Newton [15] assumed that the resistance is entirely due to turbulent resistance and deduced as $F_R = 0.055\pi d^2(V_m)^2\rho_f$. After substitution in equation 11.3

$$0.055\pi d^2 (V_m)^2 \rho_f = \frac{\pi}{6} g d^3 (\rho_p - \rho_f)$$

On computation $V_m = \sqrt{\dfrac{3gd(\rho_p - \rho_f)}{\rho_f}}$ (11.14)

This is a modified form of Newton's equation. This is applicable for the particle Reynolds number from 1000 to 200,000 and particle sizes of more than 2000 microns in general. Beyond 200,000, the coefficient of resistance decreases sharply. This is due to the formation of an eddy in the fluid behind the particle which travels with particle, resulting in sharp decrease in the drag on the particle.

For the most common sizes of particles encountered in mineral processing, the Reynolds number for settling under gravity in water falls into the transitional region, where the flow is neither fully turbulent nor fully laminar and the resistance to flow changes non-linearly with velocity. For the transition region ($1.0 < N_{Re} < 1000$) it is difficult to propose a hard and fast correlation between C_D and N_{Re}. A large number of empirical equations based on experimental data have been proposed in the literature for estimating the drag coefficient in the transition region. Newton's equation can be applied for this transition region to arrive at approximate value of settling velocity.

The following is an example for calculation of terminal settling velocity:

Let a spherical silica particle of density 2.65 gm/cm^3 and diameter of 0.1 cm is settling in water.

First estimate: Calculate velocity using Stokes' equation.

$$V_m = \frac{d^2 g (\rho_p - \rho_f)}{18\mu_f} = \frac{(0.1)^2 (981)(2.65 - 1.0)}{18(0.01)} = 89.9 \text{ cm/sec}$$

Calculate Reynolds number.

$$N_{Re} = \frac{d V_m \rho_f}{\mu_f} = \frac{(0.1)(89.9)(1.0)}{0.01} = 899 \text{ which is much greater than 1.0.}$$

Therefore, Stokes' law is certainly not valid for such a large, fast-settling particle. However, the Reynolds number is approaching 1000, so it may be in the validity range for Newton's Law.

Second estimate: Calculate velocity using modified form of Newton's equation:

$$V_m = \sqrt{\frac{3gd(\rho_p - \rho_f)}{\rho_f}} = \sqrt{\frac{3(981)(0.1)(2.65 - 1.0)}{1.0}} = 22.0 \text{ cm/sec}$$

Recalculate Reynolds number.

$$N_{Re} = \frac{dV_m \rho_f}{\mu_f} = \frac{(0.1)(22.0)(1.0)}{0.01} = 220, \text{ which is much less than } 1000.$$

So Newton's equation can be used. Referring to Figure 11.3.1, for a Reynolds number of 220, the value of C_D is approximately 0.7.

$$V_m = \sqrt{\frac{4}{3} \frac{gd}{C_D} \frac{(\rho_p - \rho_f)}{\rho_f}} = \sqrt{\frac{4(981)(0.1)(2.65 - 1.0)}{3(0.7)(1.0)}} = 17.56 \text{ cm/sec}$$

Recalculate Reynolds number. $N_{Re} = \dfrac{dV_m \rho_f}{\mu_f} = \dfrac{(0.1)(17.56)(1.0)}{0.01} = 175$

From figure 11.3.1, for N_{Re} of 175, the value of C_D is 0.8.

$$V_m = \sqrt{\frac{4}{3} \frac{gd}{C_D} \frac{(\rho_p - \rho_f)}{\rho_f}} = \sqrt{\frac{4(981)(0.1)(2.65 - 1.0)}{3(0.8)(1.0)}} = 16.4 \text{ cm/sec}$$

Recalculate Reynolds number. $N_{Re} = \dfrac{dV_m \rho_f}{\mu_f} = \dfrac{(0.1)(16.4)(1.0)}{0.01} = 164$

From figure 11.3.1, for N_{Re} of 164, the value of C_D is close to 0.8.
Hence the terminal settling velocity is approximately **16.4 cm/sec.**

Terminal settling velocity can be calculated through an equation formed between C_D and N_{Re}. Equation can be formed as follows:

Equation 11.9 is $\qquad C_D = \dfrac{4}{3} \dfrac{gd}{V_m^2} \dfrac{(\rho_p - \rho_f)}{\rho_f}$

Taking logarithm on both sides

$$\log C_D = \log \left[\frac{4gd(\rho_p - \rho_f)}{3\rho_f} \right] - 2 \log V_m \qquad (11.15)$$

Reynolds number at the terminal velocity is $N_{Re} = \dfrac{dV_m \rho_f}{\mu_f}$

Taking logarithm on both sides

$$\log N_{Re} = \log \frac{d\rho_f}{\mu_f} + \log V_m \qquad (11.16)$$

Elimination of V_m from equations 11.15 and 11.16 gives

$$\log C_D = -2 \log N_{Re} + \log \left[\frac{4gd^3 \rho_f (\rho_p - \rho_f)}{3\mu_f^2} \right] \qquad (11.17)$$

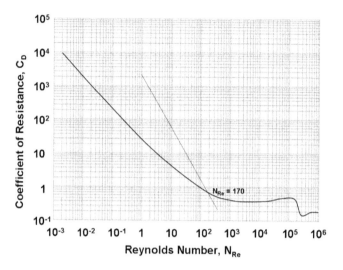

Figure 11.3.1.1 Determination of Reynolds number.

At $N_{Re} = 1$, equation 11.17 becomes

$$C_D = \frac{4gd^3 \rho_f(\rho_p - \rho_f)}{3\mu_f^2} \tag{11.18}$$

For the above example, C_D is to be calculated first

$$C_D = \frac{4gd^3 \rho_f(\rho_p - \rho_f)}{3\mu_f^2} = \frac{4 \times 981 \times (0.1)^3 \times 1.0(2.65 - 1.0)}{3(0.01)^2} = 21,852$$

At $N_{Re} = 1$ and $C_D = 21,852$, a straight line is to be drawn with slope of -2 on Figure 11.3.1. This straight line intersects the C_D vs N_{Re} curve. Corresponding to this intersection point $N_{Re} = 170$ (Fig. 11.3.1.1)
Therefore

$$\text{Terminal settling velocity} = V_m = \frac{N_{Re}\mu_f}{d\rho_f} = \frac{170 \times 0.01}{0.1 \times 1.0} = 17 \, \text{cm/sec}$$

A chart of $\log C_D$ vs. $\log N_{Re}$ is given in Figure 11.3.2 for different values of sphericity and can be used for estimation of terminal settling velocity values for non spherical particles.

Examples 11.3.1 to 11.3.5 illustrates the calculations for determination of settling velocities.

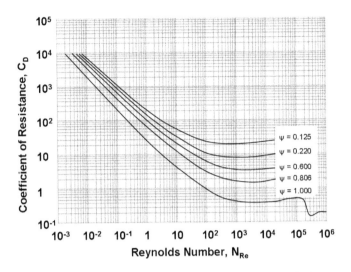

Figure 11.3.2 Relationship of C_D to N_{Re} for different values of sphericity.

Example 11.3.1: *Calculate the terminal settling velocity of coal particle having 1.4 specific gravity and 20 microns in size settling in water.*

Solution:

Given

Density of the galena particle $= \rho_p = 1.4\,\text{gm/cm}^3$
Diameter of the galena particle $= d = 20\,\text{microns} = 0.002\,\text{cm}$
Density of water $= \rho_w = 1.0\,\text{gm/cm}^3$
Viscosity of water $= \mu_w = 0.01\,\text{poise}$

Stokes' law can be applied as the size of the particle is <100 microns

$$\text{Terminal settling velocity} = v_m = \frac{d^2 g(\rho_p - \rho_w)}{18\mu_w}$$

$$= \frac{(0.002)^2(981)(1.4 - 1.0)}{18(0.01)} = 0.0087\,\text{cm/sec}$$

$$\text{Reynolds number } N_{Re} = \frac{dV_m\rho_w}{\mu_w} = \frac{(0.002)(0.0087)(1.0)}{0.01} = 0.00174$$

As $N_{Re} < 1$, the particle is settling in Stokes' law region.

Example 11.3.2: *Calculate the maximum velocity at which a spherical particle of galena having specific gravity of 7.5 and 6 mm in diameter will fall in water. Also find the type of flow. Density and viscosity of water are 1 gm/cc and 0.82 cp respectively.*

Solution:

Given

Density of the galena particle $= \rho_p = 7.5\,\text{gm/cm}^3$
Diameter of the galena particle $= d = 6\,\text{mm} = 0.6\,\text{cm}$
Density of water $= \rho_w = 1.0\,\text{gm/cm}^3$
Viscosity of water $= \mu_w = 0.82\,\text{centi poise} = 0.0082\,\text{poise}$

As the size of the galena particle is 0.6 cm, Newton's law is applicable

$$\text{Maximum settling velocity} = v_m = \sqrt{\frac{3gd(\rho_p - \rho_w)}{\rho_w}}$$

$$= \sqrt{\frac{3 \times 981 \times 0.6(7.5 - 1.0)}{1.0}} = 107.1\,\text{cm/sec}$$

$$\text{Reynolds number} = N_{Re} = \frac{dv\rho_w}{\mu_w} = \frac{0.6 \times 107.1 \times 1.0}{0.0082} = 7836.6$$

As $N_{Re} > 1000$, the flow is turbulent.

Example 11.3.3: *Calculate the size of a spherical silica particle settling in water from rest at 20° C with a terminal settling velocity of 0.5 cm/sec. Specific gravity of silica is 2.65.*

Solution:

Given

Density of the silica particle $= \rho_p = 2.65\,\text{gm/cm}^3$
Density of water $= \rho_w = 1.0\,\text{gm/cm}^3$
Viscosity of water $= \mu_w = 0.01\,\text{poise}$
Terminal settling velocity $= v_m = 0.5\,\text{cm/sec}$

As per Stokes' law $v_m = \dfrac{d^2 g(\rho_p - \rho_w)}{18\mu_w}$

$$d = \sqrt{\frac{v_m 18\mu_w}{(\rho_p - \rho_w)g}} = \sqrt{\frac{0.5 \times 18 \times 0.01}{(2.65 - 1.0)981}} = 0.0078\,\text{cm}$$

$$\text{Reynolds Number} = N_{Re} = \frac{dV_m\rho_w}{\mu_w} = \frac{(0.0078)(0.5)(1.0)}{0.01} = 0.39 < 1.0$$

Hence Stokes' law is applicable

Diameter of silica particle of 0.5 cm/sec terminal velocity = 78 microns

Example 11.3.4: *If the value of Reynolds number is 0.153 at which quartz spheres of 2.65 density settle in water at 68°F of viscosity of 0.01 poise at a gravitational constant of 981 cm/sec², calculate the size of the spheres and their terminal velocity.*

Solution:

Given

$$
\begin{aligned}
\text{Reynolds number} &= N_{Re} = 0.153 \\
\text{Density of the quartz} &= \rho_q = 2.65\,\text{gm/cm}^3 \\
\text{Viscosity of water} &= \mu_w = 0.01\,\text{poise} \\
\text{Gravitational constant} &= g = 981\,\text{cm/sec}^2 \\
\text{Density of water} &= \rho_w = 1.0\,\text{gm/cm}^3
\end{aligned}
$$

Reynolds number $\qquad N_{Re} = \dfrac{d v_m \rho_w}{\mu_w}$

$\Rightarrow \qquad v_m = \dfrac{N_{Re}\mu_w}{d\rho_w} = \dfrac{0.153 \times 0.01}{d \times 1} = \dfrac{0.00153}{d}$

$\Rightarrow \qquad v_m = \dfrac{0.00153}{d} \qquad\qquad$ (I)

As $N_{Re} < 1.0$, Stokes' law is applicable

Terminal velocity $= v_m = \dfrac{d^2 g(\rho_q - \rho_w)}{18\mu_w}$

$$= \dfrac{2.65 - 1.0}{18 \times 0.01} \times 981 \times d^2 = 8992.5 \times d^2$$

$\Rightarrow \qquad v_m = 8992.5 \times d^2 \qquad\qquad$ (II)

Equating (I) and (II)

$$8992.5 \times d^2 = \dfrac{0.00153}{d} \qquad \Rightarrow \quad d = 5.54\,\text{microns}$$

Terminal velocity $= v_m = \dfrac{0.00153}{d} = \dfrac{0.00153}{0.00554} = 0.276\,\text{cm/sec}$

Example 11.3.5: *Calculate the settling velocity of galena particle of 0.225 mm size and 7.5 specific gravity in water. Assume the sphericity is 0.806.*

Solution:

Given

$$
\begin{aligned}
\text{Size of Sphalerite} \quad &= d = 0.0225\,\text{cm} \\
\text{Density of sphalerite} \quad &= \rho_s = 7.5\,\text{gm/cm}^3 \\
\text{Sphericity of sphalerite} &= \psi = 0.806 \\
\text{Density of water} \quad &= \rho_w = 1.0\,\text{gm/cm}^3 \\
\text{Viscosity of water} \quad &= \mu_w = 0.01\,\text{poise}
\end{aligned}
$$

$$
C_D = \frac{4gd^3\rho_f(\rho_p - \rho_w)}{3\mu_w^2} = \frac{4 \times 981 \times (0.0225)^3 \times 1.0(7.5 - 1.0)}{3(0.01)^2} = 968
$$

At $N_{Re} = 1$ and $C_D = 968$, a straight line is to be drawn with slope of -2 on Figure 11.3.2. This straight line intersects the C_D vs N_{Re} curve of 0.806 sphericity. Corresponding to this intersection point $N_{Re} = 9$ (Fig. 11.3.5).

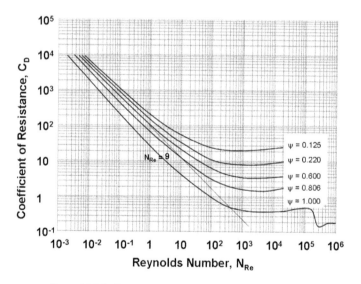

Figure 11.3.5 Determination of N_{Re} for example 11.3.5.

Therefore

$$
\text{Terminal settling velocity} = V_m = \frac{N_{Re}\mu_w}{d\rho_w} = \frac{9 \times 0.01}{0.0225 \times 1.0} = 4\,\text{cm/sec}
$$

The terminal settling velocity of a spherical particle is a function of size and specific gravity (density) of the particle. If two particles have the same specific gravity, then the larger diameter particle has higher terminal velocity and if two particles have the same diameter, then the heavier particle has higher terminal velocity. The velocity of an irregularly shaped particle with which it is settling in a fluid medium also depends on its shape. As almost all natural particles are irregular in shape, principle can be stated as

The coarser, heavier and rounder particles settle faster than the finer, lighter and more angular particles

11.4   FREE SETTLING

In all the foregoing discussions, the particle is settling in a large volume of fluid by its own specific gravity, size and shape and uninfluenced by the surrounding particles as particles are not crowded. Such settling process is called **Free settling**. Free settling predominates in a well dispersed pulps where the percent solids by weight is less than 10.

11.5   HINDERED SETTLING

When the particles settle in relatively small volume of fluid, they are crowded in the pulp. Particles collide each other during their settling and this collision affects their settling velocities. Such settling process is called **Hindered settling**. Hindered settling predominates when the percent solids by weight is more than 15. Under hindered settling conditions, the settling velocity or rate of settling of each individual particle will be considerably less than that for the free settling conditions. While the motion of large particles gets hindered, the small particles tend to get dragged downwards by the large ones and thus get accelerated. It must be noted that each particle is in fact settling through a suspension of other particles in the liquid rather than through the simple liquid itself.

The effective density and viscosity of a concentrated suspension or slurry are much larger than those of clear liquid. The settling medium therefore offers high resistance and this resistance to fall is mainly due to turbulence created. Hindered settling velocity can be approximately estimated from Stokes' equation or Newton's equation after replacing ρ_f and μ_f by ρ_{sl} and μ_{sl}.

Density of the slurry, ρ_{sl}, can be determined by the following formula knowing the volume fraction of the solids, C_v, in the slurry

$$\rho_{sl} = C_v \rho_p + (1 - C_v)\rho_f \tag{11.19}$$

To be precise, slurries are non-Newtonian fluids. The apparent viscosity of the slurry is a function of the shear rate and depends on its rheological characteristics. The viscosity (μ_{sl}) of the slurry containing spherical particles is related to the volume fraction of the solids in the slurry by the formula [16]

$$\frac{\mu_{sl}}{\mu_f} = \frac{10^{1.82C_v}}{1 - C_v} \tag{11.20}$$

By substituting ρ_{sl} and μ_{sl} in Stokes' equation (11.12), an expression for hindered settling velocity V_H can be obtained as follows for the fine particles:

$$V_H = \frac{d^2 g(\rho_p - \rho_{sl})}{18\mu_{sl}} \tag{11.21}$$

$$= \frac{d^2 g(\rho_p - \rho_f)}{18\mu_f}\left[\frac{(1-C_v)^2}{10^{1.82C_v}}\right] = V_m F_s$$

$$\Rightarrow \quad \frac{V_H}{V_m} = F_s \tag{11.22}$$

where $F_s = (1 - C_V)^2 / 10^{1.82C_V}$

F_s is therefore the ratio of hindered settling velocity (V_H) of the particle to its free settling velocity (V_m) and its magnitude will evidently be less than 1.0. Standard plots of F_s versus C_v and the viscosity ratio (μ_{sl}/μ_f) versus C_v are available in the literature [16].

Similarly, an expression for hindered settling velocity V_H for coarse particles can be obtained by replacing ρ_f by ρ_{sl} in Newton's equation (11.10):

$$V_H = \sqrt{\frac{4}{3}\frac{gd}{C_D}\frac{(\rho_p - \rho_{sl})}{\rho_{sl}}} \tag{11.23}$$

The lower the density of the particle, the more marked is the effect of reduction of the effective density, $(\rho_p - \rho_{sl})$, and the greater is the reduction in falling velocity. Similarly, the larger the particle, the greater is the reduction in falling rate as the pulp density increases.

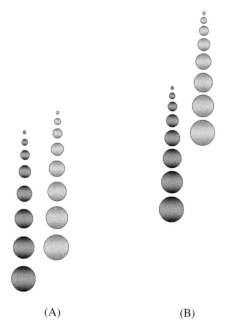

(A)   (B)

Figure 11.5 Settling of particles under (A) Free settling (B) Hindered settling.

Figure 11.5 shows how the particles of different sizes and two specific gravities settle under free and hindered settling conditions. By comparing figure 11.5 (A) & (B), it is evident that more heavier (or lighter) particles can be separated when they settle by hindered settling. This is possible because hindered settling reduces the effect of size and increases the effect of specific gravity.

Example 11.5.1: *Calculate the settling velocity of glass spheres having a diameter of 0.01554 cm in water. The slurry contains 60% by weight of solids. The density of the glass spheres is 2.467 gm/cm³.*

Solution:

Given

$$\text{Diameter of the glass spheres} = d = 0.01554 \, \text{cm}$$
$$\text{Density of the glass spheres} = \rho_{gl} = 2.467 \, \text{gm/cm}^3$$
$$\text{Fraction of solids by weight} = C_w = 0.6$$
$$\text{Density of water} = \rho_w = 1.0 \, \text{gm/cm}^3$$
$$\text{Viscosity of water} = \mu_w = 0.01 \, \text{poise}$$

$$\text{Fraction of solids by volume} = C_v = \frac{0.6/2.467}{(0.4/1.0) + (0.6/2.467)} = 0.378$$

$$\text{Density of the slurry} = \rho_{sl} = C_v \rho_{gl} + (1 - C_v)\rho_w$$

$$= 0.378 \times 2.467 + 0.622 = 1.554 \, \text{gm/cm}^3$$

$$V_H = \frac{d^2 g(\rho_{gl} - \rho_w)}{18\mu_w}\left[\frac{(1 - C_v)^2}{10^{1.82 C_v}}\right]$$

$$= \frac{(0.01554)^2 \times 981 \times (2.467 - 1.0)}{18 \times 0.01}\frac{(1 - 0.378)^2}{10^{1.82 \times 0.378}} = 0.153 \, \text{cm/sec}$$

$$N_{Re} = \frac{dV_H \rho_{sl}}{\mu_{sl}} = \frac{dV_H \rho_{sl}}{\mu_w \dfrac{10^{1.82 C_v}}{1 - C_v}} = \frac{dV_H \rho_{sl}(1 - C_V)}{\mu_w \times 10^{1.82 C_V}}$$

$$= \frac{0.01554 \times 0.153 \times 1.554 \times (1 - 0.378)}{0.01 \times 10^{1.82 \times 0.378}} = 0.047$$

Hence, the settling is in the laminar range.

11.6 EQUAL SETTLING PARTICLES

Particles are said to be equal settling if they have the same terminal velocities in the same fluid and in the same field of force.

11.7 SETTLING RATIO

Settling ratio is the ratio of the sizes of two particles of different specific gravities fall at equal rates.

Under free settling conditions, settling ratio is known as free settling ratio and can be obtained by equating the terminal velocities of lighter and heavier particles of different sizes. Let d_1 and d_2 are the diameters of lighter and heavier particles and ρ_{p1} and ρ_{p2} are the densities of lighter and heavier particles. When the terminal velocities of these two particles are same and the particles are fine obeying Stokes' law of settling, the equation for the terminal settling velocity (equation 11.12) can be written as

$$V_m = \frac{d_1^2 g(\rho_{p1} - \rho_f)}{18\mu_f} = \frac{d_2^2 g(\rho_{p2} - \rho_f)}{18\mu_f} \tag{11.24}$$

$$\Rightarrow \quad \text{Free settling ratio} = \frac{d_1}{d_2} = \left(\frac{\rho_{p2} - \rho_f}{\rho_{p1} - \rho_f}\right)^{1/2} \tag{11.25}$$

When the terminal velocities of the two particles are same and the particles are coarse obeying Newton's law of settling, equation 11.10 can be written as

$$V_m = \sqrt{\frac{4}{3}\frac{gd_1}{C_D}\frac{(\rho_{p1} - \rho_f)}{\rho_f}} = \sqrt{\frac{4}{3}\frac{gd_2}{C_D}\frac{(\rho_{p2} - \rho_f)}{\rho_f}} \tag{11.26}$$

$$\Rightarrow \quad \text{Free settling ratio} = \frac{d_1}{d_2} = \frac{\rho_{p2} - \rho_f}{\rho_{p1} - \rho_f} \tag{11.27}$$

The general expression for free-settling ratio can be deduced from equations 11.25 and 11.27 as

$$\text{Free settling ratio} = \frac{d_1}{d_2} = \left(\frac{\rho_{p2} - \rho_f}{\rho_{p1} - \rho_f}\right)^n \tag{11.28}$$

where $n = 0.5$ for fine particles obeying Stokes' law and $n = 1$ for coarse particles obeying Newton's law. The value of n lies in the range 0.5–1 for particles in the intermediate size range where neither Stokes' law nor Newton's law is applicable.

Consider a mixture of galena (density 7.5 gm/cc) and quartz (density 2.65 gm/cc) particles settling in water. For fine particles, obeying Stoke's law, the free settling ratio is

$$\sqrt{\frac{7.5 - 1}{2.65 - 1}} = 1.98$$

i.e., a fine galena particle will settle at the same rate as fine quartz particle of diameter 1.98 times larger than galena particle. Figure 11.7(A) shows the settling of fine particles.

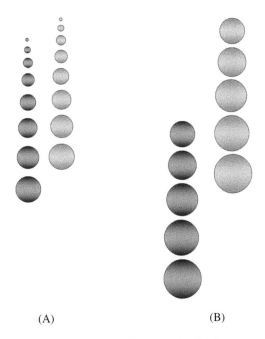

Figure 11.7 Free settling of (A) Fine particles (B) Coarse particles.

For coarse particles, obeying Newton's law, the free settling ratio is

$$\frac{7.5 - 1}{2.65 - 1} = 3.94$$

i.e., a coarse galena particle will settle at the same rate as coarse quartz particle of diameter 3.94 times larger than galena particle. Figure 11.7(B) shows the settling of coarser particles.

Therefore the free settling ratio for coarse particles is larger than for fine particles. This means that density difference between the particles has more effect at coarser size ranges when they settle.

Hindered settling ratio can be obtained by equating the terminal velocities of lighter and heavier particles of different sizes in Stokes' equation (equation 11.21)

$$V_m = \frac{d_1^2 g(\rho_{p1} - \rho_{sl})}{18\mu_{sl}} = \frac{d_2^2 g(\rho_{p2} - \rho_{sl})}{18\mu_{sl}} \tag{11.29}$$

$$\Rightarrow \qquad \text{Hindered settling ratio} = \frac{d_1}{d_2} = \left(\frac{\rho_{p2} - \rho_{sl}}{\rho_{p1} - \rho_{sl}}\right)^{1/2} \tag{11.30}$$

Similarly hindered settling ratio is obtained by equating the terminal velocities of lighter and heavier particles of different sizes in Newton's equation (equation 11.23)

$$V_m = \sqrt{\frac{4}{3} \frac{gd_1}{C_D} \frac{(\rho_{p1} - \rho_{sl})}{\rho_{sl}}} = \sqrt{\frac{4}{3} \frac{gd_2}{C_D} \frac{(\rho_{p2} - \rho_{sl})}{\rho_{sl}}} \tag{11.31}$$

$$\Rightarrow \qquad \text{Hindered settling ratio} = \frac{d_1}{d_2} = \frac{\rho_{p2} - \rho_{sl}}{\rho_{p1} - \rho_{sl}} \tag{11.32}$$

For mixture of galena and quartz particles settling in a pulp of density 1.5, the hindered settling ratio is

for fine particles $\left(\dfrac{7.5 - 1.5}{2.65 - 1.5}\right)^{1/2} = 2.28$

for coarse particles $\dfrac{7.5 - 1.5}{2.65 - 1.5} = 5.22$

i.e., a galena particle will settle at the same rate as quartz particle of diameter 2.28 times larger than galena particle if the particles are fine obeying Stokes' law and it is 5.22 times larger than galena particle if the particles are coarse obeying Newton's law.

When the hindered settling ratio of 5.22 is compared with the free settling ratio of 3.94, it is evident that hindered settling reduces the effect of size, while increasing the effect of density, which means that more heavy (or light) particles can be separated in hindered settling. Hindered settling ratio is always greater than the free settling ratio. As the pulp density increases, this ratio also increases.

Free settling conditions are used in classifiers, in which case they are called Free settling classifiers (Mechanical classifiers or Horizontal current classifiers), to increase the effect of size on separation. Hindered settling conditions are used in classifiers, in which case they are called Hindered settling classifiers (Hydraulic classifiers or Vertical current classifiers), to increase the effect of density on separation.

Almost all gravity concentration operations and many sizing devices make use of the hindered settling phenomenon during the separation of particles.

In example 11.7.1, settling ratios are calculated.

Example 11.7.1: *Determine the ratio of diameters for spherical particles of spha-lerite of specific gravity of 4.0 and quartz of specific gravity of 2.65 that have the same terminal settling velocities in water at 68°F considering free settling conditions.*

Solution:

Given

Density of the quartz particle $\quad = \rho_q = 2.65\,\text{gm/cc}$
Density of the sphalerite particle $= \rho_s = 4.00\,\text{gm/cc}$
Density of water $\quad\quad\quad\quad\quad = \rho_w = 1.0\,\text{gm/cc}$
Viscosity of water $\quad\quad\quad\quad = \mu_w = 0.01\,\text{poise}$

Let the diameters of the quartz and sphalerite particles be d_q and d_s

By Stokes' law Free settling ratio $= \dfrac{d_q}{d_s} = \left(\dfrac{\rho_s - \rho_w}{\rho_q - \rho_w}\right)^{1/2} = \left(\dfrac{4.0 - 1.0}{2.65 - 1.0}\right)^{1/2} = 1.35$

By Newton's law   Free settling ratio $= \left(\dfrac{\rho_s - \rho_w}{\rho_q - \rho_w}\right) = \left(\dfrac{4.0 - 1.0}{2.65 - 1.0}\right) = 1.82$

The motion of a particle, when it starts settling in the fluid and move through the fluid, can be divided into two stages viz., acceleration period and terminal velocity period. Initially the velocity of the particle is zero with respect to the fluid and increases to the terminal velocity during a short period, usually of the order of one tenth of a second or less. During this first stage of short period, there are initial-acceleration effects. When once particle reaches its terminal velocity, second stage starts and continued as long as the particle continues to settle. Classification and thickening processes make use of the terminal velocity period, whereas jigging makes use of differences in particle behavior during the acceleration period for separation.

11.8 SETTLING OF LARGE SPHERES IN A SUSPENSION OF FINE SPHERES

Large spheres settle in a suspension of fine spheres as if the fine spheres were a part of the fluid. When a water-solid suspension is made, the system begins to behave as a heavy liquid whose density is that of the pulp rather than that of carrier liquid. This is the principle used in Heavy Medium Separation operations in cleaning coal with a suspension of sand in water. Thus a large sphere of quartz falling in a suspension of quartz sand having an average density of ρ_{sl} falls at a speed determined by the equation 11.23.

$$V = \sqrt{\frac{4}{3} \frac{gd}{C_D} \frac{(\rho_p - \rho_{sl})}{\rho_{sl}}} \qquad (11.33)$$

11.9 PROBLEMS FOR PRACTICE

11.9.1: *Compute the maximum velocity at which particles of silica of 0.01 cm in diameter and 2.65 in specific gravity will fall from rest through quiet water.*
        [0.9 cm/sec]

11.9.2: *Calculate the particle size of a spherical galena particle of specific gravity 7.5 settling in water from rest with a terminal velocity of 0.57 cm/sec.*
        [40 μm]

11.9.3:   *Calculate the free settling ratio of magnetite and quartz particles having specific gravity of 5.6 and 2.65 respectively.*

[1.67, 2.79]

11.9.4:   *Calculate the diameter of the galena particle that settle equally with a quartz particle of 25 micron size in water. The densities of galena and quartz are 7500 kg/m³ and 2600 kg/m³ respectively.*

[12.41 μm]

Chapter 12

Classification

Classification is a method of separating mixtures of particles of different sizes, shapes and specific gravities into two or more products on the basis of the velocity with which the particles fall through a fluid medium i.e., settling velocity. Generally classification is employed for those particles which are considered too fine to be separated efficiently by screening.

The basic principle of classification is:

The coarser, heavier and rounder particles settle faster than the finer, lighter and more angular particles

In classification, certain particles are only allowed to settle in the fluid medium in order to separate the particles into two fractions.

12.1 CLASSIFIERS

The units in which the separation of solids in fluid medium is carried out are known as classifiers. These classifiers may be grouped into three broad classes as

1 Sizing classifiers
2 Sorting classifiers
3 Centrifugal classifiers

Let us reconsider Figure 11.5 by assigning the numbers in the order of increasing size as shown in the Figure 12.1.1 to explain how the particles can be separated.

Under free settling conditions, if sufficient time for the light particle 5 is not given for settling, all the light particles of 1 to 5 and heavy particles of 1 to 4 will be at the top of the classifier, and light particles of 6 to 8 and heavy particles of 5 to 8 will be at the bottom of the classifier. Top and bottom fractions are removed by suitable means as overflow and underflow product.

This separation is illustrated in Figure 12.1.2(A). Each fraction contains both light and heavy particles and almost of the same size or closely sized particles. It means that all the particles are separated into two size fractions. Hence this type of classification is a sizing classification. Classifier used is called **sizing classifier**.

A typical sizing classifier consists of a sloping rectangular trough. The feed slurry is fed into a pool at feed point as shown in Figure 12.1.2(B) at such a rate that the coarser

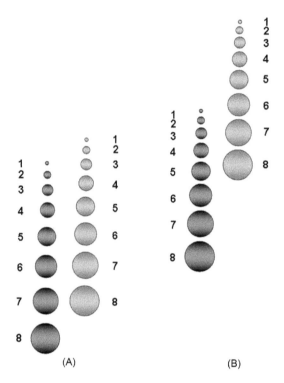

Figure 12.1.1 (A) Free settling and (B) Hindered settling.

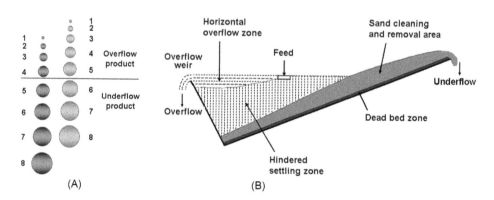

Figure 12.1.2 Separation in sizing classifier.

and faster settling particles fall to the bottom of the trough and conveyed by spiral or rake to the other end of the classifier for discharging. A stream of pulp consists of fines having low settling velocities flow horizontally from the feed inlet to the overflow weir and get discharged. These sizing classifiers are mechanical classifiers. These are also called free settling classifiers, horizontal current classifiers and pool classifiers. These

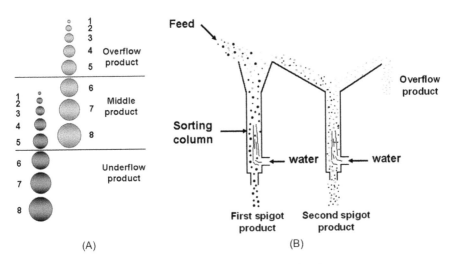

Figure 12.1.3 Separation in sorting classifier.

classifiers are extensively used in closed circuit with a ball mill where underflow coarse product is fed to the ball mill.

In case of hindered settling conditions, if top and bottom fractions are removed without allowing light particle 8 to settle, the bottom fraction contains solely of heavy particles 6 to 8. As the settling velocity of all the light particles is less than the settling velocity of light particle 8, they remain at the top along with the heavy particles 1 to 5 and get discharged as overflow product. If these particles are again classified without allowing light particle 5 to settle, the bottom fraction contains heavy particles 1 to 5 and light particles 6 to 8. Only light particles 1 to 5 remain at the top and get discharged as overflow product. This separation is illustrated in Figure 12.1.3(A). Top and bottom fractions contain all light and heavy particles respectively. However, middle fraction containing heavy particles of 1 to 5 and light particles of 6 to 8 are together under hindered settling conditions which means that they cannot be separated. Classifier used for this kind of separation is called **sorting classifier.**

A typical sorting classifier consists of a series of sorting columns. Sorting classifier with two sorting columns is shown in Figure 12.1.3(B). The feed slurry is introduced centrally near the top of the first sorting column. A current of water known as **hydraulic water** is introduced at the bottom of the sorting column at a velocity slightly less than the smallest heavy particle among the particles required to be discharged in the first sorting column. All those particles having settling velocity less than that of rising water velocity will not settle and rise to the top of the column and fed to the second column. The particles having settling velocity more than that of the rising water velocity settle to the bottom of the first sorting column and get discharged through the spigot. The velocity of the hydraulic water in the second sorting column is less than that of the velocity in the first sorting column. The particles having settling velocity more than that of rising water velocity will settle to the bottom and get discharged through the spigot. Remaining particles have settling velocity less than that of rising water velocity will rise to the top of the column and get discharged as overflow product. These sorting

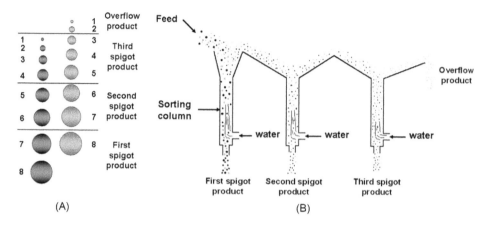

Figure 12.1.4 Sizing in hydraulic classifier.

classifiers are also called hindered settling classifiers, hydraulic classifier, and vertical current classifiers.

It is very interesting to note that, under free settling conditions, only heavy particle 8 and light particles 1 and 2 can be separated whereas in hindered settling conditions, heavy particles of 6 to 8 and light particles of 1 to 5 can be separated.

It can be reiterated that free settling conditions are to be maintained to separate the mixed density and mixed size particles according to their sizes which increases the effect of size and decreases the effect of density on separation. Hindered settling conditions are to be maintained to separate the mixed density and mixed size particles according to their densities which increases the effect of density and decreases the effect of size on separation.

Even though sorting classifiers are not truly sizing classifiers, they are sometimes used to sort out the particles in close size range by maintaining the free settling conditions as shown in Figure 12.1.4(A). A four product separation is shown in Figure 12.1.4(B) by using three sorting columns.

The sorting classifier commonly called hydrosizer is used to sort the feed into different size fractions which are further fed to separate concentrating tables as practiced in chrome ore beneficiation plants.

Different calculations are illustrated in examples 12.1.1 to 12.1.5.

Example 12.1.1: *An ore consists of valuables of 5.8 specific gravity and gangue of 2.6 specific gravity having size range of 10–25 microns is separated in a mechanical classifier. Can all valuables be separated in underflow? If so, will they gangue free? Explain.*

Solution:

Given

Density of gangue $= \rho_g = 2.6 \, \text{gm/cc}$
Density of valuables $= \rho_v = 5.8 \, \text{gm/cc}$
Density of water $= \rho_w = 1.0 \, \text{gm/cc}$
Size range of the particles $= 10 \text{ to } 25 \text{ microns}$

Settling velocity of coarsest heavy particle by Stokes' law

$$= v_m = \frac{d_v^2 g(\rho_v - \rho_w)}{18\mu_w} = \frac{(0.0025)^2 981(5.8 - 1.0)}{18 \times 0.01} = 0.164 \, \text{cm/sec}$$

$$N_{Re} = \frac{d_v v_m \rho_w}{\mu_w} = \frac{0.0025 \times 0.164 \times 1.0}{0.01} = 0.041$$

As Reynolds number (N_{Re}) for coarsest heavy particle is less than 1, all particles settle under laminar conditions. Hence Stokes' law is applicable.

Size of the gangue particle that settles equally with the smallest valuable particle

$$= d_g = \left(\frac{\rho_v - \rho_w}{\rho_g - \rho_w}\right)^{1/2} d_v = \left(\frac{5.8 - 1.0}{2.6 - 1.0}\right)^{1/2} \times 10 = 17.32 \, \text{microns}$$

It means that gangue particles of size 17.32 to 25 microns will also settle by the time smallest valuable particle settle. So underflow of classifier contains all valuable particles and gangue particles of size 17.32 to 25 microns. Hence underflow product is not gangue free. Evidently, overflow product contains only gangue particles of size 10 to 17.32 microns. Hence overflow product is valuable free.

Example 12.1.2: *The size distribution of galena ore is as shown in Table 12.1.2.*

Table 12.1.2 Particle size distribution for example 12.1.2.

Particle size μm	20	30	40	50	60	70	80	90	100
Cum wt fraction	0.33	0.53	0.67	0.77	0.83	0.88	0.91	0.93	0.95

The ore contains 30% galena by weight (remaining is silica) is to be classified with water flowing at 0.6 cm/sec. If the flow zone is essentially laminar, what fraction of galena feed will be in the overflow and underflow products and what will be the weight fraction of galena in these products? The specific gravities of galena and silica are 7.5 and 2.65 respectively.

Solution:

Given

Density of silica particles $= \rho_l = 2.65 \, \text{gm/cc}$
Density of galena particles $= \rho_g = 7.5 \, \text{gm/cc}$
Density of water $= \rho_w = 1.0 \, \text{gm/cc}$
Viscosity of water $= \mu_w = 0.01 \, \text{poise}$

Since the flow zone is laminar, Stokes' law is applicable.

Diameter of the galena particle whose terminal velocity is equal to the upward velocity of water can be determined by Stokes' equation.

$$d_g = \sqrt{\frac{v_m 18 \mu_w}{(\rho_g - \rho_w)g}} = \sqrt{\frac{0.6 \times 18 \times 0.01}{(7.5 - 1.0) \times 981}} = 0.00412 \, \text{cm} = 41.2 \, \text{microns}$$

Graph is drawn between particle size and cumulative weight fraction and shown in Figure 12.1.2.1.

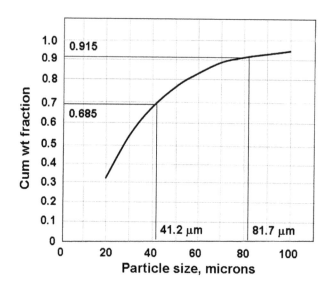

Figure 12.1.2.1 Particle size versus cumulative weight fraction graph.

From the graph, cumulative weight fraction corresponding to a particle size of 41.2 microns is 0.685. This means 68.5% of galena particles have a size less than 41.2 microns. Since all galena particles having size less than 41.2 microns will have a settling velocity less than 0.6 cm/sec, we can conclude that 68.5% of galena fed will be in the overflow. Obviously the remaining 31.5% will be in the underflow product.

The diameter of the silica particle having terminal velocity equal to upward velocity of water of 0.6 cm/sec will be

$$d_s = \sqrt{\frac{v_m 18\mu_w}{(\rho_s - \rho_w)g}} = \sqrt{\frac{0.6 \times 18 \times 0.01}{(2.65 - 1.0) \times 981}} = 0.00817 \, cm = 81.7 \, microns$$

From the graph, cumulative weight fraction corresponding to a particle size of 81.7 microns is 0.915. This means 91.5% of silica particles fed will be in the overflow. Obviously the remaining 8.5% will be in the underflow product.

Let the total feed   = 100 kg
Weight of galena fed   = 100 × 0.30 = 30 kg
Weight of galena in overflow = 30 × 0.685 = 20.55 kg
Weight of silica in overflow   = 70 × 0.915 = 64.05 kg

Weight fraction of galena in overflow = $\dfrac{20.55}{20.55 + 64.05}$ = 0.243

Weight of galena in underflow = 30 − 20.55 = 9.45 kg

Weight of silica in underflow $= 70 - 64.05 = 5.95\,\text{kg}$

Weight fraction of galena in underflow $= \dfrac{9.45}{9.45 + 5.95} = 0.614$

Example 12.1.3: *An ore contains valuables of specific gravity 5.1 and gangue of specific gravity 2.0 has size range of 10 microns to 100 microns. When the ore is classified in a free settling hydraulic classifier, find the upward water velocity in order to remove all gangue particles in overflow. Will this overflow fraction contains any valuables? If so find the size range of valuables collected in overflow.*

Solution:

Given

$$
\begin{aligned}
\text{Density of gangue particles} \quad &= \rho_g = 2.0\,\text{gm/cc}\\
\text{Density of valuable particles} &= \rho_v = 5.1\,\text{gm/cc}\\
\text{Density of water} \qquad\qquad &= \rho_w = 1.0\,\text{gm/cc}\\
\text{Size range of the particles} \quad &= 0.001 - 0.01\,\text{cm}
\end{aligned}
$$

Settling velocity of coarsest gangue particle

$$
= v_m = \frac{d_g^2 g(\rho_g - \rho_w)}{18\mu_w} = \frac{(0.01)^2 981(2.0 - 1.0)}{18 \times 0.01} = 0.545\,\text{cm/sec}
$$

$$
N_{Re} = \frac{d_g v_m \rho_w}{\mu_w} = \frac{0.01 \times 0.545 \times 1.0}{0.01} = 0.545
$$

As $N_{Re} < 1$, flow zone is laminar. Hence Stokes, law is applicable

If upward velocity is 0.545 cm/sec, then all the gangue particles will be collected in overflow.

The size of valuable particle having 0.545 cm/sec settling velocity is

$$
d_v = \sqrt{\frac{v_m 18\mu_w}{(\rho_v - \rho_w)g}} = \sqrt{\frac{0.545 \times 18 \times 0.01}{(5.1 - 1.0)981}} = 0.00494\,\text{cm} = 49.4\,\text{microns}
$$

All valuable particles of size less than 49.4 microns will have settling velocity less than 0.545 cm/sec

Hence – 49.4 microns size valuables will be collected in overflow.

$\therefore$ Size range of valuables collected in overflow is 10 to 49.4 microns.

Example 12.1.4: *Galena ore of the following size distribution of galena of specific gravity 7.5 and gangue of specific gravity 2.65 is to be separated in a hydraulic classifier. Assuming Newton's conditions with a coefficient of resistance of 0.4, calculate the upward velocity of hydraulic water required in the classifier in order to obtain the galena particles completely. Will this fraction be gangue free?*

Particle size (mm)	Mass fraction
$-0.58 + 0.49$	0.62
$-0.49 + 0.40$	0.21
$-0.40 + 0.36$	0.17

Solution:

Given

Density of gangue particles	$= \rho_g$	$= 2.65$ gm/cc
Density of galena particles	$= \rho_G$	$= 7.5$ gm/cc
Density of water	$= \rho_w$	$= 1.0$ gm/cc
Diameter of coarsest gangue particle	$= d_g$	$= 0.058$ cm
Diameter of smallest galena particle	$= d_G$	$= 0.036$ cm
Coefficient of resistance	$= C_D$	$= 0.4$

Under Newton's conditions, settling velocity of the coarsest gangue particle is

$$v_m = \sqrt{\frac{4}{3C_D}\frac{\rho_g - \rho_w}{\rho_w}gd_g} = \sqrt{\frac{4}{3 \times 0.4} \times \frac{2.65 - 1.0}{1.0} \times 981 \times 0.058} = 17.69 \text{ cm/sec}$$

Settling velocity of the smallest valuable particle is

$$v_m = \sqrt{\frac{4}{3C_D}\frac{\rho_G - \rho_w}{\rho_w}gd_G} = \sqrt{\frac{4}{3 \times 0.4} \times \frac{7.5 - 1.0}{1.0} \times 981 \times 0.036} = 27.66 \text{ cm/sec}$$

Velocity of upward hydraulic water must be greater than 17.69 cm/sec and less than 27.66 cm/sec to obtain all galena particles at underflow. No single particle of gangue will settle to bottom as the hydraulic water velocity is greater than the coarsest gangue particle. Hence galena particles are gangue free.

Example 12.1.5: *Quartz and pyrite are to be separated in a hydraulic free settling classifier. The feed to the classifier ranges in size between 10 microns and 300 microns. The specific gravity of quartz is 2.65 and that of pyrite is 5.1. If the mixture is best separated into three fractions such as one containing only quartz, another containing only pyrite and the third a mixture of quartz and pyrite, estimate the size ranges of the two materials in these fractions. Assume the flow zone to be essentially laminar.*

Solution:

Given

Density of quartz	$= \rho_q$	$= 2.65$ gm/cc
Density of pyrite	$= \rho_{py}$	$= 5.1$ gm/cc
Density of water	$= \rho_w$	$= 1.0$ gm/cc
Viscosity of water	$= \mu_w$	$= 0.01$ poise
Size range of the particles	$= 10-300$ microns	

Size of the quartz particle that settles equally with the smallest pyrite particle

$$= d_q = \left(\frac{\rho_{py} - \rho_w}{\rho_q - \rho_w} \right)^{1/2} d_{py} = \left(\frac{5.1 - 1.0}{2.65 - 1.0} \right)^{1/2} \times 10 = 15.76 \text{ microns}$$

15.76 microns size of quartz particles settle with the same velocity as the smallest pyrite particles. Quartz particles smaller than 15.76 microns size will settle only after the smallest pyrite particle of 10 microns has settled. Hence all quartz particles of size less than 15.76 microns size will constitute an overflow fraction of the classifier.

Size of the pyrite particle that settles equally with the largest quartz particle

$$= d_{py} = \left(\frac{\rho_q - \rho_w}{\rho_{py} - \rho_w} \right)^{1/2} d_q = \left(\frac{2.65 - 1.0}{5.1 - 1.0} \right)^{1/2} \times 300 = 190.314 \text{ microns}$$

It means that largest quartz particle will settle only after all the pyrite particles of size more than 190.314 microns have settled. Hence all pyrite particles of size more than 190.314 microns will constitute an underflow fraction.

∴ Size ranges of

a) Overflow fraction contains only quartz – 10 to 15.76 microns
b) Underflow fraction contains only pyrite – 190.314 to 300 microns
c) Middle fraction contains pyrite – 10 to 190.314 microns and
        quartz – 15.76 to 300 microns

The efficiency of classifiers is difficult to quantify. The usual method consists of screen sizing the classifier overflow and underflow, and of calculating efficiency from those data. The following formula similar to formula 5.5.4 used in screening can be used to express the efficiency η as a percentage

$$\eta = \frac{p(f - u)}{f(p - u)} \times 100 \tag{12.1.1}$$

where f, p and u are the percent of minus material in the feed, overflow and underflow respectively, the minus material being any size such that neither f, p, nor u is nil. Since some feed is by-passed into the overflow, the efficiency given by the equation 12.1.1 is increased. To dispose of this objection, it has been proposed to use the following equation [2]

$$\eta = \frac{10,000(p - f)(f - u)}{f(100 - f)(p - u)} \tag{12.1.2}$$

This formula expresses efficiency as the ratio, on percentage basis, of the classified material in the overflow to classifiable material in the feed. The efficiency of classifiers is usually in the range of 50 to 80 percent, but it would be appreciably greater if gauged by sedimentation analyses rather than by screen analyses. This is because of partly flocculated condition during classification.

Examples 12.1.6 and 12.1.7 illustrates the calculations of classifier efficiency by using the formula 12.1.2.

Example 12.1.6: *A mill in closed circuit with a classifier discharges the material containing 21.1% by weight of −200 mesh size. If the overflow and underflow of the classifier contains 60.2% and 8.2% of −200 mesh material, what is the efficiency of a classifier?*

Solution:

Given

% of −200 mesh material in feed $= f = 21.1\%$
% of −200 mesh material in overflow $= p = 60.2\%$
% of −200 mesh material in underflow $= u = 8.2\%$

$$\text{Efficiency of a classifier} = \eta = \frac{10,000(p-f)(f-u)}{f(100-f)(p-u)}$$

$$= \frac{10,000 \times (60.2 - 21.1)(21.1 - 8.2)}{21.1 \times (100 - 21.1)(60.2 - 8.2)} = 58.3\%$$

Example 12.1.7: *It is intended to produce −200 mesh material using ball mill classifier closed circuit operation. Screen analyses data is given in the Table 12.1.7. Calculate the efficiency of a classifier.*

Table 12.1.7 Screen analyses data for example 12.1.7.

Tyler screen mesh	Ball mill discharge wt% retained this size	Classifier	
		Overflow wt% retained this size	Underflow wt% retained this size
20	1.5	0.5	2.7
28	4.8	0.9	7.6
35	6.8	2.3	16.2
48	15.6	4.7	22.7
65	15.8	11.5	18.5
100	17.1	18.6	15.6
150	10.2	15.3	6.8
200	5.7	8.2	3.8
−200	22.5	38.0	6.1

Solution:

From the data given

% of −200 mesh material in feed $= f = 22.5\%$
% of −200 mesh material in overflow $= p = 38.0\%$
% of −200 mesh material in underflow $= u = 6.1\%$

$$\text{Efficiency of a classifier} = \eta = \frac{10,000(p-f)(f-u)}{f(100-f)(p-u)}$$

$$= \frac{10,000 \times (38.0 - 22.5)(22.5 - 6.1)}{22.5 \times (100 - 22.5)(38.0 - 6.1)} = 45.7\%$$

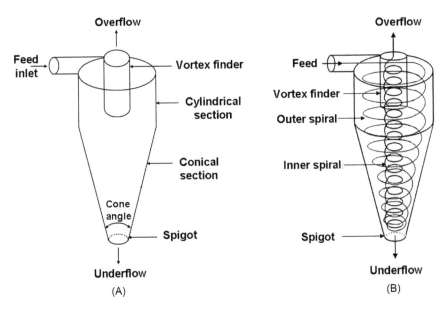

Feed inlet

Overflow

Vortex finder

Cylindrical section

Conical section

Cone angle

Spigot

Underflow

(A)

Overflow

Feed

Vortex finder

Outer spiral

Inner spiral

Spigot

Underflow

(B)

Figure 12.1.5 Hydrocyclone.

Under gravity force, the settling rate of a particle varies as its effective mass. If centrifugal force is applied, the effective mass increases and therefore settling rate increases. As particles are ground smaller they reach a size where the surface drag against the surrounding fluid almost neutralizes the gravitational pull, with the result that the particle may need hours, or even days, to fall a few inches through still water. This slowing down of settling rate reduces the tonnage that can be handled and increases the quantity of machinery and plant required. By superimposing centrifugal force, the gravitational pull can be increased from 50 to 500 times depending on the pressure at which the pulp is fed and the size of the vessel. The hydrocyclone is one which utilizes centrifugal force to accelerate the settling rate of particles.

Hydrocyclone (Fig. 12.1.5A) has no moving parts. It consists of a cylindrical section with a tangential feed inlet. A conical section, connected to it, is open at the bottom, variously called underflow nozzle, discharge orifice, apex or spigot. The top of the cylindrical section is closed with a plate through which passes an axially mounted central overflow pipe. The pipe is extended into the body of the cyclone by a short, removable section known as vortex finder, which prevents short-circuiting of feed directly into the overflow.

When a slurry is fed tangentially into a cyclone under pressure, a vortex is generated about the longitudinal axis. The accompanying centrifugal acceleration increases the settling rates of the particles, the coarser of which reach the cone's wall. Here they enter a zone of reduced pressure and flow downward to the apex, through which they are discharged.

At the center of the cyclone is a zone of low pressure and low centrifugal force which surrounds an air-filled vortex. Part of the pulp, carrying the finer particles with major portion of feed water, moves inward toward this vortex and reaches the gathering zone surrounding the air pocket. Here it is picked up by the vortex finder, and removed through a central overflow orifice (Fig. 12.1.5B).

Main use of hydrocyclone in mineral processing is as a classifier, which has proved extremely efficient at fine separation sizes (between 150 and 5 microns). It is used increasingly in closed-circuit grinding operations. It is also used for many other purposes such as de-sliming, de-gritting and thickening. It has also found wide acceptance for the washing of fine coal in the form of Heavy Medium Cyclone and Water Only Cyclone.

In open circuit operation the solids content of the slurry is about 30% and in closed circuits, it could be as high as 60%. For most operations the feed pressure ranges between 345 kPa to 700 kPa and in actual practice depends on cyclone diameter. The minimum pressure for a stable air core is around 30–35 kPa. The feed velocity is about 3.7 to 6.1 meters/sec and its acceleration in the feed chamber is inversely proportional to the hydrocyclone diameter.

Material balance equations over hydrocyclone for total material as well as for particular size of the material as percent or fraction of solids can be written as given in equations 5.4.1 and 5.4.2 to analyze or to know the performance of the hydrocyclone. Examples 12.1.8 to 12.1.12 illustrates the material balance calculations for hydrocyclone.

Example 12.1.8: *In a continuous operation, 40.8% of the feed solids fed into a hydrocyclone get discharged through the underflow. The sizing analyses of the overflow and the underflow as determined are given in Table 12.1.8.*

Table 12.1.8 Size analyses of hydrocyclone overflow and underflow.

Mesh (Tyler)	+100	100/150	150/270	−270
Overflow	6.1	14.9	42.4	36.6
Underflow	42.7	19.7	23.9	13.7

Calculate the sizing analysis of the feed to hydrocyclone.

Solution:

$$\text{Let} \quad F = 100 \text{ tons};$$
$$\text{Underflow} = U = 40.8\% = 40.8 \text{ tons}$$
$$\text{Overflow} = P = 100 - 40.8 = 59.2 \text{ tons}$$
$$\% +100 \text{ material in feed} = \frac{6.1}{100} \times 59.2 + \frac{42.7}{100} \times 40.8 = 21.0328$$
$$\% \ 100/150 \text{ material in feed} = \frac{14.9}{100} \times 59.2 + \frac{19.7}{100} \times 40.8 = 16.8584$$
$$\% \ 150/270 \text{ material in feed} = \frac{42.4}{100} \times 59.2 + \frac{23.9}{100} \times 40.8 = 34.8520$$
$$\% -270 \text{ material in feed} = \frac{36.6}{100} \times 59.2 + \frac{13.7}{100} \times 40.8 = 27.2568$$

Size analysis of the feed

Size	+100	100/150	150/270	−270
Wt%	21.0328	16.8584	34.8520	27.2568

Example 12.1.9: *A hydrocyclone is fed with a slurry to produce two products. Samples from feed, underflow and overflow are collected and determined slurry densities as 1140 kg/m³, 1290 kg/m³ and 1030 kg/m³. Calculate weight flow rate of feed solids to the cyclone, when 3-litre sample of underflow takes 4 seconds. Density of dry solid is 3000 kg/m³.*

Solution:

Given

Density of the solids $= \rho_p = 3000\,\text{kg/m}^3 = 3.00\,\text{gm/cm}^3$

Density of the feed slurry $= \rho_{\text{slF}} = 1140\,\text{kg/m}^3 = 1.14\,\text{gm/cm}^3$

Density of the underflow slurry $= \rho_{\text{slU}} = 1290\,\text{kg/m}^3 = 1.29\,\text{gm/cm}^3$

Density of the overflow slurry $= \rho_{\text{slP}} = 1030\,\text{kg/m}^3 = 1.03\,\text{gm/cm}^3$

% solids in feed $= \dfrac{100\rho_p(\rho_{\text{slF}} - 1)}{\rho_{\text{slF}}(\rho_p - 1)} = \dfrac{100 \times 3(1.14 - 1)}{1.14(3 - 1)} = 18.42\%$

% solids in underflow $= \dfrac{100\rho_p(\rho_{\text{slU}} - 1)}{\rho_{\text{slU}}(\rho_p - 1)} = \dfrac{100 \times 3(1.29 - 1)}{1.29(3 - 1)} = 33.72\%$

% solids in overflow $= \dfrac{100\rho_p(\rho_{\text{slP}} - 1)}{\rho_{\text{slP}}(\rho_p - 1)\cdot} = \dfrac{100 \times 3(1.03 - 1)}{1.03(3 - 1)} = 4.37\%$

Volumetric flow rate of underflow slurry $= \dfrac{3}{4}\text{litres/sec} = \dfrac{3000}{4} \times \dfrac{3600}{10^6} = 2.7\,\text{m}^3/\text{hr}$

Weight flow rate of underflow slurry $= U = 2.7 \times 1290 = 3483\,\text{kg/hr}$

Let F, U and P are the weight flow rates of feed, underflow and overflow slurry, f, u and p are the percent solids in feed, underflow and overflow.

Slurry balance $\Rightarrow F = U + P \Rightarrow P = F - U \Rightarrow P = F - 3483$

Solids balance $\Rightarrow Ff = Uu + Pp$

$\Rightarrow F \times 18.42 = 3483 \times 33.72 + (F - 3483)4.37$

Solving above two equations $F = 7275.88\,\text{kg/hr}$

Weight flow rate of feed slurry to cyclone $= 7275.88\,\text{kg/hr}$

Weight flow rate of feed solids to cyclone $= 7275.88 \times 0.1842$

$= 1340\,\text{kg/hr} = 1.34\,\text{tons/hr}$

Alternately

dilution ratios of three streams can be calculated. From the dilution ratios, weight flow rate of feed solids to cyclone can be calculated through water balance

$$\text{Dilution ratio of feed} \qquad = DR_F = \frac{100 - 18.42}{18.42} = 4.43$$

$$\text{Dilution ratio of underflow} = DR_U = \frac{100 - 33.72}{33.72} = 1.97$$

$$\text{Dilution ratio of overflow} \quad = DR_P = \frac{100 - 4.37}{4.37} = 21.88$$

Weight flow rate of underflow solids $= 3483 \times 0.3372 = 1174.5 \, \text{kg/hr}$

If F, U and P are the weight flow rates of solids of Feed, Underflow and overflow

Solids balance $\Rightarrow \quad F = U + P \quad \Rightarrow \quad P = F - U \quad \Rightarrow \quad P = F - 1174.5$

Water balance $\Rightarrow \quad F \times DR_F = U \times DR_U + P \times DR_P$

$\Rightarrow \qquad\qquad F \times 4.43 = 1174.5 \times 1.97 + (F - 1174.5) \, 21.88$

Solving above two equations $F = 1340 \, \text{kg/hr} = 1.34 \, \text{tons/hr}$

Alternately

From the solids flow rates, slurry balance gives

$$\frac{F}{\text{Fraction of solids in feed}}$$

$$= \frac{U}{\text{Fraction of solids in underflow}} + \frac{P}{\text{Fraction of solids in overflow}}$$

If ρ_{slF}, ρ_{slU}, and ρ_{slP} are the slurry densities of feed, underflow, and overflow respectively, then from equation 6.5

$$\left[\because C_w = \frac{\rho_p(\rho_{sl} - 1)}{\rho_{sl}(\rho_p - 1)} \qquad\qquad (6.5) \right]$$

$$\Rightarrow \quad \frac{F\rho_{slF}(\rho_p - 1)}{\rho_p(\rho_{slF} - 1)} = \frac{U\rho_{slU}(\rho_p - 1)}{\rho_p(\rho_{slU} - 1)} + \frac{P\rho_{slP}(\rho_p - 1)}{\rho_p(\rho_{slP} - 1)}$$

$$\Rightarrow \quad \frac{F\rho_{slF}}{(\rho_{slF} - 1)} = \frac{U\rho_{slU}}{(\rho_{slU} - 1)} + \frac{P\rho_{slP}}{(\rho_{slP} - 1)} = \frac{U\rho_{slU}}{(\rho_{slU} - 1)} + \frac{(F - U)\rho_{slP}}{(\rho_{slP} - 1)}$$

$$\Rightarrow \quad \frac{U}{F} = \frac{(\rho_{slF} - \rho_{slP})(\rho_{slU} - 1)}{(\rho_{slU} - \rho_{slP})(\rho_{slF} - 1)} = \frac{(1.14 - 1.03)(1.29 - 1)}{(1.29 - 1.03)(1.14 - 1)} = 0.88$$

$$\Rightarrow \quad F = \frac{U}{0.88} = \frac{1174.5}{0.88} = 1334.7 \, \text{kg/hr} = 1.334 \, \text{tons/hr}$$

Example 12.1.10: *A hydrocyclone operated in closed circuit with ball mill is fed at the rate of 800 tons/hr of dry solids and the underflow is 75% of the feed solids by weight. The feed slurry stream contains 40% solids by volume and 45% of the water is recycled, calculate the % circulating load and the volume concentration of solids in hydrocyclone products. Density of the solids is 3.2 tons/m³.*

Solution:

Given

Flow rate of feed $= 800$ tons/hr solids
Underflow $\quad = 75\%$ of feed by weight
% solids in feed $\: = 40\%$ by volume
Water recycled $\quad = 45\%$
Density of solids $= 3.2$ tons/m³

As the underflow of the cyclone is 75% of feed solids,
Weight flow rate of underflow solids $= 800 \times 0.75 = 600$ tons/hr
Weight flow rate of overflow solids $\: = 800 - 600 = 200$ tons/hr
Weight flow rate of solids fed to ball mill
$\qquad = $ Weight flow rate of overflow solids $= 200$ tons/hr

$$\% \text{ circulating load} = \frac{600}{200} \times 100 = 300\%$$

Solids balance by weight
Weight flow rate of solids in feed to hydrocyclone
$\qquad\qquad = $ Weight flow rate of solids in underflow
$\qquad\qquad\qquad + $ Weight flow rate of solids in overflow
$\Rightarrow \qquad\qquad 800 = 600 + 200$
Solids balance by volume
Volume of solids in feed $= $ Volume of solids in underflow
$\qquad\qquad\qquad + $ Volume of solids in overflow
$$\Rightarrow \qquad\qquad \frac{800}{3.2} = \frac{600}{3.2} + \frac{200}{3.2}$$
$$\Rightarrow \qquad\qquad 250\,\text{m}^3 = 187.5\,\text{m}^3 + 62.5\,\text{m}^3$$
$$\text{Volume of water in feed} = 250 \times \frac{60}{40} = 375\,\text{m}^3$$
$$\text{Volume of water in underflow} = 0.45 \times 375 = 168.75\,\text{m}^3$$
$$\% \text{ solids in underflow by volume} = \frac{187.5}{187.5 + 168.75} \times 100 = 52.6\%$$
$$\text{Volume of water in overflow} = 375 - 168.75 = 206.25\,\text{m}^3$$
$$\% \text{ solids in overflow by volume} = \frac{62.5}{62.5 + 206.25} \times 100 = 23.3\%$$

Example 12.1.11: *In a ball mill hydrocyclone closed circuit, new feed solids to the ball mill is 250 MTPH. Overflow from the hydrocyclone is 40% solids by weight. Specific gravity of the solids is 2.9. When it is desired to have 225% circulating load, calculate the following in overflow, underflow and feed to hydrocyclone:*

tonnage of solids, slurry and water,
% solids by weight,
density of the slurry,
flowrate of the slurry in litres per second,
% solids by volume

Assume the % solids by weight in underflow is 75.

Solution:

Overflow

MTPH solids	$= 250$
MTPH slurry	$= \dfrac{250}{0.40} = 625$
MTPH water	$= 625 - 250 = 375$
% solids by wt	$= 40$
Volume of solids	$= 250 \times \dfrac{1000}{2900} = 86.2 \, m^3/hr$
Volume of water	$= 375 \, m^3/hr$
Volume of slurry	$= 375 + 86.2 = 461.2 \, m^3/hr$
Slurry density	$= \dfrac{625}{461.2} = 1.355$
Flow rate of slurry	$= 461.2 \times \dfrac{1000}{3600} = 128.1 \, litres/sec$
% solids by volume	$= \dfrac{86.2}{461.2} \times 100 = 18.7\%$

Underflow

MTPH solids	$= 250 \times 2.25 = 562.5$
MTPH slurry	$= \dfrac{562.5}{0.75} = 750$
MTPH water	$= 750 - 562.5 = 187.5$
% solids by wt	$= 75$
Volume of solids	$= 562.5 \times \dfrac{1000}{2900} = 194 \, m^3/hr$
Volume of water	$= 187.5 \, m^3/hr$
Volume of slurry	$= 194 + 187.5 = 381.5 \, m^3/hr$
Slurry density	$= 750/381.5 = 1.966$
Flow rate of slurry	$= 381.5 \times \dfrac{1000}{3600} = 106 \, litres/sec$
% solids by volume	$= \dfrac{194}{381.5} \times 100 = 50.9\%$

Feed

$$\begin{aligned}
\text{MTPH solids} &= 250 + 562.5 = 812.5 \\
\text{MTPH slurry} &= 625 + 750 = 1375 \\
\text{MTPH water} &= 1375 - 812.5 = 562.5 \\
\text{\% solids by wt} &= \frac{812.5}{1375} = 0.591 = 59.1\% \\
\text{Volume of solids} &= 812.5 \times \frac{1000}{2900} = 280.2 \, \text{m}^3/\text{hr} \\
\text{Volume of water} &= 562.5 \, \text{m}^3/\text{hr} \\
\text{Volume of slurry} &= 280.2 + 562.5 = 842.7 \, \text{m}^3/\text{hr} \\
\text{Slurry density} &= \frac{1375}{842.7} = 1.632 \\
\text{Flow rate of slurry} &= 842.7 \times \frac{1000}{3600} = 234 \, \text{litres/sec} \\
\text{\% solids by volume} &= \frac{280.2}{842.7} \times 100 = 33.3\%
\end{aligned}$$

Example 12.1.12: *Rougher spiral tailings of an Iron Ore Beneficiation plant has been fed to Hydrocyclone at the rate of 350 tons/hr for further recovery of Iron from the fines by using HGMS. Hydrocyclone feed, overflow and underflow are sampled. The results of analyses of samples are given in Table 12.1.12.*

Table 12.1.12 Analyses results of feed, overflow and underflow.

	% solids	% Fe
Feed to hydrocyclone	16.24	59.55
Hydrocyclone overflow	09.88	60.13
Hydrocyclone underflow	40.02	53.36

What is the percent recovery of Fe and percent recovery of solids in both overflow and underflow of hydrocyclone?

Calculate the quantity of Fe in Overflow and Underflow per each ton of feed solids.

Solution:

Given

%Fe in Feed	$= f = 59.55$	%solids in Feed	$= 16.24$
%Fe in Overflow	$= p = 60.13$	%solids in Overflow	$= 9.88$
%Fe in Underflow	$= u = 53.36$	%solids in Underflow	$= 40.02$

Let $F = \text{Feed}$; $P = \text{Overflow}$; $U = \text{Underflow of Hydrocyclone}$

For determination of Iron recovery

Solid balance $\quad F = P + U$

Iron balance $\quad Ff = Pp + Uu \quad \Rightarrow 59.55F = 60.13P + 53.36U$

$\Rightarrow \qquad \dfrac{U}{F} = 0.08567 \quad \& \quad \dfrac{P}{F} = 0.9143$

Recovery of Fe in Underflow $= \dfrac{Uu}{Ff} \times 100 = \dfrac{0.08567 \times 53.36}{59.55} \times 100 = 7.7\%$

Recovery of Fe in Overflow $= \dfrac{Pp}{Ff} \times 100 = \dfrac{0.9143 \times 60.13}{59.55} \times 100 = 92.3\%$

Fe in underflow $= 0.08567 \times 0.5336 = 0.0457$ ton

$= 45.70$ kg per ton of feed solids

Fe in overflow $= 0.9143 \times 0.6013 = 0.5498$ ton

$= 549.80$ kg per ton of feed solids

For determination of Solids Recovery

Slurry Balance $\quad F = P + U$

Solids Balance $\quad Ff = Pp + Uu \quad \Rightarrow \quad 16.24F = 9.88P + 40.02U$

$\Rightarrow \qquad \dfrac{U}{F} = 0.211 \quad \& \quad \dfrac{P}{F} = 0.789$

Recovery of solids in Underflow $= \dfrac{Uu}{Ff} \times 100 = \dfrac{0.211 \times 40.02}{16.24} \times 100 = 52\%$

Recovery of solids in Overflow $= \dfrac{Pp}{Ff} \times 100 = \dfrac{0.789 \times 9.88}{16.24} \times 100 = 48\%$

Summary

	Overflow	Underflow
% Fe	60.13	53.36
% recovery of Fe	92.30	07.70
% solids	09.88	40.02
% recovery of solids	48.00	52.00
Quantity of Fe per ton of feed solids	549.80 kg	45.70 kg

12.2 EFFICIENCY OF SEPARATION IN HYDROCYCLONE

The commonest method of representing the efficiency of operation and separation of hydrocyclone is by a performance or partition or efficiency curve. This curve is drawn between weight fraction or percentage of each particle size in feed which reports to the spigot or underflow and the particle size.

Table 12.2.1 Size analyses data of underflow and overflow of hydrocyclone.

Size μm	Weight%	
	underflow	overflow
+1180	00.3	–
−1180 + 850	03.6	–
−850 + 600	20.7	00.1
−600 + 425	21.3	00.6
−425 + 300	13.1	02.2
−300 + 212	07.0	04.7
−212 + 150	03.7	05.9
−150 + 106	03.4	06.7
−106 + 75	02.7	07.0
−75 + 53	01.7	04.7
−53	22.5	68.1

To determine the efficiency of separation of a sample of known size distribution, pulp density and flow rate, a hydrocyclone of known geometry, including the inlet, overflow and underflow diameters, is operated in closed circuit until a steady state is reached. Simultaneous samples of the feed, overflow and underflow streams are collected, dried and analyzed for size distribution. The calculations involved to determine the efficiency, the method of construction of the efficiency curve can best be illustrated with an example.

The slurry is being classified in a hydrocyclone. The percent solids in feed, underflow and overflow are 70.4, 83.5 and 64.2 respectively. The size analyses of underflow and overflow solids are given in Table 12.2.1.

$$\text{Dilution ratio of feed} \quad = \frac{100 - C_w}{C_w} = \frac{100 - 70.4}{70.4} = 0.42$$

$$\text{Dilution ratio of underflow} = \frac{100 - C_w}{C_w} = \frac{100 - 83.5}{83.5} = 0.20$$

$$\text{Dilution ratio of overflow} \quad = \frac{100 - C_w}{C_w} = \frac{100 - 64.2}{64.2} = 0.56$$

Let F be the rate of dry solids fed to the hydrocyclone, U and P be the rate of dry solids of underflow and overflow from hydrocyclone respectively. Material balance equations are written over hydrocyclone as under:

Solid balance $\qquad\qquad\qquad F = U + P$

Water balance $\qquad\qquad 0.42F = 0.20U + 0.56P$

Solving above two equations $\quad \dfrac{U}{F} = 0.389$

i.e. the underflow is 38.9% of the feed and the overflow is 61.1% of the feed.

The following are the details of calculations (with reference to the Table 12.2.2 that follows) necessary to draw performance or partition or efficiency curves.

Table 12.2.2 Partition coefficient calculations.

A	B	C	D	E	F	G	H	I	J
	Weight%		Wt% of feed		Recons-tituted Feed	Nominal size	Partition coefficient	Corrected Partition coefficient	
Size μm	U/F	O/F	U/F B × 0.389	O/F C × 0.611	D + E		$y_a = \frac{D}{F} \times 100$	$Y_c = \frac{y_a - R_f}{100 - R_f} \times 100$	$\frac{d}{d_{50c}}$
+1180	00.3	–	0.12	–	0.12	–	100.0	100.0	–
−1180 + 850	03.6	–	1.40	–	1.40	1015	100.0	100.0	3.4
−850 + 600	20.7	00.1	8.05	0.06	8.11	725	99.3	99.1	2.5
−600 + 425	21.3	00.6	8.29	0.37	8.66	512.5	95.7	94.7	1.7
−425 + 300	13.1	02.2	5.10	1.34	6.44	362.5	79.2	74.5	1.2
−300 + 212	07.0	04.7	2.72	2.87	5.59	256	48.7	37.1	0.9
−212 + 150	03.7	05.9	1.44	3.60	5.04	181	28.6	12.4	0.6
−150 + 106	03.4	06.7	1.32	4.09	5.41	128	24.4	07.2	0.4
−106 + 75	02.7	07.0	1.05	4.28	5.33	90.5	19.7	01.5	0.3
−75 + 53	01.7	04.7	0.66	2.87	3.53	64.0	18.7	00.2	0.2
−53	22.5	68.1	8.75	41.61	50.36	–	–	–	–
			38.90	61.10	100.00				

Column D & E are the size analyses of underflow and overflow in relation to the feed material. Column D is to be calculated by multiplying the values of Column B by 0.389 as this is the fraction of feed reporting to underflow. Similarly Column E is to be calculated by multiplying the values of Column C by 0.611 as this is the fraction of feed reporting to overflow. Column F is the reconstituted size analysis of the feed material and is calculated by adding the corresponding values of Column D and Column E. Column G is the arithmetic mean of the corresponding sieve size ranges. Column H, weight% of the feed reporting to underflow, called **Partition coefficient**, Y_a, is to be calculated by dividing each weight% in Column D by the corresponding weight% in Column F multiplied by 100.

A graph is to be plotted between nominal size (Column G) and the partition coefficient (Column H). This is the efficiency curve of hydrocyclone and is shown in the Figure 12.2.1.1 for this example.

From this graph, d_{50} is determined corresponding to 50% of feed reporting to the underflow. d_{50} is a cut point or separation size, and is defined as the point on the curve for which 50% of particles in the feed of that size report to the underflow. It means that the particles of that size have an equal chance of going either with the overflow or underflow. d_{50} in this example is 255 microns. It is to be noted that the curve does not pass through the origin. It is due to a fraction of the slurry bypassing the cyclone without classification called **dead flux**. Thus if 5% of the feed slurry bypassed or short-circuited the unit, then, only 95% of the slurry would be subjected to the classification. Thus the d_{50} obtained by the above method has to be corrected. It is suggested that the fraction of the solids in each size fraction that is bypassed from the feed to the underflow is in the same ratio as the fraction of feed water that reported to the underflow, i.e. **flow ratio** (or) fluid flow ratio, R_f.

In this case, fraction of the feed water that reported to underflow is

$$R_f = \frac{38.9 \times 0.20}{100 \times 0.42} = 0.185 \quad \Rightarrow \quad 18.5\%$$

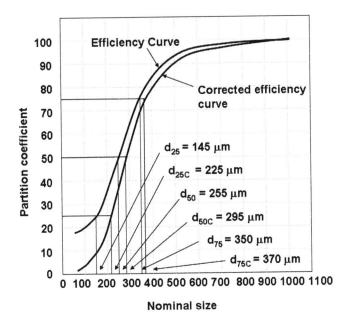

Figure 12.2.1.1 Efficiency curves.

The corrected weight% of each size particles reporting to underflow called **corrected partition coefficient** is determined by the following equation (column I)

$$y_c = \frac{y_a - R_f}{100 - R_f} \times 100$$

where y_a is the actual weight% of a particular size particles reporting to underflow and R_f is the percent of the feed water that reported to underflow.

The corrected efficiency curve is plotted between nominal size and corrected partition coefficient. This curve represents the efficiency of separation of that portion of slurry which is subjected to classification. From this curve d_{50C} (corrected d_{50}) can be read. The value of d_{50C} is $295\,\mu m$ in this example.

The sharpness of the cut depends on the slope of the central portion of the efficiency curve. If the slope is closer to vertical, the efficiency is higher. The slope of the curve can be expressed by taking the points at which 75% and 25% of the feed particles report to the underflow. These are the d_{75} and d_{25} sizes respectively.

Probable Error (or Ecart Terra) is defined as $\dfrac{d_{75C} - d_{25C}}{2}$

The efficiency of separation, or **Imperfection**, I, is expressed as the ratio of Probable Error to the cut size

$$I = \frac{\text{Probable error (Ecart Terra)}}{\text{Cut size}} = \frac{(d_{75C} - d_{25C})/2}{d_{50C}} = \frac{d_{75C} - d_{25C}}{2d_{50C}}$$

The smaller the imperfection, the better the classification. For hydrocyclones, the range is from 0.2 to 0.8 with an average of about 0.3–0.4. The advantage of imperfection is that it is independent of the particle size. Even so, it is not a good measure of sharpness at extremely fine cut sizes. The **Grade Efficiency** is defined as the ratio of d_{75C} to d_{25C}. The **Sharpness Index** (or **Selectivity Index**) S_I, the reciprocal of grade efficiency, is the ratio of d_{25C} to d_{75C}. It is a measure of the sharpness of classification.

The following are various values in this example.

		Uncorrected	*Corrected*
d_{75}		350	370
d_{25}		145	225
d_{50}		255	295
Imperfection $= I = \dfrac{d_{75C} - d_{25C}}{2d_{50C}} =$		---	0.25
Sharpness Index $= S_I = \dfrac{d_{25C}}{d_{75C}} =$		---	0.61

The corrected efficiency curve can be normalized (made dimensionless) by dividing each particle size, d, by d_{50C} (column J). Plotting d/d_{50C} against the % feed to underflow gives the curve known as Reduced Efficiency Curve. Lynch and Rao found that reduced efficiency curve describes the performance of a hydrocyclone. It is concluded that for geometrically similar cyclones, the reduced efficiency curve is the function of the material classified and is independent of size of hydrocyclone. The curve shown in Figure 12.2.1.2 is the reduced efficiency curve drawn for this example.

The advantage of plotting reduced efficiency curve is that the results can be translated to any larger size cyclone. The reduced efficiency curves for different minerals

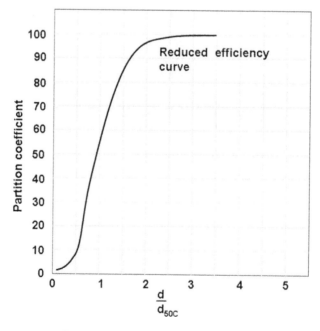

Figure 12.2.1.2 Reduced efficiency curve.

of different density and shape are different but as the size d is simply divided by a constant (d_{50C}), the nature of the curve remains unaltered.

Lynch and Rao [17] have given the following equation for the reduced efficiency curve

$$y_c = \frac{e^{\alpha \frac{d}{d_{50C}}} - 1}{e^{\alpha \frac{d}{d_{50C}}} + e^{\alpha} - 2}$$

where y_c = the corrected partition coefficient; α = efficiency parameter which completely describes a change in the shape of the curve.

The value of α is typically 3–4 for a single stage cyclone but can be as high as 6. A closed circuit grinding operation can have values around 2.5.

Example 12.2.1:   *The input and output streams of an operating cyclone were sampled simultaneously for the same period of time. The dried samples were analysed. The size distributions of underflow and overflow streams are shown in Table 12.2.1.1.*

Table 12.2.1.1 Size distributions of underflow and overflow.

Size μm	Underflow	Overflow
+425	2.0	0.0
−425 + 300	6.3	0.0
−300 + 212	12.9	0.3
−212 + 150	21.2	1.8
−150 + 106	28.0	15.2
−106 + 75	10.0	26.2
−75 + 53	5.0	38.4
−53	14.6	18.1

% solids in feed = 35.0%
% solids in underflow = 70.2%
% solids in overflow = 17.2%

Draw partition curves and evaluate the performance of hydrocyclone.

Solution:

Given

% solids in feed   = 35.0%
% solids in underflow = 70.2%
% solids in overflow = 17.2%

$$\text{Dilution ratio of feed} = \frac{100 - 35.0}{35.0} = 1.86$$

$$\text{Dilution ratio of underflow} = \frac{100 - 70.2}{70.2} = 0.42$$

$$\text{Dilution ratio of overflow} = \frac{100 - 17.2}{17.2} = 4.81$$

Let F be the rate of dry solids fed to the hydrocyclone, U and P be the rate of dry solids of underflow and overflow from hydrocyclone. The following are the material balance equations over hydrocyclone

Solid balance $$F = U + P$$

Water balance $$1.86F = 0.42U + 4.81P$$

Solving above two equations $$\frac{U}{F} = 0.672$$

i.e. the underflow is 67.2% of the feed and the overflow is 32.8% of the feed. The percent of the feed water that reported to underflow

$$= R_f = \frac{67.2 \times 0.42}{100 \times 1.86} \times 100 = 15.2\%$$

The values of partition coefficients are calculated as detailed in article 12.2 and tabulated in Table 12.2.1.2.

Table 12.2.1.2 Partition coefficient calculations.

A	B	C	D	E	F	G	H	I
	Weight%		Wt% of feed		Recons-tituted Feed	Nominal size	Partition coefficient	Corrected Partition coefficient
Size μm	U/F	O/F	U/F B $\times$ 0.672	O/F C $\times$ 0.328	D + E		$y_a = \frac{D}{F} \times 100$	$Y_c = \frac{y_a - R_f}{100 - R_f} \times 100$
+425	2.0	0.0	1.34	0.0	1.34	–	100.0	100.0
−425 + 300	6.3	0.0	4.23	0.0	4.23	362.5	100.0	100.0
−300 + 212	12.9	0.3	8.67	0.10	8.77	256.0	98.9	98.7
−212 + 150	21.2	1.8	14.25	0.59	14.84	181.0	96.0	95.3
−150 + 106	28.0	15.2	18.82	4.99	23.81	128.0	79.0	75.2
−106 + 75	10.0	26.2	6.72	8.59	15.31	90.5	43.9	33.8
−75 + 53	5.0	38.4	3.36	12.60	15.96	64.0	21.1	7.0
−53	14.6	18.1	9.81	5.93	15.74	26.2	62.3	–
	100.0	100.0	67.20	32.80	100.00			

Efficiency and corrected efficiency curves are drawn and shown in Figure 12.2.1.3. Values of d_{50}, d_{75} and d_{25} are read from both the curves and shown below. Calculated values of imperfection and sharpness index are also shown.

	Uncorrected	Corrected
d_{75}	125	128
d_{25}	73	85
d_{50}	100	110

$$\text{Imperfection} = I = \frac{d_{75C} - d_{25C}}{2d_{50C}} = \quad \text{---} \quad 0.20$$

$$\text{Sharpness Index} = S_I = \frac{d_{25C}}{d_{75C}} = \quad \text{---} \quad 0.66$$

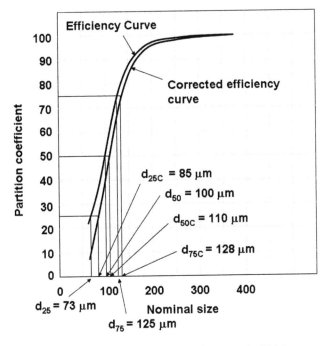

Figure 12.2.1.3 Efficiency curves for example 12.2.1.

12.3 PROBLEMS FOR PRACTICE

12.3.1:   *An ore contains valuables of specific gravity of 7.0 and gangue of specific gravity of 2.0 has to be separated in a hydraulic free settling classifier. If the mixture has the size range between 40 microns and 90 microns, estimate the maximum upward velocity of hydraulic water to be used so that overflow does not contain any valuables. Will underflow be gangue free? Consider the flow zone is laminar.*

[0.52 cm/sec, gangue free]

12.3.2:   *It is desired to separate quartz particles of specific gravity of 2.6 from galena of specific gravity 7.5 in a free settling classifier. The size range is from 5.2 to 25 microns. Three fractions are obtained. One of pure quartz, another of pure galena, and the third a mixture. Calculate the size range of the substances in the mixture.*

[Quartz 10.4 – 25 microns, Galena 5.2 – 12.4 microns]

12.3.3:   *It is required to classify the ore of −90 + 5 microns size consists of heavies of specific gravity 7.0 and light particles of specific gravity 2.5 into three fractions one contains pure heavies, one contains pure lights and another contains a mixture of both heavies and lights, in a free settling hydraulic classifier. What is the upward velocity of hydraulic water to be maintained in (a) first column to get heavies in the pocket (b) the second column to get mixture in the pocket and lights in overflow.*

[0.662 cm/sec, 0.0327 cm/sec]

12.3.4:   The Table 12.3.4 shows the sizing analyses of feed, overflow and underflow solids of a hydrocyclone.

Table 12.3.4 Size analyses for problem 12.3.4.

Mesh (Tyler)	+100	100/150	150/270	−270
Feed	21.0	16.9	34.8	27.3
Overflow	6.1	14.9	42.4	36.6
Underflow	42.7	19.7	23.9	13.7

If the separation is aimed at 150 mesh, calculate the efficiency of the hydrocyclone

[42.5%]

12.3.5:   A classifier was fed with ground silica at the rate of 50 tph. The screen analyses of feed, product and tailings from a classifier are as shown in Table 12.3.5.

Table 12.3.5 Screen analyses for problem 12.3.5.

Size (microns)	Feed%	Overflow%	Underflow%
850	15.2	0.0	18.3
500	8.5	0.0	10.2
250	33.6	2.9	39.9
125	34.2	65.4	27.8
75	8.5	31.8	3.7
	100.0	100.0	100.0

Estimate the tons of dry ore per day in classifier underflow.

[243.7 tons/day]

12.3.6:   30 dry tons of solids per hour is fed to a cyclone. The percent solids in cyclone feed, underflow and overflow are 30%, 55%, and 20% by weight respectively. Calculate the tonnage of solids per hour in the underflow.

[15.7 tph]

12.3.7:   Quartz slurry at a slurry density of 1.13 tons/m³ is fed to a laboratory hydrocyclone. Pulp densities of hydrocylone underflow and overflow are 1.28 tons/m³ and 1.04 tons/m³ respectively. A 2 litre sample of underflow was taken in 3.1 seconds. Calculate the mass flowrate of feed solids to the cyclone. The density of quartz is 2.65 tons/m³.

[1.29 tph]

12.3.8: *The slurry of quartz is being classified in a hydrocyclone. The slurry densities of feed, underflow and overflow of hydrocyclone is measured online as 1670, 1890 and 1460 kg/m³ respectively. The density of quartz is 2700 kg/m³. The size analyses of underflow and overflow solids are given in Table 12.3.8.*

Table 12.3.8 Size analyses of underflow and overflow.

	Weight%	
Size μm	U/F	O/F
+1168	14.7	–
−1168 + 589	21.8	–
−589 + 295	25.0	5.9
−295 + 208	7.4	9.0
−208 + 147	6.3	11.7
−147 + 104	4.8	11.2
−104 + 74	2.9	7.9
−74	17.1	54.3

Draw the performance curves.
    And determine d_{50}, d_{50C}, imperfection and sharpness index.
                          [177.5 microns, 320 microns, 0.31, 0.52]

12.3.9: *Evaluate the efficiency of separation of a hydrocyclone operated in closed circuit with ball mill. The percent solids and flowrates of feed to the hydrocyclone, underflow and overflow from the hydrocyclone are 55.0% solids at 206.5 tons/hr, 78.25 solids at 177.1 tons/hr and 19.6% solids at 29.4 tons/hr respectively. The size analyses of samples collected from underflow and overflow are given in Table 12.3.9.*

Table 12.3.9 Size analyses of underflow and overflow from hydrocyclone.

	Weight%	
Size μm	U/F	O/F
−600 + 425	68.3	0.0
−425 + 300	13.6	2.0
−300 + 250	6.2	6.8
−250 + 150	4.6	16.0
−150 + 106	2.4	15.6
−106 + 75	1.2	10.9
−75	3.7	48.7

[I = 0.38; S_I = 0.44]

12.3.10:   *A hydrocyclone is operated in closed circuit with ball mill. The size distributions of feed, underflow and overflow for the hydrocyclone is shown in Table 12.3.10.*

Table 12.3.10   Size distributions of feed, underflow and overflow of hydrocyclone.

Size μm	Feed	Underflow	Overflow
+500	9.6	14.7	0.0
−500 + 300	14.2	21.8	0.0
−300 + 250	18.4	25.0	5.9
−250 + 180	7.9	7.4	9.0
−180 + 130	8.2	6.3	11.7
−130 + 90	7.0	4.8	11.2
−90 + 63	4.6	2.9	7.9
−63	30.1	17.1	54.3
		100.0	100.0

If 70% of the feed goes to underflow, draw a partition curve for the hydrocyclone and determine the d_{50} size.

[110 microns]

Chapter 13

Beneficiation operations

Beneficiation operations are physical or mechanical unit operations where valuable mineral particles are separated from the mixture of liberated valuable mineral particles and gangue mineral particles. Every beneficiation operation is based on one or more physical properties in which valuable and gangue minerals differ in that property.

The following are the major beneficiation operations

- Gravity concentration
    - Heavy Medium Separation
    - Jigging
    - Flowing Film Concentration
        - Spiraling
        - Tabling
- Flotation
- Magnetic Separation
- Electrical Separation

13.1 GRAVITY CONCENTRATION

Gravity Concentration is a method of separating the mineral particles based on their specific gravities when they are allowed to settle in a fluid medium.

The motion of a particle in a fluid depends not only on its specific gravity but also on its size; large particles will be affected more than smaller ones. The efficiency of gravity concentration operation, therefore, increases with particle size. Smaller particles respond poorly because their movement is dominated mainly by surface friction. In practice, close size control of feeds to gravity processes is required in order to reduce the size effect and make the particles to move depending on their specific gravities.

Sink and Float (also called heavy liquid separation) is an operation where particles of different specific gravities are separated by using suitable heavy liquid. Solutions of required specific gravity are prepared by adding two or more liquids at different proportions. When the ore particles are introduced in this solution, mineral particles of less specific gravity will float and mineral particles of more specific gravity will sink. Then two products are removed from the solution.

Separation of minerals, Ilmenite, Monazite, Rutile, Zircon, Garnet from Beach Sands, and separation of raw coal into different specific gravity fractions are the two important examples for the application of sink and float.

The buoyant forces acting on the light particles in a dense medium make them to rise to the surface and the dense particles sink to the bottom because of gravity force. In a static bath, force balance equation is written as

$$F_g = (m_p - m_f)g \qquad (13.1.1)$$

where

F_g = gravitational force
m_p = mass of the solid particle
m_f = mass of the fluid displaced by the particle

For the particles which float, F_g will have a negative value and for the particles which sink, F_g value is positive. In a centrifugal separator, equation 13.1.1 becomes

$$F_C = (m_p - m_f)\frac{v^2}{R} \qquad (13.1.2)$$

where

F_C = centrifugal force
v = tangential velocity
R = radius of the centrifugal separator

From these two equations, it is clear that the forces causing the separation of the particles in a static bath are proportional to g whereas in a centrifugal separator, separating forces are proportional to $\dfrac{v^2}{R}$ which is much greater. Hence particles down to 0.5 mm size can be separated by centrifugal separators.

Heavy Medium Separation or Dense Medium Separation is a process similar to Heavy Liquid Separation but instead of heavy liquid, a pseudo liquid is used. A pseudo liquid is a suspension of water and solids which behaves like a true liquid. The solids used are known as medium solids. Sand, loess, shale, barites, magnetite, ferrosilicon, galena and some types of clays are the medium solids used in heavy medium separation.

Heavy Medium Separation is applicable to any ore. It is widely used for washing coal at coarser sizes. Chance Cone process, Barvoys Process, Dutch State Mines Process, Tromp Process, Drewboy Process, and Wemco Drum Separation Process are some of the heavy medium separation processes employed in Coal Washeries.

Heavy medium cyclone is a centrifugal separator similar to the conventional hydro-cyclone in principle of operation. It provides a high centrifugal force and a low viscosity of the medium. The ore or coal is suspended in a very fine medium of ferrosilicon or magnetite, and is fed tangentially to the Heavy medium cyclone under pressure. The sink product leaves the cyclone at the apex and the float product is discharged through the vortex finder. These cyclones are commonly installed with axes at 10–15°

to the horizontal thereby enabling the unit to be fed at comparatively low inlet pressure, preferably from a steady head tank. Heavy medium cyclones are widely used for beneficiation of coal in the size range of 13–0.5 mm in coal washeries.

By giving a special shape to the cone and to the cylindrical part of the cyclone, it is possible to affect separation without using heavy medium. Such cyclone is known as water-only cyclone or water washing cyclone. Vorsyl separator, LARCODEMS, Dyna Whirlpool and Tri-Flo separator are some other centrifugal separators.

Jigging is the process of separating the particles of different specific gravities, size and shape by introducing them on a perforated surface (or screen) through which the water is made to flow by pulsion and suction strokes alternatively.

In jigging, the particles are allowed to settle only for short period and the particles will never attain their terminal velocities. It means that separation will depend on the initial settling velocities of the particles. The particles will settle during their accelerating period. The initial settling velocity is extremely low and the drag force due to frictional effects is not developed. Under these circumstances, the drag force is practically zero and the two principal forces acting on the particle are the gravitational and the buoyant forces. Then the force balance equation becomes

$$m_p \frac{dv}{dt} = (m_p - m_f)g \qquad (13.1.3)$$

$$\frac{dv}{dt} = \frac{m_p - m_f}{m_p}g = \left[1 - \frac{\rho_f}{\rho_p}\right]g \qquad (13.1.4)$$

Equation 13.1.4 shows that the initial acceleration of the particles during settling depends on the force of gravity, density of the particle and the fluid and does not depend on the size or shape of the particle. Initial acceleration is maximum for the most dense particles. This situation indicates that light and heavy particles can be separated by providing short duration of settling.

When a group of both light and heavy particles, all of same size, are taken in equal number and do the jigging, after several pulsion and suction strokes, all the particles are rearranged in such a way that all light particles are at the top and heavy particles are at the bottom. Rearrangement of particles is called **stratification**.

If a group of both light and heavy particles of different sizes are taken and do the jigging, all the particles are rearranged in similar fashion. But small particles pass through the interstices between coarse particles. This phenomenon is called **consolidation trickling**.

During the pulsion stroke, solid bed is opened and expanded. When pulsion ceases, the particles settle into more homogeneous layers under the influence of gravity during the suction stroke. Stratification during the stage that the bed is open is essentially controlled by hindered settling classification as modified by differential acceleration, and during the stage that the bed is tight, it is controlled by consolidation trickling. The first process puts the coarse-heavy particles on the bottom, the fine-light particles at the top, and the coarse-light and fine-heavy particles in the middle. The second process puts the fine-heavy particles at the bottom, the coarse-light particles at the top, and the coarse-heavy and fine-light particles in the middle. By varying the relative importance of the two, and by varying the importance of differential acceleration, an almost perfect stratification according to density alone can be obtained.

Flowing film concentration is a sorting of mineral particles on flat surfaces in accordance with the size, shape and specific gravity of the particles moved by a flowing film of water.

When water is made to flow over a bare sloping deck, the velocity of water adjacent to the deck is zero and increases as the distance from the deck increases reaching maximum at the top surface of water. However, velocity at the top surface of water is slightly less than the maximum due to air friction.

If number of spheres, composed of two kinds of minerals, one heavy another light, and are of different sizes, are introduced into a thick layer of water, they will be separated during their fall through this layer. The coarsest heavy sphere falls faster on to the deck through water and least effected by the current and lies nearest to the point of entry. The smallest light sphere will be drifted farthest downstream. The others will be drifted to different distances.

The flowing water presses the sphere and make to move downstream. The differential rate at which the water is flowing over the deck causes low pressure on the bottom of the sphere tending to slide on the deck and causes high pressure at the top of the sphere tending to roll on the deck.

Since small particles are submerged in the slower-moving portion of the film, they will not move as rapidly as coarse particles. If the combined influence of deck slope and streaming velocity is sufficient to keep all the spheres in rolling movement, they rearrange themselves in the following downslope sequence

1   Small-heavy particles
2   Coarse-heavy and small-light particles
3   Coarse-light particles

It is to be noted that in flowing film concentration coarse-heavy particles are placed with small-light particles which is reverse of the stratification takes place in classification.

Humphrey Spiral is a typical unit of stationary flowing film concentrator. It is effective for particles in the range of 3 mm to 75 μm and for minerals with specific gravity differences greater than 1.0. Shaking table is a moving flowing film concentrator which utilizes reciprocating motion and cross riffles in addition to flowing film. Shaking tables are normally operate on feed sizes in the range of 3 mm to 100 microns. The particles of size less than 100 microns are treated in slimes tables whose decks have a series of planes rather than riffles.

13.2   FROTH FLOTATION

Froth flotation utilises the differences in physico-chemical surface properties of particles of various minerals. Froth flotation is a process of separating fine particles of different minerals from each other by floating certain minerals on to a water surface. In froth flotation, the ore particles are maintained in suspension and treated with chemical reagents. The valuable mineral particles are made to adhere or attach with air-bubbles. The air-bubbles, after the valuable mineral particles adhered to them, will float on the top of the pulp in the form of froth because the density of mineral adhered air-bubbles is less than that of water. The froth is then separated and dewatered to get valuable mineral particles. This process can only be applied to relatively fine particles (less than 150 microns).

Flotation reagents are substances added to the ore pulp prior to or during flotation in order to make possible to float valuable mineral particles and not to float the gangue mineral particles. Important flotation reagents are collectors, frothers, depressants, activators and pH regulators.

Collector is a chemical reagent, either an acid, base or salt, and is hetero-polar in nature; polar part of it has an affinity towards a specific mineral and non-polar part has an affinity towards air bubble. Small amount of collector is added to the pulp and agitated long enough so that the polar part is adsorbed on to the mineral to be floated while the non-polar part is oriented outwards and makes the surface of mineral particles hydrophobic. Collector increases the contact angle of the valuable mineral particles. Contact angle is the angle of contact of an air bubble with the surface of a solid measured across the water. Xanthates are most widely used collectors for flotation of sulphide minerals. Carboxylic collectors are used for flotation of non-sulphides and non-silicates. Cationic collectors are used for oxide and silicate minerals including quartz. Amines are the most commonly used cationic collectors. Oily collectors normally used are petroleum products, blast furnace oils, coal-tar and wood-tar creosotes. They are used in flotation of oxidized metalliferous ores and gold ores.

Frother is a chemical reagent and is heteropolar in nature; polar part of it has an affinity for water and non-polar part has an affinity for air or repulsion for water. Frother acts upon the air water interface. The addition of a frother decreases the surface tension of water and increases the life of bubbles produced. The main objective of a frother is to permit the production of a sufficiently stable froth to hold the mineral particles that form a network around the bubbles until they are removed from the flotation unit. As a result of the addition of a frother, the air bubbles, formed under the surface of water, are more or less completely lined with monomolecular sheath of frother molecules which allows each bubble to come in contact with other bubbles. This forms a froth. Thus a froth is simply a collection of bubbles. Cresylic acid and pine oil are the most widely used frothers. A wide range of synthetic frothers are now in use in many plants. Methyl Iso-Butyl Carbinol (MIBC) is most important among the synthetic frothers. Eucalyptus oil, camphor oil and sagebrush oil are used when they are more cheaply available than the common frothing agents.

Other chemical reagents, depressants, activators and pH regulators, called **modifiers**, are used extensively in flotation to modify the action of the collector, either by intensifying or reducing its water–repellent effect on the mineral surface. Thus they make collector action more selective towards certain minerals. **Depressants** are inorganic chemicals. They react chemically with the mineral particle surfaces to produce insoluble protective coatings of a wettable nature making them non-floatable even in the presence of a proper collector. Thus formed protective coatings prevent the formation of collector film. Sodium or potassium cyanide is a powerful depressant for sphalerite and pyrite. **Activators**, generally inorganic compounds, can modify the surface of non-floatable or poorly floatable mineral particles by adsorption on particle surface so that the collector may film the particle and induce flotation. An example of this is the use of copper sulphate in the flotation of sphalerite. **pH regulators** are used to modify the alkalinity or acidity of a flotation circuit or in otherwords to control

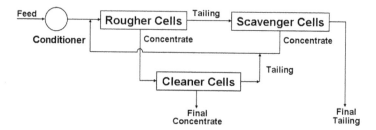

Figure 13.2.1 Typical flotation circuit.

the pH of the pulp. The pH of the pulp has an important and sometimes very critical controlling effect on the action of the flotation reagents. Common pH regulators are lime and soda ash for creating alkaline conditions, sulphuric and hydrochloric acids for creating acidic conditions.

The pulp is treated in a conditioner, called **conditioning**, with necessary reagents like depressant and collector prior to flotation in order to convert the mineral particles to respond readily in a flotation cell. The flotation operation is generally carried out in three stages namely roughing, scavenging, and cleaning called flotation circuit (Fig. 13.2.1).

Each stage in the circuit consists of a bank of cells and the number of cells in a bank is primarily depends upon the residence time of the pulp in the cells and the required throughput. The reagent conditioned feed pulp is treated in a first bank of cells called roughers. The tailing from the rougher cells, which may still contain some valuable mineral particles, is treated in another bank of cells called scavengers. The concentrate from the rougher cells is further treated in a bank of cells called cleaners to obtain high grade final concentrate. The scavenger concentrate and cleaner tailing are re-fed to rougher bank to be treated with the fresh feed pulp. The scavenger tailing is the final tailing.

13.3   MAGNETIC SEPARATION

Magnetic separation is a physical separation of particles based on the magnetic property of the mineral particles. The property of a material that determines its response to a magnetic field is the magnetic susceptibility.

Initially, magnetic separation was employed to separate strongly magnetic iron ores such as magnetite from gangue or other less magnetic minerals. With the advancement of technology and design of machines, it was adopted in separation of ores containing iron or manganese, which are only feebly magnetic.

Magnetic separators are classified mainly into two groups:

Low intensity magnetic separators
High intensity magnetic separators

Low intensity magnetic separators are used primarily for ferromagnetic minerals. These are also used for paramagnetic minerals of high magnetic susceptibility. High intensity magnetic separators are used for paramagnetic minerals of low magnetic susceptibility. Both the separations may be carried out either wet or dry. Wet processing predominates in low intensity operations, although large tonnage dry plants do exist. High intensity separators are usually dry and of low capacity. Wet magnetic separators are generally used for the particles of below 0.5 cm size.

13.4 ELECTRICAL SEPARATION

Electrical separation is a physical separation of particles based on the electrical properties of the mineral particles. The following are the three important electrical separation processes:

1 Electrostatic separation
2 Electrodynamic (High Tension) separation
3 Dielectric separation

Electrostatic and Electrodynamic separation processes utilize the difference in electrical conductivity between various minerals in the ore feed whereas dielectric separation utilizes the differences in dielectric constant of mineral particles.

Electrical conductivity is defined as current density (flow of electric charge per unit area of cross section) per unit applied electric field. It is the reciprocal of the resistivity or specific resistance of a conductor. It is measured in $ohm^{-1} cm^{-1}$.

The basis of any electrostatic separation is the interaction between an external electric field and the electric charges acquired by the various particles. Particles can be charged by

1 contacting dissimilar particles
2 conductive induction
3 ion bombardment

On beneficiation of an ore, normally two products (concentrate and tailing) are obtained.

Concentrate: The valuable mineral product obtained from the beneficiation operation

Tailing: The waste or gangue mineral product obtained from the beneficiation operation

Practically all the valuable mineral particles can not be separated due to

- Incomplete liberation (degree of liberation <100%)
- Inefficiency of beneficiation operation (efficiency of operation <100%)

Hence both the concentrate and the tailing contain locked particles of valuable and gangue minerals. And the concentrate contains little amount of gangue mineral particles and the tailing contains little amount of valuable mineral particles.

If the valuable mineral particles are not completely liberated during the size reduction operations, (i.e. degree of liberation is less than 100%) some particles contain both valuable and gangue minerals. Such particles can be separated in a beneficiation operation as a third product. This third product is called middling. Usually, this middling product is comminuted for further liberation and then beneficiated.

Chapter 14

Sink and float

Gravity concentration remained the dominant mineral processing method for two thousand years, and it is only in the twentieth century that its importance has declined, with the development of such processes flotation, magnetic separation and leaching. Gravity concentration works best for rich ores, for those showing a coarse size of liberation, for placer deposits, for pre-concentration, and for processing in remote areas or where the situation dictates the least expenditure of money. Today, gravity concentration is used for the treatment, not of one or two minerals, but for a diverse range; from andalusite to zircon, from coal to diamonds, from mineral sands to metal oxides and from industrial minerals to precious metals.

Sink and float is a method used for specific gravity fractionation using heavy liquids. Hence it is called heavy liquid separation. The separation of mineral mixtures into fractions of varying specific gravities is one of the most useful laboratory techniques serving a variety of objectives. Liberation study is one of the important objectives of conducting sink and float analysis for an ore. There are different ways of conducting sink and float analysis by using heavy liquids. Simple sink and float analysis is a method wherein samples are subjected to a single sink and float separation and the two products are weighed and analysed.

A simple example is the sink and float studies of closed size samples of manganese ore using bromoform of specific gravity of 2.86. Table 14.1 shows the results of sink and float studies of manganese ore together with the chemical analysis of both sink and float fractions for Mn and Fe.

It may be seen from the table that Mn content in sink is increasing with decrease in size, which indicates that better liberation is achieved at finer fractions. It is interesting to note that weight percentages of sinks are almost same in all the size fractions. It is also observed that Mn content in float fractions is only 1.25 to 2.67% by weight indicating that most of the gangue minerals are present in the liberated form.

Similarly a representative sample of iron ore fines is subjected to wet sieving cum washing studies in auto sieve shaker and the individual sieve fractions are subjected to sink and float separation using tetra bromoethane to elicit information on liberation characteristics of the sample and the results are presented in Table 14.2.

For the Mineral sands (Beach sands), it is customary to conduct sink and float separation by using bromoform to remove quartz gangue; the sink fraction later analysed by microscope to estimate the different minerals such as ilmenite, rutile, zircon, monazite, garnet, sillimanite etc. Table 14.3 shows the results of sink and float study of beach sands followed by microscopic examination of sink fractions.

Table 14.1 Results of sink and float studies of manganese ore [18].

Size in microns	Nature	Wt%	Mn%	Fe%	Distribution of Mn
−1000 + 500	Float	35.38	2.51	0.69	3.6
	Sink	64.62	36.48	4.17	96.4
−500 + 250	Float	35.66	2.20	1.11	2.9
	Sink	64.34	40.25	4.45	97.1
−250 + 150	Float	35.28	1.25	1.67	1.6
	Sink	64.72	42.77	3.89	98.4
−150 + 75	Float	34.22	2.67	1.39	3.0
	Sink	65.78	44.97	4.45	97.0

Table 14.2 Results of sink and float studies of iron ore fines [19].

Size μm	Wt%			Assay %			Distribution %		
				Fe	SiO_2	Al_2O_3	Fe	SiO_2	Al_2O_3
−150 + 75	3.48	Sink	2.54	64.67	2.68	1.99	2.73	1.08	1.38
			(73.04)				(90.1)	(14.9)	(33.5)
		Float	0.94	19.21	41.26	10.69	0.30	6.18	2.75
			(26.96)				(9.9)	(85.1)	(66.5)
−75 + 45	9.05	Sink	7.57	66.91	1.56	1.10	8.42	1.88	2.27
			(83.68)				(88.7)	(27.3)	(40.2)
		Float	1.48	43.67	21.25	8.38	1.07	5.01	3.39
			(16.32)				(11.3)	(72.7)	(59.8)
−45 + 25	12.99	Sink	10.49	66.24	2.45	1.02	11.54	4.09	2.92
			(80.72)				(82.6)	(55.6)	(47.6)
		Float	2.50	58.71	8.22	4.71	2.44	3.27	3.22
			(19.28)				(17.4)	(44.4)	(52.4)
−25	74.48	Sink	71.59	60.99	5.96	2.75	72.52	67.92	53.81
			(96.12)				(98.7)	(86.5)	(64.0)
		Float	2.89	20.53	22.97	38.32	0.98	10.57	30.26
			(3.88)				(1.3)	(13.5)	(36.0)
	100.00	Head		60.21	6.28	3.66			

Figures in parentheses indicate % contribution of each entity (Sink & Float) in the respective size fraction.

The data of the Table 14.3 indicates that the sample contain 7.7% total heavy minerals (THM). It can be seen from the data on size analysis that accumulation of mass fraction is more significant at −600 + 210 μm size fraction. Sink float data on close size fractions indicates that most of the heavy minerals are concentrated below 150 micron size fractions.

In another way of conducting sink and float analysis, two liquids are used. One is at a density slightly above that of the bulk of the gangue mineral and the other one is at a density known to produce a saleable grade of sink products of valuable mineral. Table 14.4 shows the sink and float analysis results of chrome ore by using heavy liquids of specific gravities of 3.6 and 3.3 and Table 14.5 shows the sink and float analysis results of manganese ore by using heavy liquids of specific gravities of 3.32 and 2.89.

Table 14.3   Results of sink and float followed by microscopic study of Beach sands [20].

Size μm	Wt%	Sink%	THM%
−1000 + 600	7.1	1.4	0.1
−600 + 420	23.6	2.54	0.6
−420 + 300	23.3	2.14	0.5
−300 + 210	36.8	6.52	2.4
−210 + 150	5.0	18.0	0.9
−150 + 100	2.3	69.5	1.6
−100 + 75	1.7	88.2	1.5
−75	0.2	50.0	0.1
Total	100.0	–	7.7

Table 14.4   Sink and float analysis results of chrome ore [21].

Size mm	Product	Wt%	%Cr_2O_3	Cr/Fe	Metal distribution%
−3.35 + 0.5	Sink at 3.6	38.10	47.70		87.97
	3.3–3.6	4.41	36.10		7.70
	Total Sink	42.51	46.45	3.52	95.67
	Float at 3.3	57.49	1.56		4.33
	Total	100.00	20.66		100.00

Table 14.5   Sink and float analysis results of manganese ore [22].

Size μm	Product	Wt%	%MnO	CaO	SiO_2
−180 + 75	Sink at 3.32	22.0	71.7	6.6	1.8
	2.89–3.32	9.4	32.1	28.1	11.6
	Float at 2.89	68.6	3.0	52.5	5.1
	Total	100.0	20.8	40.1	5.0

From the data of Table 14.4 and 14.5, sink at highest specific gravity is of higher grade. It infers that maximum valuables have been liberated. Sink at next low specific gravity is of good grade in chrome ore. Eventhough there is considerable variation between the grades of sink at 3.6 and sink at 3.3, if both are combined the grade is almost same. Hence in the Table 14.4, both sinks are combined as total sink. In case of manganese ore there is much variation between two sinks. Sink at 3.32 is of higher grade and sink at 2.89 is of poor grade. Hence both sinks cannot be combined for any further use. The float at lowest specific gravity containing very little quantity of valuables infers maximum liberation of gangue minerals in both cases.

In a sequential sink and float analysis, the ore is separated into different specific gravity fractions. There are two methods depending upon the nature of the values. The sequence may be one of decreasing or increasing specific gravity. In the case of decreasing specific gravity, the sample is first separated by using highest specific gravity

Table 14.6 Sequential sink and float analysis of −10+2 mm manganese ore [23].

| | | | | Sinks | | | | Floats | | |
Density	Wt%	%MnO₂	MnO₂ points	Cum wt%	Cum MnO₂ points	Cum % MnO₂	Cum wt%	Cum MnO₂ points	Cum% MnO₂
	A	B	$C \frac{AB}{100}$	D	E	$\frac{E}{D} \times 100$ G	H		$\frac{H}{G} \times 100$
Sinks 3.33	14.0	81.6	11.42	14.0	11.42	81.60	86.0	29.92	34.79
−3.33+2.97	20.0	67.3	13.46	34.0	24.88	73.18	66.0	16.46	24.94
−2.97+2.88	5.0	54.5	2.73	39.0	27.61	70.79	61.0	13.73	22.51
−2.88+2.81	4.0	51.5	2.06	43.0	29.67	69.00	57.0	11.67	20.47
−2.81+2.76	3.0	52.8	1.58	46.0	31.25	67.93	54.0	10.09	18.68
−2.76+2.72	2.5	38.2	0.96	48.5	32.21	66.41	51.5	9.13	17.73
−2.72+2.59	32.5	16.6	5.39	81.0	37.60	46.62	19.0	3.74	19.68
−2.59 Floats	19.0	19.7	3.74	100.0	41.34	41.34	–	–	–
	100	41.34							

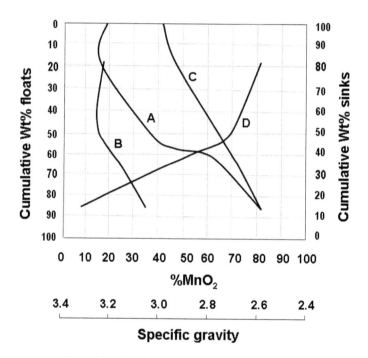

Figure 14.6 Separability curves for Manganese ore.

liquid. Then floats are separated in the next lower specific gravity liquid and so on. The sink fractions from each liquid and the float fraction of the lowest specific gravity liquid are collected, dried and analysed for required component. Table 14.6 shows the sequential sink and float analysis data for manganese ore of −10+2 mm size fraction.

With the data of Table 14.6, the following four separability curves are drawn as shown in Fig. 14.6.

A.   Characteristic or Elementary Assay curve
B.   Cumulative percent assay of floats curve
C.   Cumulative percent assay of sinks curve
D.   Density curve

Characteristic or Elementary Assay curve (curve A) is drawn between actual determined assays of each fraction and the cumulative weight percent sinks for each density. This curve indicates general character of the material in relation to separation. Cumulative percent assay of floats curve (curve B) is drawn between cumulative percent assays of floats and the cumulative weight percent of floats for each density of separation. Cumulative percent assay of sinks curve (curve C) is drawn between cumulative percent assays of sinks and the cumulative weight percent of sinks for each density of separation. These two curves indicate both the assay and the quantity of floats or sinks that would result at a particular density of separation. Density curve (curve D) is drawn between cumulative weight percent sinks and density of separation. This curve shows the cumulative weight percent of either sinks or floats to be expected at any density of separation.

The process of sequential sink and float analysis is used extensively in the coal industry but relatively rarely in the mineral industry. Sequential sink and float for coal is dealt in detail in the chapter 15 float and sink.

Sink and float analysis is extremely useful tool for the evaluation of gravity separation equipment and gravity plant performance. When material is treated in a gravity concentration process, sequential sink and float tests are carried out on each of the two products i.e. concentrate and tailing, separately and used for the evaluation of the plant performance. Evaluation of plant performance is discussed in detail in chapter 17.

Chapter 15

Float and sink

The procedure for float and sink analysis is reverse to that of the sink and float analysis. The sample is first separated by using lowest specific gravity liquid. Then sinks are separated in the next higher specific gravity liquid and so on. The float fractions from each liquid and the sink fraction of the highest specific gravity liquid are collected, dried and analysed for required component.

Sink and float analysis is adopted in most cases where the objective is the production of a concentrate of the heaviest component (i.e. sink). Sink and float analysis is started with heaviest liquid and goes down to lightest liquid. If the objective is to produce a float product, then float and sink analysis is adopted and started with lightest liquid and goes up to heaviest liquid. Float and sink analysis is extremely useful for assessing the washability characteristics of coal.

Washability of coal means amenability of coal to improvement in quality by beneficiation techniques. Washability characteristics of coal vary from locality to locality and from seam to seam in the same locality due to variance in the extent and nature of impurities associated with coal. It is, therefore, necessary to assess the cleaning (beneficiation) potentiality of a coal before sending it to a processing plant.

Ordinary mechanical cleaning processes can only separate coal particles of lower ash content from those of higher ash content. The degree of cleaning possible with any particular coal therefore depends upon the distribution of its ash forming constituents. The ash in the coal can be divided as **Free ash** and **Fixed ash**. Impurities which exist as individual discrete particles when the coal has been broken to the size at which it is cleaned contribute to the free ash content of the coal. The term fixed or inherent ash is used to designate the impurities that are structurally a part of the coal. The former type of impurities can be removed by mechanical means while the latter are difficult to remove. As coal is broken to finer sizes, more and more impurities occur as discrete individual particles and can be separated.

The normal procedure to assess the cleaning potentialities of a coal is to carry out float and sink analysis on the representative sample of desired sizes. In the case of coking coals containing a large proportion of middlings or sinks, separate float and sink analyses are carried out on middlings/sinks crushed to various sizes to find out the possible recovery of additional clean coal. For the evaluation of efficiency of washing baths, float and sink analyses are separately carried out on the individual washed products. Thus, the specific objectives of float and sink analyses of raw and/or washed products are as follows:

1   To know the cleaning possibilities of a coal including recovery of products, ash content, gravity of cut etc.
2   To study the feasibility of a washery project including selection of washing units and flow scheme for treatment
3   To assess the efficiency of different washing baths in operation
4   To predict the yield of products achievable in practice

15.1   FLOAT AND SINK TEST

Float and sink, commercially called Heavy Liquid Separation (HLS), is an operation where particles of different specific gravities are separated by using suitable heavy liquid. The principle of float and sink is:

> When two particles of different specific gravities are immersed in a liquid having specific gravity intermediate between that of two particles, lighter particle would float and heavier particle would sink

Float and sink test is based on the difference in specific gravity of coal particles. Specific gravity of clean coal (free from shale, clay, sandstone, etc.) is about 1.20 (1.12 to 1.35) while specific gravity of impure coal varies from 1.60 and 2.60.

The usual organic liquids employed for float and sink test are Benzene (sp. gr. 0.86), Tetra Chloro Ethylene (sp. gr. 1.62) and Bromoform (sp. gr. 2.80). By mixing benzene with tetra chloro ethylene in right proportions it is possible to prepare a range of liquid baths with specific gravities lying between 1.25 and 1.62. For testing of samples at specific gravity more than 1.62, mixtures of bromoform and tetra chloro ethylene are used in different proportions. Since tetra chloro ethylene and bromoform are both toxic, all tests using these liquids are necessarily conducted in well ventilated places and where possible, under an open shed out of doors.

Chlorinated salt solutions, employing specially zinc chloride, are often recommended for float and sink testing of coal samples above 6 mm in size. Zinc chloride can be conveniently used to prepare test solutions ranging in specific gravity from 1.25 to 1.70.

In carrying out float and sink tests of coking coals and mature types of non-coking coals, it has also been found advantageous to employ sulphuric acid solutions (specific gravity range 1.25 to 1.80) for size fractions generally above 3 mm but preferably 13 mm, provided carbonate contents in such coals are found to be reasonably low.

Laboratory float and sink test is performed on coal to know the washability characteristics of coal thereby to determine the economic separating density. Suitability of heavy medium separation can also be assessed from this test.

Liquids covering a range of specific gravities in incremental steps are prepared and the representative sample of coal is introduced into the liquid of lowest specific gravity. The sink product is removed, washed and placed in a liquid of next higher specific gravity, whose sink product is then transferred to the next higher specific

Table 15.1.1 Laboratory observed values of float and sink analysis.

Specific gravity of liquid used	Weight of watch glass gm	Weight of watch glass with coal floated gm	Weight of empty crucible gm	Weight of crucible with coal gm	Weight of crucible with coal heated at 750°C till constant weight gm
1.30	18.62	31.62	16.843	17.792	16.881
1.35	18.75	38.25	17.346	18.262	17.401
1.40	18.56	44.56	19.982	20.956	20.060
1.45	18.46	50.96	16.827	17.784	16.961
1.50	17.32	30.32	17.924	18.907	18.160
1.55	18.48	24.98	17.465	18.430	17.774
1.60	18.32	24.82	17.638	18.619	18.070
1.75	18.62	25.12	17.239	18.196	17.775
2.00	18.35	21.60	16.347	17.333	16.939
2.00 (sink)	17.52	20.77	16.466	17.439	17.060

gravity and so on. All the float products are collected, drained, washed, dried and then weighed together with the final sink product, to give the specific gravity distribution of the sample by weight.

A sample of less than 1 gm is prepared separately from each float fraction as well as final sink fraction and ash percentages are determined as detailed in article 1.3.1.3.

It is desirable to size the coal and make separate float and sink tests on the separate sizes. Conducting float and sink tests on separate sizes not only assists separation but also gives information regarding the distribution of the impurities according to the size which is essential to select suitable beneficiation process.

The laboratory observations of float and sink analysis on an individual size fraction are to be tabulated. Table 15.1.1 shows such a tabulated values for a hypothetical coal. Hypothetical data is considered for the convenience in discussion.

From the laboratory observed values of Table 15.1.1, weight of the coal floated at each specific gravity, its weight percentages, cumulative weight percentages and ash percentages in floated coal at each specific gravity are calculated and shown in Table 15.1.2.

In Table 15.1.2, first four columns are the results of float and sink test. Column 5 is the calculated ash% in floated coal at each specific gravity. Column 6 represents the coal floated on percentage basis, also known as differential weight percentage or fractional yield. Column 7 represents cumulative yield percentages of cleans (floats) for each gravity fraction and obtained by simply adding fractional yields of column 6 from topmost value to the value at respective specific gravity.

From the values of Table 15.1.2, it is evident that every coal particle in the sample tested has its own ash and their ash vary from 4% to 61% ash.

From the above results ash percentages corresponding to each cumulative yield percent of floats as well as sinks are calculated. Table 15.1.3 shows values after such calculations.

In Table 15.1.3, the values in column 4 are the product of values in column 2 and 3 and called Ash product or Ash point. Column 5 and 6 are the simple addition of values in respective columns 2 and 4. The values in column 7 (cumulative ash percentages or

Table 15.1.2 Calculated values of float and sink analysis.

Sp. gr. of liquid used	Wt of coal floated gm	Wt of coal taken for ash determination gm	Wt of ash gm	Ash% in floated coal	Wt% of floated coal	Cum wt% of floated coal
1	2	3	4	5	6	7
1.30	13.00	0.949	0.038	4.0	10.0	10.0
1.35	19.50	0.916	0.055	6.0	15.0	25.0
1.40	26.00	0.974	0.078	8.0	20.0	45.0
1.45	32.50	0.957	0.134	14.0	25.0	70.0
1.50	13.00	0.983	0.236	24.0	10.0	80.0
1.55	6.50	0.965	0.309	32.0	5.0	85.0
1.60	6.50	0.981	0.432	44.0	5.0	90.0
1.75	6.50	0.957	0.536	56.0	5.0	95.0
2.00	3.25	0.986	0.592	60.0	2.5	97.5
2.00 (sink)	3.25	0.973	0.594	61.0	2.5	100.0
	130.00				100.0	

Table 15.1.3 Results of cumulative yields of floats and sinks.

Sp. gr.	Wt%	Ash%	Ash product	Cumulative floats			Cumulative sinks		
				Wt%	Ash product	Ash%	Wt%	Ash product	Ash%
1	2	3	4	5	6	7	8	9	10
1.30	10.0	4.0	40.0	10.0	40.0	4.0	90.0	1802.5	20.0
1.35	15.0	6.0	90.0	25.0	130.0	5.2	75.0	1712.5	22.8
1.40	20.0	8.0	160.0	45.0	290.0	6.4	55.0	1552.5	28.2
1.45	25.0	14.0	350.0	70.0	640.0	9.1	30.0	1202.5	40.1
1.50	10.0	24.0	240.0	80.0	880.0	11.0	20.0	962.5	48.1
1.55	5.0	32.0	160.0	85.0	1040.0	12.2	15.0	802.5	53.5
1.60	5.0	44.0	220.0	90.0	1260.0	14.0	10.0	582.5	58.3
1.75	5.0	56.0	280.0	95.0	1540.0	16.2	5.0	302.5	60.5
2.00	2.5	60.0	150.0	97.5	1690.0	17.3	2.5	152.5	61.0
2.00 (sink)	2.5	61.0	152.5	100.0	1842.5	18.4	0.0	0.0	0.0

ash percentages in total yields at respective specific gravities) is obtained by dividing values of column 6 by values in column 5. The last value 18.4% of column 7 is the ash percent of the coal sample taken for float and sink test. Ash percentages in cumulative sinks is obtained essentially the same manner but the values are cumulated from bottom to top to get cumulative sinks percent, cumulative ash product and cumulative ash percentages.

All the above results are confined to one particular size fraction. Similar tables are to be prepared for all size fractions of coal separately by similar calculations.

Finally the combined float and sink results of the overall coal can be obtained by adding weights of floated coal in all size fractions at respective specific gravities and computing ash percentages in each floated fraction and then carrying out similar calculations to get cumulative ash percentages.

In addition to expressing the float and sink results of a coal in the form of table as shown in Table 15.1.3, they are represented in the form of curves, familiarly known as Washability Curves developed in 1902 [24].

The graphical representation of float and sink results helps to understand more clearly the washing or cleaning possibilities of coal at any desired ash level of floats or at any desired specific gravity of cut and is therefore more suitable for studying the implications of these results for practical use. The Four different kinds of standard washability curves which are conveniently drawn for this purpose are:

1   The Characteristic curve, also known as elementary ash curve, or fractional yield ash curve
2   Total floats-ash curve or cumulative floats curve
3   Total sinks-ash curve or cumulative sinks curve
4   The yield gravity curve

The characteristic curve is a primary washability curve, also called Henry-Reinhard plot, and is constructed by plotting the cumulative yield of floats expressed in percentages of the total coal against the ash percentages of the individual fractions. At any yield level of the total floats, it shows directly the ash contents of those particles of the floats which have the highest ash content.

The total floats-ash curve is obtained by plotting the cumulative yield of floats expressed in percentage at each specific gravity against the cumulative ash percentage of floats at that specific gravity. As the total floats ash curve directly tells us the recovery of clean or washed coal obtainable for any desired ash content of the floated coal, this curve finds more frequent use in practice than the characteristic curve.

Total sinks-ash curve is drawn between cumulative yield of sinks expressed in percentage at each specific gravity and the cumulative ash percentage of sinks at that specific gravity. This curve is complementary to the total floats-ash curve and shows directly the ash content of the total sinks at any yield level of sinks or floats.

The yield gravity curve is drawn between cumulative yield of floats expressed in percentage at each specific gravity and the specific gravity. The desired specific gravity of separation corresponding to any yield level of floated coal can be obtained by this curve.

All these four curves are usually drawn on a common diagram as shown in Figure 15.1.

A.   Characteristic Curve
B.   Total floats-ash Curve
C.   Total sinks-ash Curve
D.   Yield gravity Curve
E.   Near gravity material Curve

The advantages of drawing all the curves on a common diagram are that all the essential information required for studying the cleaning possibilities of a coal can be readily obtained by cross projections. For example, if one is interested to recover 10% ash clean coal, he has to read first from the total floats ash curve the percent yield

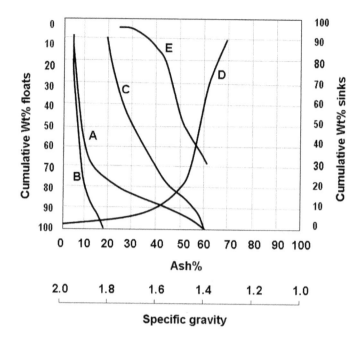

Figure 15.1 Washability Curves.

of cleans corresponding to 10% ash. Then from this yield point, a horizontal line is drawn to cut the total sinks ash curve and the yield gravity curve. At the cut point of the total sinks ash curve, the ash content of the sinks is read from the ash-axis and at the cut point of the yield-gravity curve, the required specific gravity of separation is read from the gravity axis.

Of these families of curves, the characteristic curve is rightly regarded as the parent curve from which other two curves (total floats-ash curve and total sinks-ash curve) may be derived either by direct calculation or by graphical plotting.

While the basic data obtained from laboratory float and sink analysis is directly used for the construction of the characteristic curve, it is important to realize that the determined ash content of any particular specific gravity fraction represents the weighted average of ash contents of all particles of different specific gravities included in that fraction which leads to an error. This error can be minimized and better accuracy in constructing characteristic curve can be achieved by using more gravity baths in closer intervals (0.025) during float and sink tests.

Both the shape and position of the characteristic curve are of much significance in assessing whether the coal is easy or difficult to wash. If the characteristic curve approaches straight line, the coal becomes more difficult to wash and this is all the more true when the curve approaches vertically i.e., becomes nearly parallel to vertical axis. This nearly vertical shape with steep slope represents relatively small ash and correspondingly small specific gravity differences, hence more difficult to wash.

Flat slope represents the reverse physical condition and correspondingly easy separation. If the characteristic curve lies mostly near the vertical axis and then takes a sudden turn towards horizontality, the coal becomes very easy to wash.

The chief limitation of the standard washability curves is that the yield of middlings of any stipulated ash content cannot be directly read from them.

These standard washability curves do not provide any common and accurate basis for the comparison and correlation of the washability characteristics of different coals. The need for such a basis is felt all the more, when there is wide divergence between the ash contents of raw coals and then the yield levels of clean coal cannot be maintained at any uniform ash or specific gravity level. So more standard criteria have been proposed from time to time to compare and correlate the washibility characteristics of different coals according to the ease or difficulty of washing.

These are:

1   Near Gravity Materials (NGM)
2   Yield Reduction Factor (YRF)
3   Washability Index (WI)
4   Optimum degree of Washability (ODW)
5   Washability Number (WN)

15.2   NEAR GRAVITY MATERIALS

In common usage, the near gravity material is defined as the percent yield of total coal, which is within plus and minus 0.10 specific gravity range of the effective density range of separation. Sometimes an additional curve known as ±0.10 specific gravity distribution curve or Near Gravity Material Curve can be constructed along with the standard washability curves to directly measure the ease or difficulty of washing at different densities of cut. The curve can be easily derived from the yield gravity curve.

The near gravity material curves may be of variable nature. For easy washing coals, the near gravity material not only show a tendency to decrease both at lower most and higher most density ranges of cut but becomes maximum in the intermediate density ranges of cut, which generally vary from 1.45 to 1.55. Incidentally, it may be mentioned that the optimum densities of cut for the majority of Indian coals fall in the specific gravity range between 1.45 and 1.55. As a general rule, coal contains more of fixed ash show high percentages of near gravity materials, and those containing more of free dirt show lower percentages of near gravity materials. But precise amount of near gravity materials always depend on the quantum of crowding of the specific gravity fractions around the desired density of separation.

In 1931, B.M. Bird [25] of America proposed the following classification based on near gravity materials to indicate roughly the degree of difficulty faced in the washing of a coal at any selected density of separation. The basis of BIRD's classification is shown in Table 15.2.1.

As per BIRD's classification, most of the Indian coals can be considered as formidable to washing, whereas most of the American or continental coals fall under simple to moderately difficult washing groups. In Indian coals (except Assam coals), the near gravity materials hardly come down to below 25 to 30% at 15 to 17% ash level. Most of the inferior coal seams of India exhibit much worse cleaning characteristics, as they often contain 60 to 80% near gravity materials at the desired density of separation. Hence major quantity of Indian coals at 75 mm–13 mm size is treated by Heavy Medium Separation.

Table 15.2.1 BIRD's classification.

±0.10 sp. gr. distribution present	Degree of difficulty	Gravity process recommended	Type
0 to 7	Simple	Almost any process High tonnages	Jigs, Tables, Spirals
7 to 10	Moderately difficult	Efficient process High tonnages	Sluices, Cones, HMS
10 to 15	Difficult	Efficient process Medium tonnages Good operation	HMS
15 to 20	Very difficult	Efficient process Low tonnages Expert operation	HMS
20 to 25	Exceedingly Difficult	Very efficient process Low tonnages Expert operation	HMS
above 25	Formidable	Limited to a few exceptionally efficient processes Expert operation	HMS with close control

Table 15.2.2 Values of ±0.10 near gravity material.

Sp. gr.	Wt% of ±0.10 near gravity material
1.35	70.0
1.40	70.0
1.45	60.0
1.50	45.0
1.55	22.0
1.60	14.0
1.65	10.0
1.70	6.0
1.75	5.0

Table 15.2.2 shows the values of ±0.10 near gravity material at respective specific gravities deduced from the yield gravity curve for the coal under consideration.

Near Gravity Material curve is also drawn along with standard washability curves in Figure 15.1.

15.3 YIELD REDUCTION FACTOR

Yield Reduction Factor is expressed as the percent reduction or sacrifice in the yield for each percent reduction or decrease in ash content at any selected ash level of the clean coal. In the present case, the ash content of the raw coal is 18.4%. Let the coal is washed at 1.50 specific gravity. The percent yield and its ash are 80.0% and 11.0% respectively. A reduction of 7.4% ash is effected by sacrificing 20.0% yield and so the Yield Reduction Factor works out to be 20.0 divided by 7.4 i.e., 2.7. It has been observed that the yield reduction factor for any particular coal seam remains reasonably constant so long the size grading of the coal and clean coal ash level do not undergo any appreciable change.

The lower the Yield Reduction Factor, the better is the washability characteristics of the coal. The comparatively easy-washing coals of the Jharia field of India often show 3 to 3.5 yield reduction factor when the raw coal is crushed to below 75 mm and separated at 16.0% ash level. Under similar conditions of treatment, the yield reduction factors of the difficult washing coals are as high as 7 to 10. Except Assam coals, the average yield reduction factor values for Indian coals lie between 4 and 7 as against 1.2 to 2.5 obtained for American and continental coals.

On the basis of yield reduction factors at 15.0% ash level of cleans from raw coals crushed to below 15 mm, coal seams of different washability characteristics may be subdivided as easy coals if yield reduction factor is less than 4, medium coals if the yield reduction factor ranges from 4 to 6, and difficult coals if yield reduction factor is more than 6.

The application of yield reduction factors assumes greater importance in estimating the expected yield of the day to day operation of a washery, receiving supplies of raw coal from varied sources.

15.4 WASHABILITY INDEX

The most important criterion in defining the washability characteristics of coals for the purpose of comparing and correlating the results is the Washability Index proposed for the first time by CIMFR [26] at the Fourth International Coal Preparation Congress held at Harrogate (UK) in 1962. This Index is expressed by a number ranging between zero and 100 and is independent of the overall ash content of the raw coal or the level of clean coal ash. When this index is low, the coal is difficult to wash and when the index is high the coal becomes easy to wash.

A graph is to be drawn between cumulative weight percent of floats and cumulative ash as percentage of the total ash in the raw coal at different yield levels. For any coal, this yield ash distribution curve originates at zero yield (corresponding to zero ash) and terminates at 100 yield (corresponding to 100 ash). For any hypothetical coal, which is absolutely non-washable, this curve has to take the form of a straight line running from zero yield (with zero ash) to 100 yield (with 100 ash). On the other hand, the greater the concavity of this curve, the better is the cleaning characteristics of the coal. When the area bounded by the curve and the diagonal through zero yield and 100 percent ash is expressed as the percent of the total area bounded by the two axes and the diagonal, Washability Index in terms of number is obtained, which has necessarily to lie between zero and 100.

The washability indices of a large number of coals from different coalfields of India, particularly from Jharia coalfield have been evaluated. It has been observed that the Indian coals, except the coals from the Assam coalfield which is of tertiary origin, gave indices between 15 and 43. For Assam coals, these indices varied between 47 and 56. The indices for British, German and American coals range between 45 and 76 indicating that they are comparatively much easier to wash. On the other hand, the indices of Japanese, Australian and South African coals range between 20 and 49 being more or less similar to those obtained for Indian coals.

Since the reserves of Indian coals having washability indices above 40 are limited, it may be convenient to adopt a suitable classification for defining the Washability Index. Thus, coals having washability indices above 40 may be termed as easy coals, those lying between 21 and 40 as medium coals and those below 20 as difficult coals.

Table 15.4 shows the values calculated for drawing the curve to determine washability index for the hypothetical coal under consideration. The last column is the cumulative ash percentages with respect to total ash present in the coal.

Figure 15.4 shows the curve drawn with the above values. The washability index for this coal is 45.5.

$$WI = \frac{\text{Area of ACBA}}{\text{Area of AOBA}} = 45.5$$

Table 15.4 Calculated values to determine WI.

Sp. gr.	Wt%	Ash%	Ash product	Cumulative floats		
				Wt%	Ash product	Ash%
1.30	10.0	4.0	40.0	10.0	40.0	2.2
1.35	15.0	6.0	90.0	25.0	130.0	7.1
1.40	20.0	8.0	160.0	45.0	290.0	15.7
1.45	25.0	14.0	350.0	70.0	640.0	34.7
1.50	10.0	24.0	240.0	80.0	880.0	47.8
1.55	5.0	32.0	160.0	85.0	1040.0	56.4
1.60	5.0	44.0	220.0	90.0	1260.0	68.4
1.75	5.0	56.0	280.0	95.0	1540.0	83.6
2.00	2.5	60.0	150.0	97.5	1690.0	91.7
2.00 (sink)	2.5	61.0	152.5	100.0	1842.5	100.0

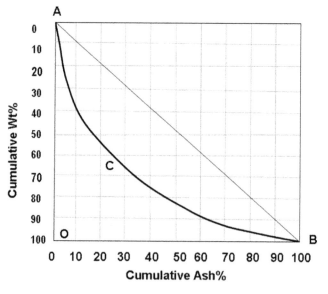

Figure 15.4 Curve to determine Washability Index.

15.5   OPTIMUM DEGREE OF WASHABILITY (ODW)

For every coal, there is an optimal cut point, where one can expect the maximum advantage in coal-dirt separation.

While the washability index gives a general impression about the overall washability characteristics of a coal, it fails to locate this optimal cut point. With respect to a coal of some specified top size, the degree of washability at any particular level of cut can be represented by the following expression:

Degree of washability at any particular level of cut

$$= \frac{\text{Recovery percent of clean coal} \times (\text{Ash\% in raw coal} - \text{Ash\% in clean coal})}{\text{Ash\% in raw coal}}$$

(15.5)

The degree of washability of any coal at zero recovery level has to be zero, as one of the factors in the numerator becomes zero. So also at the recovery level of 100%, where there is no difference between the ash level of raw coal and that of clean coal, the product of the two factors representing the numerator becomes zero, signifying that the degree of washability at that level is again zero. In between these two recovery levels there is a value that represents the optimum degree of washability (ODW).

By plotting the degree of washability against recovery percent of cleans at different specific gravity levels, a peak point is observed for each coal. The degree of washability corresponding to this peak point is designated as the optimum degree of washability. This optimum degree of washability has its values lying between 0 and 100 and generally bears a rectilinear relationship with the washability index. A coal can be considered to have better washability characteristics, when it has higher ODW value and the lower corresponding clean coal ash.

Table 15.5 shows the values calculated for drawing the curve to determine optimum degree of washability for the coal under consideration.

The curve drawn between degree of washability and cumulative weight percent is shown in Figure 15.5 and the value of optimum degree of washability is obtained as 36.

Table 15.5   Calculated values to determine ODW.

Sp. gr.	Wt%	Ash%	Cum Wt%	Cum Ash%	Degree of Washability
1.30	10.0	4.0	10.0	4.0	7.8
1.35	15.0	6.0	25.0	5.2	17.9
1.40	20.0	8.0	45.0	6.4	29.4
1.45	25.0	14.0	70.0	9.1	35.4
1.50	10.0	24.0	80.0	11.0	32.2
1.55	5.0	32.0	85.0	12.2	28.6
1.60	5.0	44.0	90.0	14.0	21.5
1.75	5.0	56.0	95.0	16.2	11.4
2.00	2.5	60.0	97.5	17.3	5.8
2.00 (sink)	2.5	61.0	100.0	18.4	0.0

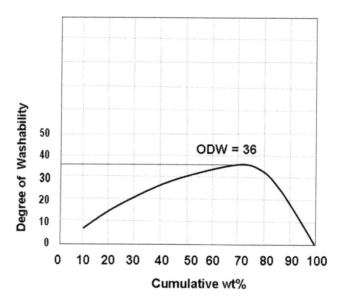

Figure 15.5 Curve to determine Optimum Degree of Washability.

15.6 WASHABILITY NUMBER

To classify coals in accordance with their washablity characteristics, it may be more convenient to express the final value as the ratio of the Optimum Degree of Washability to the clean coal ash at this optimum level.

Since in coals having extremely intractable cleaning characteristics, the value of this ratio is often less than 1, a magnified scale is used by multiplying the resultant value by 10 and rounding up the decimal (if any) to the nearest whole number. This number calculated as

$$\frac{ODW}{Clean\ coal\ ash} \times 10 \tag{15.6}$$

has been designated as Washability Number of a coal.

For the coal under consideration, the clean coal ash corresponding to optimum degree of washability is 8.8%.

$$\text{Hence the washability number is } \frac{36}{8.8} \times 10 = 41$$

15.7 EFFECT OF SIZING ON WASHABILITY CHARACTERISTICS OF A COAL

It is well known that the size of a coal generally influences its washability characteristics. It is, therefore, expected that the various parameters used to define the washability characteristics of any coal should vary with the size of screening and size of crushing the coal.

Table 15.7.1 Size-wise ash analysis of −38 + 0.5 mm coal [27].

Size of the Coal fraction	Weight percent	Ash percent
−38 + 25 mm	18.04	39.19
−25 + 13 mm	51.46	38.28
−13 + 4 mm	21.93	33.35
−4 + 2 mm	04.05	30.69
−2 + 0.5 mm	04.52	25.76
	100.00	36.48

Table 15.7.2 Size-wise float and sink data of raw coal crushed to −38 mm [27].

Sp. gr.	−38 + 25 mm		−25 + 13 mm		−13 + 4 mm		−4 + 2 mm		−2 + 0.5 mm	
	Wt%	Ash%	Wt%	Ash%	Wt%	Ash%	Wt%	Ash%	Wt%	Ash%
Floats 1.40	–	–	–	–	0.11	4.95	0.06	2.68	0.06	2.65
1.40–1.50	5.88	26.56	15.44	24.01	2.22	10.23	0.13	4.34	0.20	4.15
1.50–1.60	5.10	36.42	11.48	34.78	5.62	19.42	0.77	9.23	1.24	7.77
1.60–1.70	3.40	45.66	15.47	44.52	7.12	34.27	1.09	23.06	0.83	17.52
1.70–1.80	0.60	51.58	4.14	48.63	4.49	47.92	0.93	35.51	1.15	30.91
1.80–1.90	3.06	58.48	4.05	62.03	1.61	57.07	0.61	48.31	0.54	47.04
1.90 sinks	–	–	0.88	66.16	0.76	63.36	0.46	63.05	0.50	60.10
	18.04	39.19	51.46	38.28	21.93	33.35	4.05	30.69	4.52	25.76

ROM non-coking coal is crushed to −38 mm size and screened at 25, 13, 4, 2 and 0.5 mm. Fines of −0.5 mm size are removed. Table 15.7.1 shows the weight and ash percent of each size fraction of coal.

Each size fraction of coal is subjected to float and sink analysis and size/specific gravity distribution is shown in the Table 15.7.2 with ash percent in each fraction.

Data of Table 15.7.2 reveals that the ash% decreases with the size of the coal particles. It infers that more and more ash forming constituents of coal got liberated from non-ash forming constituents at finer sizes.

Table 15.7.3 shows size wise washability characteristics in terms of Washability Index (WI), Optimum Degree of Washability (ODW), Optimum Recovery% (OR), Optimum Specific Gravity (OSG), Optimum Ash % (OA), Near Gravity Material (NGM) at optimum cut point and Washability Number (WN).

From Table 15.7.3, it can be readily observed that screening results in yielding coal fractions having widely different washability characteristics caused mainly by the differential segregation of petrographic components in the different size fractions.

Calculations on washability character of coal are illustrated in examples 15.7.1 to 15.7.5.

Table 15.7.3 Washability Characteristics of coal [27].

Size	Ash%	WI	ODW	OR	OSG	OA	NGM	WN
−38 + 25 mm	39.19	16.2	12.5	60	1.60	30	47	4
−25 + 13 mm	38.28	17.5	13.2	52	1.60	29	52	5
−13 + 4 mm	33.35	22.5	19.4	52	1.65	22	58	9
−2 + 2 mm	30.69	32.8	24.2	54	1.70	17	50	14
−2 + 0.5 mm	25.76	40.1	30.0	53	1.70	10	44	30

Example 15.7.1:   *When a sample of coal is subjected to float and sink analysis, the values recorded are shown in Table 15.7.1.1.*

Table 15.7.1.1   Float and sink analysis data for example 15.7.1.

Sp. gr. of liquid used	Weight of watch glass gm	Weight of watch glass with coal floated gm	Weight of empty crucible gm	Weight of crucible with coal gm	Weight of crucible with coal heated at 750°C till constant weight gm
1.30	18.48	27.07	16.842	17.707	16.885
1.40	18.25	38.96	17.468	18.405	17.554
1.50	17.87	49.37	17.354	18.199	17.520
1.60	17.54	26.03	17.637	18.457	17.877
1.70	18.36	21.37	16.987	17.767	17.299
1.80	17.65	20.13	17.123	17.918	17.492
1.80 (sink)	18.53	33.75	17.234	18.069	17.865

Calculate the ash percentage of the coal sample taken.

Solution:

Values of wt%, Ash% & Ash product are calculated and tabulated in the Table 15.7.1.2.

Table 15.7.1.2   Calculated values for example 15.7.1.

Sp. gr.	Wt of coal	Wt%	Weight of Coal taken	Weight of Ash	Ash%	Ash product
1.30	8.59	9.55	0.865	0.043	4.97	47.4635
1.40	20.71	23.01	0.937	0.086	9.18	211.2318
1.50	31.50	35.00	0.845	0.166	19.64	687.4000
1.60	8.49	9.43	0.820	0.240	29.27	276.0161
1.70	3.01	3.34	0.780	0.312	40.00	133.6000
1.80	2.48	2.76	0.795	0.369	46.42	128.1192
1.80 (sink)	15.22	16.91	0.835	0.631	75.57	1277.8887
	90.00	100.00				2761.7193

$$\text{Ash \% of coal sample} = \frac{2761.7193}{100} = 27.62\%$$

Example 15.7.2: *Float and sink test data of a coal of −38 + 25 mm size is shown in Table 15.7.2.1. Evaluate the washability characteristics.*

Table 15.7.2.1 Float and sink test data for example 15.7.2.

Specific gravity	Wt% of floats	Floats Ash%
1.40	12.35	15.28
1.50	34.96	22.08
1.60	25.04	33.17
1.70	3.45	42.37
1.80	13.06	48.69
1.90	2.61	53.15
1.90 (sink)	8.53	76.03

Solution:

Cumulative wt% of floats and their ash% and cumulative wt% of sinks and their ash% are calculated and tabulated as shown in Table 15.7.2.2.

Table 15.7.2.2 Float and sink test results for example 15.7.2.

Specific gravity	Wt% of floats	Floats Ash%	Ash product	Cumulative floats			Cumulative sinks		
				Wt%	Ash product	Ash%	Wt%	Ash product	Ash%
1.40	12.35	15.28	188.71	12.35	188.71	15.28	87.65	3171.83	36.19
1.50	34.96	22.08	771.92	47.31	960.63	20.31	52.69	2399.91	45.55
1.60	25.04	33.17	830.58	72.35	1791.21	24.76	27.65	1569.33	56.76
1.70	3.45	42.37	146.18	75.80	1937.39	25.56	24.20	1423.15	58.81
1.80	13.06	48.69	635.89	88.86	2573.28	28.96	11.14	787.26	70.67
1.90	2.61	53.15	138.72	91.47	2712.00	29.65	8.53	648.54	76.03
1.90 (sink)	8.53	76.03	648.54	100.0	3360.54	33.61	–	–	–

Percent ±0.1 near gravity material (NGM) at each specific gravity, cumulative ash percentages with respect to total ash present in the coal and degree of washabilty at each specific gravity by using the expression

$$\frac{\text{Recovery percent of clean coal} \times (\text{Ash\% in raw coal} - \text{Ash\% in clean coal})}{\text{Ash\% in raw coal}}$$

are calculated and tabulated as shown in Table 15.7.2.3.

With these values, washability curves are drawn as shown in Figure 15.7.2.1.

1   Characteristic Curve
2   Total floats-ash Curve
3   Total sinks-ash Curve
4   Yield gravity Curve
5   Near gravity material Curve

Table 15.7.2.3 Calculated values of NGM & DW for example 15.7.2.

Sp. gr.	Wt%	NGM	Ash%	Ash product	Cum Ash%	DW
1.40	12.35	47.31	15.28	188.71	15.28	6.74
1.50	34.96	60.00	22.08	771.92	20.31	18.72
1.60	25.04	28.49	33.17	830.58	24.76	19.05
1.70	3.45	16.51	42.37	146.18	25.56	18.16
1.80	13.06	15.67	48.69	635.89	28.96	12.29
1.90	2.61	–	53.15	138.72	29.65	10.78
1.90 (sink)	8.53	–	76.03	648.54	33.61	–

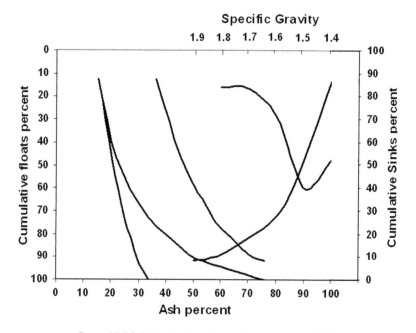

Figure 15.7.2.1 Washability Curves for example 15.7.2.

Another two graphs, one to determine washability index and another to determine optimum degree of washability, are drawn as shown in Figure 15.7.2.2 & 15.7.2.3.

From the shape of characteristic curve, it is difficult to wash the coal. The near gravity material at 1.5 and 1.6 specific gravities is more than 25%. Hence the coal is exceedingly difficult to wash as per the BIRD'S classification. At 1.7 specific gravity, the coal is difficult to wash.

The value of Washability Index (WI) obtained from the graph is 26 and the value of Optimum Degree of Washability (ODW) obtained from the graph is 21. The yield corresponding to ODW of 21 is 62% from the graph. From washability curves, ash in

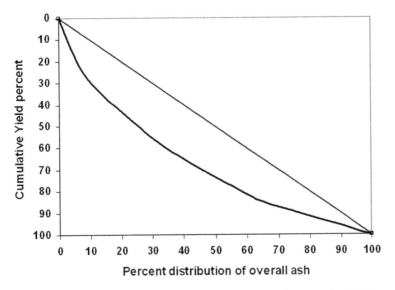

Figure 15.7.2.2 Graph to determine Washability Index for example 15.7.2.

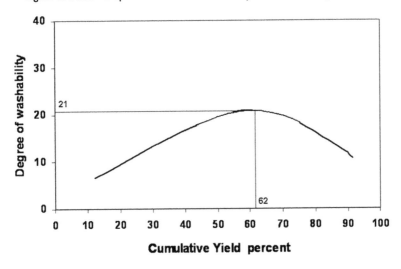

Figure 15.7.2.3 Graph to determine ODW for example 15.7.2.

clean coal corresponding to this yield is 22, corresponding specific gravity of separation is 1.55, near gravity material (NGM) at this optimum cut point is 47.

$$\text{Washability number (WN)} = \frac{21}{22} \times 10 = 9.5$$

From the values of NGM, WI and WN, it can be concluded that the coal is difficult to wash.

Example 15.7.3:   *A student, who wants to study the washability characteristics of coal, has collected a sample of coal and carried out the float and sink analysis. The observations he recorded are shown in Table 15.7.3.1.*

Table 15.7.3.1   Float and sink analysis for example 15.7.3.

Specific gravity of liquid used	Weight of watch glass gm	Weight of watch glass with coal floated gm	Weight of empty crucible gm	Weight of crucible with coal gm	Weight of crucible with coal heated at 750°C till constant weight gm
1.30	18.48	27.07	16.842	17.707	16.885
1.40	18.25	38.96	17.468	18.405	17.554
1.50	17.87	49.37	17.354	18.199	17.523
1.60	17.54	26.03	17.637	18.457	17.877
1.70	18.36	21.37	16.987	17.767	17.299
1.80	17.65	20.13	17.123	17.918	17.492
1.80 (sink)	18.53	33.75	17.234	18.069	17.865

By using the above data, draw the washability curves.
If this coal is to be washed to give the clean coal of 15% Ash, determine the percent yield and the specific gravity of separation.
Find out the percent rejects and its Ash percent.
Determine percent ±0.1 near gravity material at the separation density.
Calculate yield reduction factor.

Solution:

Calculations of weight% and ash% from float and sink analysis data are done and shown in the Table 15.7.3.2.

Table 15.7.3.2   Calculation of Weight% and Ash%.

Specific gravity of liquid used	Weight of coal floated gm	Weight% of coal floated gm	Weight of coal in crucible gm	Weight of ash in crucible gm	Ash%
1.30	8.59	9.55	0.865	0.043	4.97
1.40	20.71	23.01	0.937	0.086	9.18
1.50	31.50	35.00	0.845	0.169	20.00
1.60	8.49	9.43	0.820	0.240	29.27
1.70	3.01	3.34	0.780	0.312	40.00
1.80	2.48	2.76	0.795	0.369	46.42
1.80 (sink)	15.22	16.91	0.835	0.631	75.57
	90.00	100.00			

Calculation of cumulative weight percentages, cumulative ash percentages and weight percent ±0.1 near gravity material are done and shown in the Table 15.7.3.3.

Table 15.7.3.3 Calculation of Cumulative Weight%, Ash% and NGM.

				Cumulative floats			Cumulative sinks			
Sp. gr.	Wt%	Ash%	Ash product	Wt%	Ash product	Ash%	Wt%	Ash product	Ash%	NGM
							100.00	2774.32		–
1.30	9.55	4.97	47.46	9.55	47.46	4.97	90.45	2726.86	30.15	–
1.40	23.01	9.18	211.23	32.56	258.69	7.95	67.44	2515.63	37.30	58.01
1.50	35.00	20.00	700.00	67.56	958.69	14.19	32.44	1815.63	55.97	44.43
1.60	9.43	29.27	276.02	76.99	1234.71	16.04	23.01	1539.61	66.91	12.77
1.70	3.34	40.00	133.60	80.33	1368.31	17.03	19.67	1406.01	71.48	6.10
1.80	2.76	46.42	128.12	83.09	1496.43	19.01	16.91	1277.89	75.57	–
1.80 (sink)	16.91	75.57	1277.89	100.00	2774.32	27.74	–	–	–	–

Ash% of the sample of coal = 27.74 (Last column of cumulative floats)

With the calculated data, 5 washability curves drawn and shown in the figure.

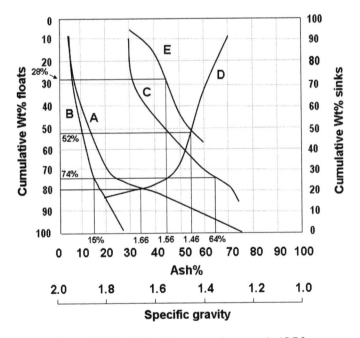

Figure 15.7.3.1 Washability curves for example 15.7.3.

From the Figure 15.7.3.1,

$$\text{yield percent corresponding to 15\% ash} = 74\%$$

$$\text{Specific gravity of separation} = 1.56$$

$$\% \text{ Rejects} = 100 - 74$$

$$= 26\%$$

$$\text{Ash\% in rejects} = 64\%$$

$$\pm 0.1 \text{ near gravity material at the separation specific gravity} = 80 - 52$$

$$= 28\%$$

$$\text{Yield reduction factor} = \frac{100 - 74}{27.74 - 15.0} = 2.04$$

Example 15.7.4: *Float and sink analysis of a sample of non coking coal is shown in Table 15.7.4.1:*

Table 15.7.4.1 Float and sink analysis for example 15.7.4.

Sp. gr. of liquid used	Weigh of watch glass gm	Weight of watch glass with coal floated gm	Weight of empty crucible gm	Weight of crucible with coal gm	Weight of crucible with coal heated at 750°C till constant weight gm
1.40	–	–	–	–	–
1.50	17.2234	66.1084	18.6735	19.6513	18.9332
1.60	16.3927	58.8427	18.2100	18.9937	18.4954
1.70	17.8531	46.1281	18.6140	19.4101	18.9775
1.80	15.5216	20.4716	22.2625	23.2770	22.7858
1.90	16.3762	41.8162	23.5930	24.2230	23.9614
1.90 (sink)	–	–	–	–	–

Calculate the ash% of the sample. It is desired to wash this coal to obtain 30% ash coal. Determine the yield for 100 tons/hr capacity plant and specific gravity of separation. What is the yield reduction factor?

Solution:

Calculations of weight% and ash% from sink and float analysis data are done and shown in Table 15.7.4.2.

Calculation of cumulative weight percentages and cumulative ash percentages are done and shown in Table 15.7.4.3.

Ash% of the sample of coal $= 40.94$

With the calculated data of Table 15.7.4.3, total floats ash curve and yield gravity curve are drawn and shown in Figure 15.7.4.1.

Table 15.7.4.2 Calculation of Weight% and Ash% for example 15.7.4.

Sp. gr. of liquid used	Weight of coal floated gm	Weight% of coal floated gm	Weight of coal in crucible gm	Weight of ash in crucible gm	Ash%
1.40	–	–	–	–	–
1.50	48.885	32.59	0.9778	0.2597	26.56
1.60	42.450	28.30	0.7837	0.2854	36.42
1.70	28.275	18.85	0.7961	0.3635	45.66
1.80	4.950	3.30	1.0145	0.5233	51.58
1.90	25.440	16.96	0.6300	0.3684	58.48
1.90 (sink)	–	–	–	–	–
	150.000				

Table 15.7.4.3 Calculation of Cumulative Weight% and Ash%.

Sp. gr.	Wt%	Ash%	Ash product	Cumulative floats Wt%	Ash product	Ash%
1.40	–	–	–	–	–	–
1.50	32.59	26.56	865.59	32.59	865.59	26.56
1.60	28.30	36.42	1030.69	60.89	1896.28	31.14
1.70	18.85	45.66	860.69	79.74	2756.97	34.57
1.80	3.30	51.58	170.21	83.04	2927.18	35.25
1.90	16.96	58.48	1167.26	100.00	4094.44	40.94
1.90 (sink)	–	–	–	–	–	–

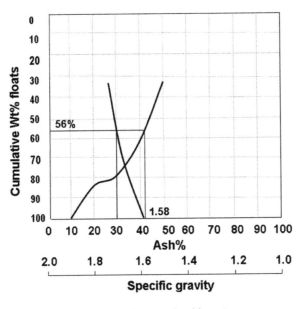

Figure 15.7.4.1 Total floats and yield gravity curves.

From the Figure 15.7.4.1,

Yield percent corresponding to 30% ash = 56%

Therefore the yield for 100 tons/hr capacity plant = 56 tons/hr

Specific gravity of separation = 1.58

$$\text{Yield reduction factor} = \frac{100 - 56}{40.94 - 30.0} = 4.02$$

Example 15.7.5: *Washability test data of a coal sample is shown in Table 15.7.5.1.*

Table 15.7.5.1 Washability test data for example 15.7.5.

Sp. gr.	Differential% of floats	Ash% in differential floats
1.30	7.6	5.9
1.40	21.3	12.7
1.50	36.7	21.8
1.60	13.8	30.5
1.70	11.6	39.6
1.80	4.8	48.3
1.80 (sink)	4.2	63.4

Determine the specific gravity at which the separation should be done for the beneficiation of this coal so that the ash content of the clean coal is not more than 17%. What is the expected yield of clean coal corresponding to this separation? Find out the ash percent in refuse and maximum ash of a particle in the yield. Calculate the yield reduction factor.

Solution:

The required values are calculated and tabulated as shown in Table 15.7.5.2.

Table 15.7.5.2 Calculated values for example 15.7.5.

				Cumulative floats			Cumulative sinks		
Sp. gr.	Wt%	Ash%	Ash product	Wt%	Ash product	Ash%	Wt%	Ash product	Ash%
1	2	3	4	5	6	7	8	9	10
1.30	7.6	5.9	44.84	7.6	44.84	5.90	92.4	2448.95	26.50
1.40	21.3	12.7	270.51	28.9	315.35	10.91	71.1	2178.44	30.64
1.50	36.7	21.8	800.06	65.6	1115.41	17.00	34.4	1378.38	40.07
1.60	13.8	30.5	420.90	79.4	1536.31	19.35	20.6	957.48	46.48
1.70	11.6	39.6	459.36	91.0	1995.67	21.93	9.0	498.12	55.35
1.80	4.8	48.3	231.84	95.8	2227.51	23.25	4.2	266.28	63.40
1.80 (sink)	4.2	63.4	266.28	100.0	2493.79	24.94	–	–	–

From the 7th column of Table 15.7.5.2, for floats of 17% ash the following are read:

a) Recovery of the clean coal   = 65.6%   (5th column)

b) Ash percent in rejects   = 40.07%   (10th column)

c) Maximum ash in any particle
in the clean coal product   = 21.8%   (3rd column)

d) Specific gravity of separation   = 1.50   (1st column)

$$\text{Reduction in yield} = 100 - 65.6 = 34.4\%$$

$$\text{Reduction in ash} = 24.94 - 17.0 = 7.94\ \%$$

$$\therefore \text{Yield Reduction Factor} = \frac{34.4}{7.94} = 4.33$$

15.8 MAYER CURVE

The Mayer curve, known as the M-curve, is a method of plotting float-and-sink analysis to predict the results of cleaning properties of coal in a three-component system. The method was first described by F.W. Mayer [28] in 1950 and applications of the method were discussed by him in 1956 and 1957. It is a useful tool to predict cleaning properties of a three-product separation wherein clean coal and middling can be predicted for the required ash percentages of clean coal and refuse. It can also be used in predicting cleaning properties of a product mixture resulting from blending a cleaned coal with un-cleaned coal, or blending two different plant clean coal products.

15.8.1 Construction of M-curve

The M-curve is constructed with the cumulative weight percent on the ordinate and cumulative ash percent on the abscissa. Diagonal lines (known as ash lines) are drawn from the origin to intersect the cumulative ash axis at each cumulative ash percent value from float-and-sink analysis. Each cumulative ash percent value is then plotted on its respective diagonal at the intersection of the corresponding cumulative weight ordinate. The M-curve is drawn through these points. Figure 15.8.1.1 shows the construction of M-curve for the float-and-sink analysis data given in Table 15.1.3.

15.8.2 M-curve for a three product system

M-curve can be used to determine yield and ash percentages of clean coal and middling in a three product separation system. The detailed procedure for use of M-curve is explained taking float-and-sink data of Table 15.1.3 as an example.

In predicting the cleaning properties of three product separation system, desired ash contents for the clean coal and refuse are to be selected first. Let these selected values of ash percentages for clean coal and refuse be 10% and 55% respectively. Line OA is drawn (Fig. 15.8.1.2) from the origin to the selected clean coal ash% value of 10%. The line OA intersects the M-curve at a point C. Weight percent corresponding to this point C is the yield of clean coal and it is 74%.

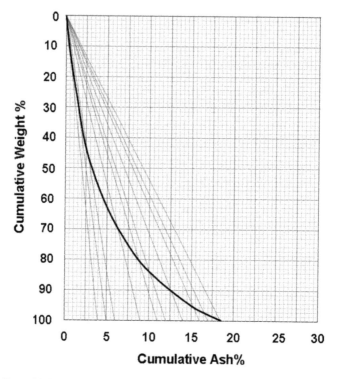

Figure 15.8.1.1 M-curve for float-and-sink analysis data of Table 15.1.3.

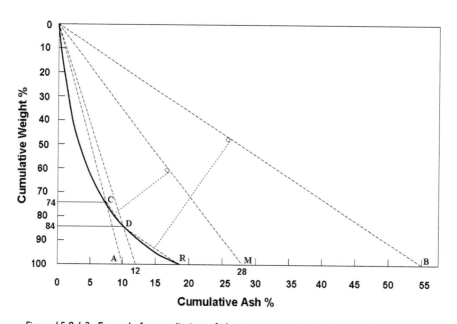

Figure 15.8.1.2 Example for prediction of cleaning properties in three product system.

Next, line OB is drawn from the origin to the selected refuse ash% value of 55%. A line parallel to OB is drawn from point R, the ash percent of raw coal. This line intersects the M-curve at a point D. Weight percent corresponding to this point D is the combined yield of clean coal and middling and it is 84%. Subtracting the yield of clean coal percent 74% from the combined yield of clean coal and middling 84%, gives the yield of middling 10%. Finally, line OM is drawn from the origin parallel to a line connecting points C and D. The intersection of this line on the ash axis at M gives the ash percent of the middling, 28%. Line OD, drawn from the origin, intersects the ash axis, when extended, at 12% ash. This is the ash% of clean coal and middling together. The specific gravities to be maintained to get three products can be read from yield gravity curve corresponding to respective yields.

This graphical procedure is one of the useful ways to carry out the optimal yield study. The accuracy that is possible depends to a large extent on the degree of curvature exhibited by the curve.

15.8.3   M-curve for blended cleaned and un-cleaned coal

In certain instances, in few of the coal washeries, due to lack of costly fine coal beneficiation facility, un-cleaned fine coal is blended with cleaned coarse coal without affecting the resulted ash percentage much. In such cases M-curve can be used to predict the properties of such blended mixture. Substantial reduction in time and labor may be saved through the use of the M-curve. The practical use of the M-curve in predicting the properties of blended clean coal and un-cleaned coal mixture at optimum conditions is described by taking float-and-sink data of Table 15.1.3 as an example.

Let us consider the blending of an un-cleaned coal of 20% ash with a clean coal product. To prepare the blended clean coal and un-cleaned coal at 12% ash, the M-curve is employed to find the ash percent of clean coal.

The ash lines of the blended coal mixture (12%) and un-cleaned coal (20%) are drawn from the origin on the M-curve (Fig. 15.8.1.3). A tangent parallel line to the 20% ash line is drawn on the M-curve at point C. This line intersects the 12% ash line at point T. This corresponds to the ash percent of the clean coal and un-cleaned coal. The ash line is drawn from the origin through point C. On extending this line, it intersects the ash axis at 8% ash. Corresponding to point C, the yield is 63% read on weight axis. This is the yield of clean coal at 8% ash without adding un-cleaned coal. Corresponding to point of intersection of 12% ash line with the M-curve, the yield is 84%. It means that an yield of 84% is achievable if raw coal alone is washed to get 12% ash. From the set of washability curves drawn for the same data in Figure 15.1, the specific gravity at which the raw coal can be washed to get 84% yield with 12% ash is 1.54. When un-cleaned coal is added to clean coal, the yield is 93% corresponding to point T. It is evident that the weight percent of un-cleaned coal added to the clean coal to obtain resulted 12% ash is 30% (93%–63%). From the set of washability curves drawn for the same data in Figure 15.1, the specific gravity at which the raw coal can be washed to get 63% yield with 8% ash is 1.42. If 30% of uncleaned coal of 20% ash is added, the ash% of blended coal is 12%.

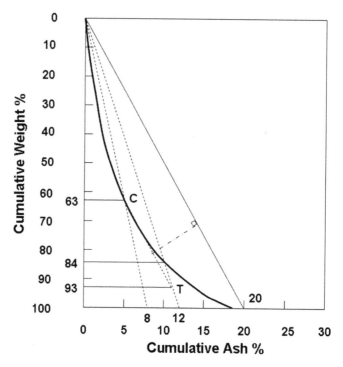

Figure 15.8.1.3 Predicting clean product by addition of un-cleaned coal to clean coal.

15.8.4 M-curve for blending clean coal from two plants

Figure 15.8.1.4 shows the M-curve for two raw coals, A and B, respectively. These two coals are cleaned separately to the required product ash percent of 12% by either an independent cleaning system in the same coal preparation plant or cleaned in two coal preparation plants separately. The 12% ash line of the product mixture is drawn on these two M-curves. This ash line intersects the two M-curves at a_1 and b_1. The intersection points indicate yields of 95% and 65% respectively. The separation specific gravities at ash percent of 12% are 1.80 and 1.48. With an assumed mixture ratio of two raw feed coals A:B = 50:50, a total yield of 80% [(95 + 65)/2] can be achieved. This is shown in the Figure 15.8.1.4 corresponding to point m_1 which is a midpoint of a_1 and b_1. It is to be determined whether 80% is maximum or some other maximum is there by combining two clean coals in the same ratio of 50:50 but washed to yield different ash percentages.

The following are the steps to be followed to determine separation specific gravity at which the yield increases at a mix ratio of 50:50 with 12% ash.

1   Connect equal specific gravity points (1.50, 1.60, and 1.70) on curves A and B.
2   Divide these connected lines according to the ratio 50:50 i.e. midpoint.

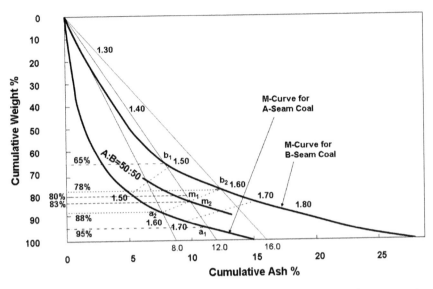

Figure 15.8.1.4 Mixing of two coals at 50:50 ratio to obtain highest yield for a given ash.

3   Connect the midpoints of the connected lines by a curve that represents the curve for a 50:50 mixture of coals A and B.
4   From the 50:50 mixture curve of A and B coals, determine the yield at 12% ash as well as the separation specific gravity at this yield.

As shown in Figure 15.8.1.4, an yield of 83% is achievable with the same mix ratio of 50:50, and both the coals are to be washed at 1.60 specific gravity. An yield of 88% at 8% ash for A coal and an yield of 78% at 16% ash for B coal will be obtained.

A comparison between the two methods shows certain essential differences. In the first case, both coals A and B produce a clean coal with an ash percent of 12% and a yield of 95% and 65% are obtained when cleaned at the different separation specific gravities of 1.80 and 1.48 respectively. The combined yield is 80%. In the second case, two raw feed coals with different ash contents are washed at the same separation specific gravity of 1.60 yielding 88% and 78% respectively. The combined yield 83% is achieved at the desired ash percent of 12%. In the second case, the yield has increased by 3% when compared to the first case.

The validity of this method is not limited to the mixing ratio of 50:50 and can be extended to the other ratios. The ratio of 50:50 has been selected to facilitate clarity of the illustration and necessary explanations.

Example 15.8.1:   *The washability test data of a coal sample is given in Table 15.8.1.1.*

Table 15.8.1.1 Washability test data for example 15.8.1.

Sp. gr.	Wt%	Ash%
1.40	14.32	10.79
1.50	31.32	21.09
1.60	11.06	29.33
1.70	10.98	34.80
1.80	20.33	46.72
1.90	01.23	53.47
1.90 (sink)	10.76	78.84

By using M-curve, determine yield and ash percentage of clean coal and middlings for the clean coal of 17% ash and refuse of 50% ash.

Solution:

Cumulative weight percentages and ash percentages are calculated necessary for drawing M-curve and shown in Table 15.8.1.2.

Table 15.8.1.2 Cumulative percentages for example 15.8.1.

Specific gravity	Wt% of floats	Floats Ash%	Ash product	Cumulative floats Wt%	Cumulative floats Ash product	Ash%
1.40	14.32	10.79	154.51	14.32	154.51	10.79
1.50	31.32	21.09	660.54	45.64	815.05	17.86
1.60	11.06	29.33	324.39	56.70	1139.44	20.10
1.70	10.98	34.80	382.10	67.68	1521.54	22.48
1.80	20.33	46.72	949.82	88.01	2471.36	28.08
1.90	1.23	53.47	65.77	89.24	2537.13	28.48
1.90 (sink)	10.76	78.84	848.32	100.00	3385.45	33.85

Yield gravity curve is drawn and shown in Figure 15.8.1.5.

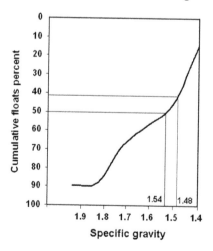

Figure 15.8.1.5 Yield gravity curve for example 15.8.1.

M-curve is drawn as per the procedure given in article 15.8.1. and yield and ash percentages are determined from M-curve as per the procedure given in article 15.8.2. and shown in Figure 15.8.1.6.

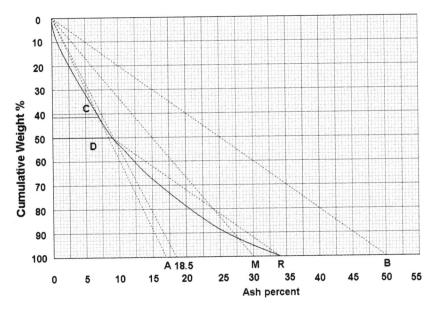

Figure 15.8.1.6 M-curve for example 15.8.1.

The following results were obtained from Figure 15.8.1.6:

Yield of clean coal corresponding to point $C = 42\%$
Yield of clean coal and middlings corresponding to point $D = 50\%$
∴ Yield of middlings $= 50 - 42 = 8\%$

Ash% of middlings at $M = 30\%$
Ash% of clean coal and middlings together $= 18.5\%$
Yield of refuse $= 100 - 50 = 50\%$

From yield gravity curve of Figure 15.8.1.5,

Specific gravity of separation to get clean coal $= 1.48$
Specific gravity of separation to get middlings $= 1.54$

Example 15.8.2: *For the problem 15.8.1, determine clean coal product to be obtained in order to blend the uncleaned coal of 35% ash to this clean coal to get blended mixture of 25% ash. Also determine ash% of clean coal.*

Solution:

M-curve is constructed following the procedure given in article 15.8.3 and is shown in Figure 15.8.2.1.

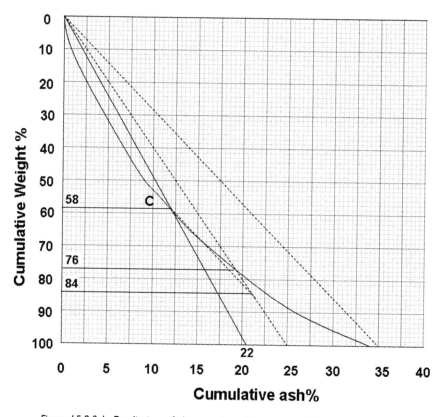

Figure 15.8.2.1 Prediction of clean coal product required for example 15.8.2.

From the Figure 15.8.2.1,

$$\text{Wt\% of uncleaned coal to be blended} = 84 - 58 = 26\%$$

Yield of clean coal of 25% ash achievable
if raw coal is washed $= 76\%$

Yield of clean coal of 22% ash achievable
if raw coal is washed $= 58\%$

Wt% of clean coal of 22% ash to be added
to uncleaned coal of 35% ash
to yield blended coal of 25% ash $= 100 - 26 = 74\%$

15.9 PROBLEMS FOR PRACTICE

15.9.1: *Float and sink test data of a coal of −2 + 0.5 mm size is shown in Table 15.9.1.*

Table 15.9.1 Float and sink test data for problem 15.9.1.

Sp. gr.	Wt% of floated coal	Ash% of floated coal
1.30	2.10	4.70
1.40	10.00	6.22
1.50	13.67	13.24
1.60	13.52	20.26
1.70	13.21	27.86
1.80	8.62	39.64
1.90	5.54	46.83
2.00	5.13	51.01
2.00 (sink)	28.21	74.23

Evaluate the washability characteristics.

15.9.2: *From the observations as shown in Table 15.9.2 obtained from Float and Sink experiment, calculate the ash percentage of the coal sample taken.*

Table 15.9.2 Float and sink analysis for problem 15.9.2.

Sp. gr. of the liquid	Weight of the coal floated, gm	Weight of the coal taken in crucible, gm	Weight of the ash in crucible, gm
1.30	15.89	0.921	0.047
1.40	34.23	0.831	0.068
1.50	7.91	0.760	0.149
1.60	3.15	0.625	0.199
1.80	3.36	0.634	0.280
1.80 (sink)	5.46	0.740	0.487

If the coal is washed at 1.5 sp. gr., what is the degree of difficulty based on near gravity material.

[very difficult to wash]

15.9.3: *By using the float and sink data of a coal sample shown in Table 15.9.3, calculate ash% in coal sample, clean coal and rejects when the coal is washed at 1.5 specific gravity.*

Table 15.9.3 Float and sink data for problem 15.9.3.

Specific gravity	Weight of the coal floated, gm	Weight of the ash in floated coal, gm
1.30	11.4	0.5
1.40	24.4	2.0
1.50	5.6	1.1
1.60	2.3	0.7
1.80	2.4	1.0
2.00	3.9	2.5

[15.6, 8.7%, 48.8%]

15.9.4: *The washability test data of a sample of coal is given in Table 15.9.4*

Table 15.9.4 Washability test data for problem 15.9.4.

Sp. gr.	Wt%	Ash%
− 1.30	36.0	2.6
+1.30 − 1.35	25.7	7.5
+1.35 − 1.40	11.0	14.1
+1.40 − 1.45	5.3	19.2
+1.45 − 1.50	3.5	23.9
+1.50 − 1.55	2.0	28.1
+1.55 − 1.60	1.2	31.7
+1.60 − 1.65	1.1	34.3
+1.65 − 1.70	0.7	37.5
+1.70	13.5	67.4
		36.28

When this coal is washed, a clean coal of 8.8% ash is obtained. Find

a) *Recovery of clean coal*
b) *Ash percent in rejects*
c) *Maximum ash in any particle in the clean coal product*
d) *Specific gravity of separation*

Calculate yield reduction factor and assess the washability character.

[85.8%, 65.9%, 34.3%, 1.65, 1.75, easy to wash]

15.9.5: *Washability test data of a coal sample is shown in Table 15.9.5.*

Table 15.9.5 Washability test data for problem 15.9.5.

Sp. gr.	Differential % of floats	Ash% in differential floats
1.40	7.5	14.1
1.50	13.1	26.6
1.60	12.6	36.6
1.70	11.8	42.7
1.80	6.8	47.9
1.90	7.2	53.8
2.00	7.0	61.1
2.00 (sink)	34.0	76.3

Determine yield and ash percentages of clean coal and middlings to get clean coal of 30% ash and refuse of 75% ash.

[43%, 9%, 45%]

Chapter 16

Metallurgical accounting

Metallurgical accounting is required in order to assess the production and metallurgical performance of processing plant accurately. This will normally consist of:

1. Sampling of feed, concentrate and tailing streams

    Samples collected should be as representative as possible of the stream that is being sampled. Samples are usually collected from conveyors (dry) and pipe lines (both wet and dry).

2. Weight flow measurement of the same streams

    Weight flow rate is usually measured on dry flows using weightometers (for conveyors), or impact weighers (for vertically falling dry flows), or wet flows using nuclear density gauges integrated with magnetic flow meters (for slurry pipelines).

3. Laboratory analysis of the samples

    Sample analysis consists of sample preparation of the samples received from the plant and then assayed.

Based on the information generated from the above, performance of mineral processing plant is evaluated through the calculations by using simple expressions. These simple expressions are similarly applied for calculating the results of laboratory testing. Any increase in the number of separations and mineral components to be accounted for, complexity of the computations greatly increases.

16.1 TWO PRODUCTS BENEFICIATION OPERATIONS

For a two products beneficiation operation

Let
F = Weight flow rate of the feed solids
C = Weight flow rate of the concentrate solids
T = Weight flow rate of the tailing solids
f = Fraction of the metal present in the feed
c = Fraction of the metal present in the concentrate
t = Fraction of the metal present in the tailing

The material balance for solids is

$$F = C + T \qquad (16.1)$$

i.e.,   *solids input = solids output*

Similarly, the material balance for valuable metal (or material) they contain is

$$Ff = Cc + Tt \qquad (16.2)$$

i.e.,   *valuable metal input = valuable metal output*

16.1.1 Ratio of concentration

Ratio of concentration or concentration ratio (K) is defined as the ratio of the weight of the ore fed to the weight of the concentrate produced.

$$\text{Ratio of concentration} = K = \frac{F}{C} \qquad (16.1.1.1)$$

It can be thought of as number of tons of feed required to produce one ton of concentrate. From the material balance equations 16.1.1 & 16.1.2, ratio of concentration can be computed in terms of assay values as

$$\text{Ratio of concentration} = K = \frac{F}{C} = \frac{c - t}{f - t} \qquad (16.1.1.2)$$

At operating plants, it is usually simpler to report the K based on assays. If more than one mineral or metal is recovered in a bulk concentrate, each will have its own K.

If the tonnage of concentrate produced is unknown it can be obtained using the assays of all streams and the tons of plant feed solids

$$\text{Weight of the concentrate} = C = \frac{F}{K} = F\frac{f - t}{c - t} \qquad (16.1.1.3)$$

16.1.2 Ratio of recovery

Ratio of recovery (R) is defined as the ratio of the weight of the valuable metal in the concentrate to the weight of the valuable metal originally present in the ore. It is usually expressed as percentage

$$\% \text{ Recovery} = R = \frac{Cc}{Ff} \times 100 \qquad (16.1.2.1)$$

From the material balance equations 16.1.1 & 16.1.2, % recovery can be computed in terms of assay values as

$$\% \text{ Recovery} = R = \frac{c(f - t)}{f(c - t)} \times 100 \qquad (16.1.2.2)$$

16.1.3 Ratio of enrichment

Ratio of enrichment is defined as the ratio of the assay value of the concentrate to that of the ore.

$$\text{Ratio of enrichment} = \frac{c}{f} \qquad (16.1.3.1)$$

This is equal to the product of the ratio of concentration and the recovery. The ratio of enrichment does not have the direct economic significance of the commonly used ratio of concentration. It gives better idea of the cleanliness of the concentrate.

16.1.4 Metallurgical efficiency

Metallurgical efficiency is defined as the arithmetical average of the recoveries of the principal constituent of each product (including tailing).

$$\text{Let} \qquad R_v = \text{recovery of valuable metal in concentrate} = \frac{Cc}{Ff}$$

$$J_w = \text{rejection of waste in tailing} = \frac{T(1-t)}{F(1-f)}$$

$$\text{Then,} \quad \text{Metallurgical efficiency} = \frac{R_v + J_w}{2} \qquad (16.1.4.1)$$

16.1.5 Economic recovery or efficiency

Economic recovery or efficiency is the ratio of the actual value of the concentrate obtained from a ton of ore to the value of the concentrate theoretically obtainable in mineralogically pure form from a ton of ore.

If 0.20 ton of lead concentrate worth Rs. 10,000/- per ton is obtained in practice from an ore that theoretically should yield 0.18 ton of mineralogically pure galena concentrate worth Rs. 15,000/- per ton,

$$\text{Economic recovery} = \frac{0.20 \times 10000}{0.18 \times 15000} = 0.741 \quad \Rightarrow \quad 74.1\%$$

To achieve the recovery to the maximum possible extent, the ore is repeatedly treated by using another unit of the same device. The following is the terminology in use in such cases:

Roughing: It is the operation of removal of a rough concentrate at the earliest stage of treatment of the ground ore. The device used is called Rougher.

Scavenging: It is the operation of removal of the last recoverable fraction of valuables before discarding the final tailing from the treatment plant. The device used for this purpose is called Scavenger.

Cleaning: It is the operation of re-treating the rough concentrate to improve its quality. The device used for this cleaning stage is called Cleaner.

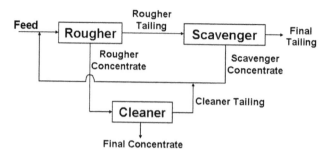

Figure 16.1 Flowsheet of three stage treatment.

Fig. 16.1 is a typical flowsheet showing three stage treatment.
Different calculations are illustrated in examples 16.1.1 to 16.1.13.

Example 16.1.1: *In an iron ore concentration operation, samples of feed, concentrate and tailing are analysed, and all stream flowrates are measured. Results are shown in Table 16.1.1.*

Table 16.1.1 Quantities and assay values.

	Flow rates tons/hr	Assay value %Fe
Feed	1390	64.77
Concentrate	1112	68.08
Tailing	278	51.53

Calculate ratio of concentration, ratio of enrichment, recovery and metallurgical efficiency. How much iron is lost in tailing?

Solution:

Given

$$\text{Rate of feed} = F = 1390 \text{ tons/hr}$$
$$\text{Rate of concentrate} = C = 1112 \text{ tons/hr}$$
$$\text{Rate of tailing} = T = 278 \text{ tons/hr}$$
$$\%\text{Fe in feed} = f = 64.77\%$$
$$\%\text{Fe in concentrate} = c = 68.08\%$$
$$\%\text{Fe in tailing} = t = 51.53\%$$

$$\text{Ratio of concentration} = \frac{F}{C} = \frac{1390}{1112} = 1.25$$

$$\text{Ratio of enrichment} = \frac{c}{f} = \frac{68.08}{64.77} = 1.05$$

$$\%\text{Concentrate recovery} = \frac{C}{F} \times 100 = \frac{1112}{1390} \times 100 = 80\%$$

$$\%\text{Metal recovery} = R_v = \frac{Cc}{Ff} \times 100 = \frac{1112 \times 68.08}{1390 \times 64.77} \times 100 = 84\%$$

$$\%\text{Rejection of waste in tailing} = J_w = \frac{T(100-t)}{F(100-f)} \times 100$$

$$= \frac{278(100-51.53)}{1390(100-64.77)} \times 100 = 27.5\%$$

$$\text{Metallurgical Efficiency} = \frac{R_v + J_w}{2} = \frac{84+27.5}{2} = 55.8\%$$

$$\text{Iron lost in tailing} = \frac{Tt}{100} = \frac{278 \times 51.53}{100} = 143.3 \text{ tons/hr}$$

Example 16.1.2: *A beneficiation plant treats 200 tons/day of lead ore. Assay analyses of samples of feed, concentrate and tailing determined as 4.4% Pb, 55% Pb and 0.05% Pb. Calculate the amount of concentrate recovered, percentage of recovery and loss of lead in tailing.*

Solution:

Given

Feed rate	$= F = 200$ tons/day	
%Pb in feed	$= f = 4.4\%$	
%Pb in concentrate	$= c = 55\%$	
%Pb in tailing	$= t = 0.05\%$	

Solids balance $\qquad\qquad F = C + T \quad \Rightarrow \quad 200 = C + T$

Lead balance $\qquad\qquad Ff = Cc + Tt \quad \Rightarrow \quad 4.4 \times 200 = 55C + 0.05T$

On solving above equations $C = 15.83$ tons/day; $\qquad T = 184.17$ tons/day

Amount of concentrate recovered $= C = 15.83$ tons/day

$$\% \text{ Recovery of lead} = \frac{Cc}{Ff}100 = \frac{15.83 \times 55}{200 \times 4.4} \times 100 = 98.94\%$$

$$\text{Loss of lead in tailing} = \frac{Tt}{100} = \frac{184.17 \times 0.05}{100} = 0.0921 \text{ ton/day}$$

Example 16.1.3: *Tin ore assaying 10% Sn is fed to a concentration plant at the rate of 100 tons/hr. If the grades of concentrate and tailing are 40% Sn and 1% Sn, calculate the recovery of tin and tin lost in tailing in tons/hr.*

Solution:

Given

Flow rate of the feed	$= F = 100$ tons/hr	
Assay value of feed	$= f = 10\%$ Sn	
Assay value of concentrate	$= c = 40\%$ Sn	
Assay value of tailing	$= t = 1\%$ Sn	

Total material balance $F = C + T$ $\Rightarrow$ $100 = C + T$

Tin balance $Ff = Cc + Tt$ $\Rightarrow$ $100 \times 10 = 40C + T$

On solving $C = 23.077$ tons/hr; $T = 76.923$ tons/hr

$$\% \text{ Recovery of tin} = \frac{Cc}{Ff} \times 100 = \frac{23.077 \times 40}{100 \times 10} \times 100 = 92.31\%$$

$$\text{Tin lost in tailing} = \frac{Tt}{100} = \frac{76.923 \times 1}{100} = 0.76923 \text{ ton/hr}$$

Example 16.1.4: *Estimate the recovery of chromite from the following data: The ore contains 18.4% Cr_2O_3. The assay of product and rejects are 52.5% and 10.8% Cr_2O_3 respectively.*

Solution:

Given

Assay value of feed	$= 18.4\%$ Cr_2O_3
Assay value of concentrate	$= 52.5\%$ Cr_2O_3
Assay value of tailing	$= 10.8\%$ Cr_2O_3

Let the flow rate of the feed $= 100$ tons/hr

Total material balance $F = C + T$ $\Rightarrow$ $100 = C + T$

Chromite balance $Ff = Cc + Tt$ $\Rightarrow$ $100 \times 18.4 = 52.5C + 10.8T$

On solving $C = 18.23$ tons/hr; $T = 81.77$ tons/hr

$$\% \text{ Recovery of chromite} = \frac{Cc}{Ff} 100 = \frac{18.23 \times 52.5}{100 \times 18.4} \times 100 = 52\%$$

Example 16.1.5: *100 tph of zinc ore containing 8.5% sphalerite and 91.5% gangue is treated in a processing plant. Recovery of the sphalerite and the gangue in the concentrate are 92% and 2.5%. Calculate flowrates of concentrate and tailing, assay of chalcopyrite and gangue in concentrate and tailing.*

Solution:

Given

Feed flow rate	$= F = 100$ tph
% sphalerite in feed	$= f = 8.5\%$
% gangue in feed	$= 91.5\%$

% recovery of sphalerite in concentrate $= 92\%$
% recovery of gangue in concentrate $= 2.5\%$

$$\text{Sphalerite recovery} = \frac{Cc}{Ff} \times 100 = 92 \quad \Rightarrow \quad \frac{Cc}{100 \times 8.5} \times 100 = 92$$

$$\Rightarrow \quad Cc = 782$$

$$\text{Gangue recovery} = \frac{C(100-c)}{F(100-f)} \times 100 = 2.5 \quad \Rightarrow \quad \frac{100C - Cc}{F(100-f)} \times 100 = 2.5$$

$$\Rightarrow \quad \frac{100C - 782}{100 \times (100 - 8.5)} \times 100 = 2.5 \quad \Rightarrow \quad C = 10.1\,\text{tph}$$

Concentrate flow rate $= C = 10.1\,\text{tph}$

Solids balance $\qquad F = C + T \qquad\qquad \Rightarrow 100 = 10.1 + T$

$$\Rightarrow T = 89.9\,\text{tph}$$

Tailing flow rate $\quad = T = 89.9\,\text{tph}$

$Cc = 782 \qquad\qquad \Rightarrow c = \dfrac{782}{10.1} = 77.4\%$

Assay of sphalerite in concentrate $= c = 77.4\%$

Assay of gangue in concentrate $\quad = 100 - 77.4 = 22.6\%$

Sphalerite balance $\quad Ff = Cc + Tt \quad \Rightarrow \quad 100 \times 8.5 = 782 + 89.9t$

$$\Rightarrow \quad t = 0.76\%$$

Assay of sphalerite in tailing $= t = 0.76\%$

Assay of gangue in tailing $\quad = 100 - 0.76 = 99.24\%$

Example 16.1.6: *A copper ore assaying 2% metal is treated in a concentrating plant with a concentration ratio of 10 and produces 200 tons of concentrate. If the loss of metal in the tailing is 1.8 tons, calculate the recovery of metal in the concentrate.*

Solution:

Given

Weight of the concentrate $= 200$ tons
Assay value of feed $\qquad = 2\%\ \text{Cu}$
Ratio of concentration $\quad = 10$
Copper lost in tailing $\qquad = 1.8$ tons
Ratio of concentration $\quad = 10 = \dfrac{F}{C} = \dfrac{F}{200}$

$$\Rightarrow \quad F \quad = 2000\,\text{tons}$$

Solids balance $\qquad\qquad F \quad = C + T$

$$\Rightarrow \quad 2000 = 200 + T \quad \Rightarrow \quad T = 1800\,\text{tons}$$

$$\text{Copper lost in tailing} = \frac{Tt}{100} = 1.8 \quad \Rightarrow \quad \frac{1800t}{100} = 1.8 \quad \Rightarrow \quad t = 0.1\%$$

$$\text{Assay of tailing} = t = 0.1\% \text{ Cu}$$

Copper balance

$$Ff = Cc + Tt \quad \Rightarrow \quad 2000 \times 2 = 200 \times c + 1800 \times 0.1$$
$$\Rightarrow \quad c = 19.1\%$$

Assay of concentrate $= c = 19.1\%$ Cu

$$\% \text{ Recovery of copper} = \frac{Cc}{Ff} 100 = \frac{200 \times 19.1}{2000 \times 2} \times 100 = 95.5\%$$

Example 16.1.7: *When 60% iron ore is fed to a concentrating plant at the rate of 200 tons/hr, 150 tons/hr of concentrate assaying 65% Fe is produced. Calculate the metallurgical efficiency.*

Solution:

Given

$$\begin{aligned}
\text{Flow rate of the feed} &= F = 200 \text{ tons/hr}\\
\text{Flow rate of the concentrate} &= C = 150 \text{ tons/hr}\\
\text{Assay value of feed} &= f = 60\% \text{ Fe}\\
\text{Assay value of concentrate} &= c = 65\% \text{ Fe}
\end{aligned}$$

Solids balance $\qquad F = C + T$

$$\Rightarrow \qquad 200 = 150 + T \quad \Rightarrow \quad T = 50 \text{ tons}$$

Iron balance $\qquad Ff = Cc + Tt$

$$\Rightarrow \qquad 200 \times 60 = 150 \times 65 + 50 \times t \quad \Rightarrow \quad t = 45\%$$

Assay of tailing $= t = 45\%$ Fe

$$\% \text{ recovery of iron} = R_v = \frac{Cc}{Ff} 100 = \frac{150 \times 65}{200 \times 60} \times 100 = 81.25\%$$

$$\% \text{ Rejection of waste in tailing} = J_w = \frac{T(100 - t)}{F(100 - f)} \times 100$$

$$= \frac{50(100 - 45)}{200(100 - 60)} \times 100 = 34.4\%$$

$$\text{Metallurgical Efficiency} = \frac{R_v + J_w}{2} = \frac{81.25 + 34.4}{2} = 57.83\%$$

Example 16.1.8: *A copper ore treated by a flotation plant produces concentrate of 28% Cu and tailing of 0.1% Cu. The ore assays 0.9% Cu. Calculate*
a) Cu recovery in concentrate
b) Fraction of feed in concentrate
c) Enrichment ratio

Solution:

Given

$$\text{Assay value of feed} = f = 0.9\% \text{ Cu}$$
$$\text{Assay value of concentrate} = c = 28\% \text{ Cu}$$
$$\text{Assay value of tailing} = t = 0.1\% \text{ Cu}$$

Let the flow rate of the feed = 100 tons/hr

$$\text{Solids balance } F = C + T \quad \Rightarrow \quad 100 = C + T$$

$$\text{Copper balance } Ff = Cc + Tt \quad \Rightarrow \quad 100 \times 0.9 = 28C + 0.1T$$

$$\text{On solving } C = 2.87 \text{ tons/hr}$$

Concentrate flow rate $= C = 2.87$ tons/hr

$$\% \text{ recovery of copper} = \frac{Cc}{Ff} 100 = \frac{2.87 \times 28}{100 \times 0.9} \times 100 = 89.3\%$$

$$\text{Fraction of feed in concentrate} = \frac{C}{F} = \frac{2.87}{100} = 0.0287$$

$$\text{Enrichment ratio} = \frac{c}{f} = \frac{28}{0.9} = 31.1$$

Example 16.1.9: *Molybdenum ore assaying 0.36% MoS$_2$ (Molybdenite) is treated in a flotation circuit consists of a bank of rougher cells and cleaner cells in series. All the streams are sampled and analysed for MoS$_2$. The results are shown in Table 16.1.9.*

Table 16.1.9 Analyses of the streams of flotation circuit.

Streams	Assay%
Rougher concentrate	7.37
Rougher tailing	0.05
Cleaner concentrate	93.65
Cleaner tailing	0.21

If the ore is treated at the rate of 100 tons/hr, calculate the flow rates of all streams, percent recovery of Molybdenite and molybdenite lost in composite tailings.

Solution:

Let F, R_C, R_T, C, C_T are the flow rates of feed, rougher concentrate, rougher tailing, concentrate, cleaner tailing and f, r_C, r_T, c, c_T are assay values in respective streams.

Material balance over rougher bank

Solids balance $F = R_C + R_T$ $\Rightarrow$ $100 = R_C + R_T$ (I)

Molybdenite balance $Ff = R_C r_C + R_T r_T$

$\Rightarrow$ $100 \times 0.36 = R_C \times 7.37 + R_T \times 0.05$ (II)

On solving (I) & (II) $R_C = 4.2$ tons/hr

$$R_T = 95.8 \text{ tons/hr}$$

Material balance over cleaner bank

Solids balance $R_C = C + C_T$ $\Rightarrow$ $4.2 = C + C_T$ (III)

Molybdenite balance $R_C r_C = Cc + C_T c_T$

$\Rightarrow$ $7.37 \times 4.2 = C \times 93.65 + C_T \times 0.05$ (IV)

On solving (III) & (IV) $C = 0.33$ ton/hr

$$C_T = 3.87 \text{ tons/hr}$$

$$\% \text{ Recovery of Molybdenite} = \frac{Cc}{Ff} \times 100 = \frac{0.33 \times 93.65}{100 \times 0.36} \times 100 = 85.8\%$$

Molybdenite lost in composite tailings $= R_T r_T + C_T c_T$

$$= 95.8 \times \frac{0.05}{100} + 3.87 \times \frac{0.21}{100}$$

$$= 0.056 \text{ ton/hr}$$

Example 16.1.10: *100 gm of graphite with 60% carbon is fed to a flotation cell in the laboratory. By using proper reagents 50 gm of concentrate assaying 90% carbon is obtained. Calculate the percentage recovery, ratio of concentration and ratio of enrichment.*

Solution:

Given

Weight of the feed	$= F = 100$ gm
Weight of concentrate	$= C = 50$ gm
Assay value of feed	$= f = 60\%$ Carbon
Assay value of concentrate	$= c = 90\%$ Carbon

Solids balance $F = C + T \Rightarrow 100 = 50 + T$

$\Rightarrow$ $T = 50$ gm

Weight of tailing $T = 50$ gm

Carbon balance $Ff = Cc + Tt$

$\Rightarrow$ $100 \times 60 = 50 \times 90 + 50t$ $\Rightarrow$ $t = 30\%$ Carbon

Assay of tailing $= t = 30\%$ Carbon

$$\text{\% recovery of carbon} = \frac{Cc}{Ff}100 = \frac{50 \times 90}{100 \times 60} \times 100 = 75\%$$

$$\text{Ratio of concentration} = \frac{F}{C} = \frac{100}{50} = 2$$

$$\text{Ratio of enrichment} = \frac{c}{f} = \frac{90}{60} = 1.5$$

Example 16.1.11: *Zinc ore of 15% ZnS is fed at the rate of 1200 tons/hr to a flotation circuit consists of rougher and cleaner cells. Rougher concentrate is treated in cleaner cells. The tailings from the rougher cells is the final tailings. The cleaner tailing of 20% ZnS is recycled to rougher and circulating load is 0.25 (recycle/fresh feed). The concentrate assays 89% ZnS and the recovery of ZnS is 98%. Calculate the flow rates and the assays of the respective streams.*

Solution:

Given

$$\begin{aligned}
\text{Assay value of fresh feed} &= 15\% \text{ ZnS}\\
\text{Assay value of concentrate} &= 89\% \text{ ZnS}\\
\text{Assay value of cleaner tailing} &= 20\% \text{ ZnS}
\end{aligned}$$

$$\text{Circulating load} = \frac{\text{cleaner tailings}}{\text{Fresh feed}} = 0.25$$

$$\text{\% Recovery of ZnS} = \frac{\text{Weight of ZnS in cleaner concentrate}}{\text{Weight of ZnS in fresh feed}} = 98$$

Let F = Flow rate of the fresh feed solids to the rougher = 1200 tons/hr
C_R = Flow rate of the concentrate solids from the rougher to cleaner
T = Flow rate of the rougher tailing solids
T_c = Flow rate of the cleaner tailing solids
C = Flow rate of the cleaner concentrate solids
f = Assay value of fresh feed = 15%
c_R = Assay value of the rougher concentrate
t = Assay value of the rougher tailings
t_c = Assay value of the cleaner tailings = 20%
c = Assay value of the cleaner concentrate = 89%

$$\text{\% Recovery of ZnS} = 98 = \frac{Cc}{Ff} \times 100 = \frac{C \times 0.89}{1200 \times 0.15} \times 100$$

$$\Rightarrow \qquad C = 198.2 \text{ tons/hr}$$

$$\frac{\text{cleaner tailings}}{\text{Fresh feed}} = 0.25$$

$$\Rightarrow \quad \text{Cleaner tailings} = T_c = 0.25 \times 1200 = 300 \text{ tons/hr}$$

Solids balance for the cleaner

$$C_R = C + T_c \quad \Rightarrow \quad C_R = 198.2 + 300 = 498.2 \text{ tons/hr}$$

Flow rate of rougher concentrate $= C_R = 498.2$ tons/hr

ZnS balance for the cleaner

$$C_R c_R = Cc + T_c t_c \quad \Rightarrow \quad 498.2 \times c_R = 198.2 \times 89 + 300 \times 20$$

$$\Rightarrow \quad c_R = 47.5\%$$

Assay of rougher concentrate $= c_R = 47.5\%$

Solids balance for the rougher

$$F + T_c = C_R + T$$

$$\Rightarrow \quad 1200 + 300 = 498.2 + T \quad \Rightarrow \quad T = 1001.8 \text{ tons/hr}$$

Flow rate of rougher tailings $= \quad T = 1001.8$ tons/hr

ZnS balance for the rougher

$$Ff + T_c t_c = C_R c_R + Tt$$

$$\Rightarrow \quad 1200 \times 15 + 300 \times 20 = 498.2 \times 47.5 + 1001.8 \times t$$

$$\Rightarrow \quad t = 0.335\%$$

Assay of ZnS in rougher tailings $= t = 0.335\%$

Example 16.1.12: *10% concentrate produced by a single bank of flotation cells assays 50% Zn. When the concentrate is again treated in another bank of flotation cells by re-circulating its tailing to first bank of cells, 8% concentrate is produced assaying 65% Zn. Calculate the increase in recovery of the zinc, if the feed assays 8% Zn.*

Solution:

Given

Assay of the feed ore	$= f = 8\%$ Zn
Quantity of concentrate without recirculation	$= C = 10\%$ of feed
Assay of concentrate without recirculation	$= c = 50\%$ Zn
Quantity of concentrate with recirculation	$= C = 8\%$ of feed
Assay of concentrate with recirculation	$= c = 65\%$ Zn

Let the ore feed $= 100$ tons

Quantity of concentrate without recirculation $= C = \dfrac{10}{100} \times 100 = 10$ tons

Recovery of zinc without recirculation $= \dfrac{Cc}{Ff} \times 100$

$$= \dfrac{10 \times 50}{100 \times 8} \times 100 = 62.5\%$$

Quantity of concentrate with recirculation $= C = \dfrac{8}{100} \times 100 = 8$ tons

Recovery of zinc with recirculation $= \dfrac{Cc}{Ff} \times 100$

$$= \dfrac{8 \times 65}{100 \times 8} \times 100 = 65\%$$

Increase in recovery of zinc $= 65.0 - 62.5 = 2.5\%$

Example 16.1.13: *When a lead ore assaying 4% Pb is treated in a bank of high grade and low grade flotation cells in series at the rate of 30 tons/hr, a high grade concentrate assaying 44% Pb and a low grade concentrate assaying 7% Pb are produced. The high grade tailings assaying 0.6% Pb is treated in low grade cells. The low grade tailings assayed 0.2% Pb. What are the weights of high and low grade concentrates produced per hour? Calculate the recovery of lead produced.*

Solution:

Given

Flow rate of lead ore	$= F = 30$ tons/hr
Assay of lead ore	$= f = 4\%$ Pb
Assay of high grade concentrate	$= c_H = 44\%$ Pb
Assay of low grade concentrate	$= c_L = 7\%$ Pb
Assay of high grade tailings	$= t_H = 0.6\%$ Pb
Assay of low grade tailings	$= t_L = 0.2\%$ Pb

Solid balance over high grade cells

$$F = C_H + T_H \quad \Rightarrow \quad T_H = 30 - C_H$$

Lead balance over high grade cells

$$Ff = C_H c_H + T_H t_H$$

$\Rightarrow \qquad 30 \times 4 = C_H \times 44 + T_H \times 0.6$

$\Rightarrow \qquad 30 \times 4 = C_H \times 44 + (30 - C_H) \times 0.6$

$\Rightarrow \qquad C_H = 2.35$ tons/hr

Flow rate of high grade concentrate $= C_H = 2.35$ tons/hr

Flow rate of high grade tailings $= T_H = 30 - 2.35 = 27.65$ tons/hr

Solid balance over low grade cells

$$T_H = C_L + T_L \quad \Rightarrow \quad T_L = 27.65 - C_L$$

Lead balance over low grade cells

$$T_H t_H = C_L c_L + T_L t_L$$

$\Rightarrow \qquad 27.65 \times 0.6 = C_L \times 7 + T_L \times 0.2$

$\Rightarrow \qquad 27.65 \times 0.6 = C_L \times 7 + (27.65 - C_L) \times 0.2$

$\Rightarrow \qquad C_L = 1.63$ tons/hr

Flow rate of low grade concentrate $= C_L = 1.63$ tons/hr

Alternately

By using equation 16.1.1.3, flow rates of the concentrates can be calculated.

$$\text{Flow rate of high grade concentrate } C_H = F\frac{f - t_H}{c_H - t_H} = 30 \times \frac{4 - 0.6}{44 - 0.6}$$

$$\Rightarrow \qquad\qquad C_H = 2.35 \text{ tons/hr}$$

$$\text{Flow rate of high grade tailings} \qquad = T_H = 30 - 2.35 = 27.65 \text{ tons/hr}$$

$$\text{Flow rate of Low grade concentrate } C_L = T_H\frac{t_H - t_L}{c_L - t_L} = 27.65 \times \frac{0.6 - 0.2}{7 - 0.2}$$

$$\Rightarrow \qquad\qquad C_L = 1.63 \text{ tons/hr}$$

$$\text{Weight of lead in concentrates} = 2.35 \times 0.44 + 1.63 \times 0.07$$

$$= 1.15 \text{ tons/hr}$$

$$\% \text{ Recovery of lead} = \frac{1.15}{30 \times 0.04} \times 100 = 95.8\%$$

16.2 THREE PRODUCTS BENEFICIATION OPERATIONS

A concentrator processing a complex ore by comminution and beneficiation operations produces two separate concentrates each is enriched with different metal or valuable mineral and a final tailing with low in both constituents. The following is the definition and notation used for three products beneficiation operations.

	Weight or Wt%	Assay of metal 1	Assay of metal 2
Feed	F	f_1	f_2
Concentrate of mineral 1	C_1	c_{11}	c_{21}
Concentrate of mineral 2	C_2	c_{12}	c_{22}
Tailing	T	t_1	t_2

Total material balance $\qquad F = C_1 + C_2 + T$ $\qquad\qquad$ (16.2.1)

Metal 1 balance $\qquad\qquad Ff_1 = C_1 c_{11} + C_2 c_{12} + T t_1$ $\qquad\qquad$ (16.2.2)

Metal 2 balance $\qquad\qquad Ff_2 = C_1 c_{21} + C_2 c_{22} + T t_2$ $\qquad\qquad$ (16.2.3)

On solving the above three equations

Weight of concentrate of mineral 1

$$= C_1 = F \times \frac{(f_1 - t_1)(c_{22} - t_2) - (f_2 - t_2)(c_{12} - t_1)}{(c_{11} - t_1)(c_{22} - t_2) - (c_{21} - t_2)(c_{12} - t_1)}$$

Weight of concentrate of mineral 2

$$= C_2 = F \times \frac{(f_2 - t_2)(c_{11} - t_1) - (f_1 - t_1)(c_{21} - t_2)}{(c_{22} - t_2)(c_{11} - t_1) - (c_{12} - t_1)(c_{21} - t_2)}$$

Ratio of concentration of metal 1 concentrate $= K_1 = \dfrac{F}{C_1}$

Ratio of concentration of metal 2 concentrate $= K_2 = \dfrac{F}{C_2}$

% Recovery of metal 1 $= R_1 = \dfrac{C_1 c_{11}}{F f_1} \times 100$

% Recovery of metal 2 $= R_2 = \dfrac{C_2 c_{22}}{F f_2} \times 100$

Different calculations are illustrated in examples 16.2.1 to 16.2.5

Example 16.2.1: *A laboratory flotation test was carried out on a sample of lead zinc ore. The results are shown in the Table 16.2.1.*

Table 16.2.1 Results of flotation test.

		Assay%	
	Weight gm	Pb	Zn
Pb concentrate	40	65	7
Zn concentrate	30	4	50
Tailing	430	0.5	1

a) What is the calculated assay of the flotation feed i.e. % Pb & % Zn.
b) What is the distribution of lead and zinc in the concentrates and tailing.

Solution:

Let F, C_L, C_Z, T are the flow rates, f_L, c_{1L}, c_{2L}, t_L are the lead assay values, f_Z, c_{1Z}, c_{2Z}, t_Z are the zinc assay values of feed, lead concentrate, zinc concentrate, and tailing respectively.

Feed for flotation test
$$F = C_L + C_Z + T$$
$$= 40 + 30 + 430 = 500 \text{ gm}$$

Lead balance
$$F f_L = C_L c_{1L} + C_Z c_{2L} + T t_L$$

$\Rightarrow$
$$500 \times f_L = 40 \times 65 + 30 \times 4 + 430 \times 0.5 = 2935$$

$\Rightarrow$
$$f_L = 5.87\% \text{ Pb}$$

Zinc balance
$$F f_Z = C_L c_{1Z} + C_Z c_{2Z} + T t_Z$$

$\Rightarrow$
$$500 \times f_Z = 40 \times 7 + 30 \times 50 + 430 \times 1.0 = 2210$$

$\Rightarrow$
$$f_Z = 4.42\% \text{ Zn}$$

Percent distribution of lead and zinc are calculated and shown in Table 16.2.1.1.

Table 16.2.1.1 Percent distribution of lead and zinc for example 16.2.1.

	Weight gm	%Pb	Pb content	Pb distribution %	%Zn	Zn content	Zn distribution %
	A	B	AB/100		C	AC/100	
Pb concentrate	40	65	26.0	88.6	7	2.8	12.7
Zn concentrate	30	4	1.2	4.1	50	15.0	67.9
Tailing	430	0.5	2.15	7.3	1	4.3	19.4
	500		29.35	100.0		22.1	100.0

Example 16.2.2: *The assay values of the feed, two concentrates and the composite tailing of a lead-zinc ore beneficiation plant are determined by analyses of samples of all streams as follows:*

	%Pb	%Zn
Feed to the plant	4.5	6.0
Lead concentrate	56.3	0.1
Zinc concentrate	0.05	55.4
Composite tailing	0.05	0.1

Calculate the yields of the concentrates and tailing.

Solution:

Let the basis of calculations is 100 tons of feed

Notations as given in article 16.2 are followed

$$\text{Weight of Lead concentrate} = C_1 = F \times \frac{(f_1 - t_1)(c_{22} - t_2) - (f_2 - t_2)(c_{12} - t_1)}{(c_{11} - t_1)(c_{22} - t_2) - (c_{21} - t_2)(c_{12} - t_1)}$$

$$= 100 \times \frac{(4.5 - 0.05)(55.4 - 0.1) - (6 - 0.1)(0.05 - 0.05)}{(56.3 - 0.05)(55.4 - 0.1) - (0.1 - 0.1)(0.05 - 0.05)} = 7.91 \text{ tons}$$

$$\text{Weight of Zinc concentrate} = C_2 = F \times \frac{(f_2 - t_2)(c_{11} - t_1) - (f_1 - t_1)(c_{21} - t_2)}{(c_{22} - t_2)(c_{11} - t_1) - (c_{12} - t_1)(c_{21} - t_2)}$$

$$= 100 \times \frac{(6 - 0.1)(56.3 - 0.05) - (4.5 - 0.05)(0.1 - 0.1)}{(55.4 - 0.1)(56.3 - 0.05) - (0.05 - 0.05)(0.1 - 0.1)} = 10.67 \text{ tons}$$

Weight of tailings $= T = 100 - 7.91 - 10.67 = 81.42$ tons

This problem can be worked out by using determinants.

Basis: 100 tons of feed

Material balance equations for total material, lead and zinc are as follows:

$$C_1 + C_2 + T = 100$$

$$56.3C_1 + 0.05C_2 + 0.05T = 100 \times 4.5 = 450$$

$$0.1C_1 + 55.4C_2 + 0.1T = 100 \times 6 = 600$$

$$\Delta = \begin{vmatrix} 1 & 1 & 1 \\ 56.3 & 0.05 & 0.05 \\ 0.1 & 55.4 & 0.1 \end{vmatrix} = \begin{vmatrix} 1 & 0 & 0 \\ 56.3 & 56.25 & 56.25 \\ 0.1 & -55.3 & 0 \end{vmatrix} = 3110.625$$

$$\Delta_1 = \begin{vmatrix} 100 & 1 & 1 \\ 450 & 0.05 & 0.05 \\ 600 & 55.4 & 0.1 \end{vmatrix} = \begin{vmatrix} 100 & 1 & 0 \\ 450 & 0.05 & 0 \\ 600 & 55.4 & 55.3 \end{vmatrix} = 24608.5$$

$$\Delta_2 = \begin{vmatrix} 1 & 100 & 1 \\ 56.3 & 450 & 0.05 \\ 0.1 & 600 & 0.1 \end{vmatrix} = \begin{vmatrix} 1 & 100 & 0 \\ 56.3 & 450 & 56.25 \\ 0.1 & 600 & 0 \end{vmatrix} = 33187.5$$

$$C_1 = \frac{\Delta_1}{\Delta} = \frac{24608.5}{3110.625} = 7.91 \text{ tons}$$

$$C_2 = \frac{\Delta_2}{\Delta} = \frac{33187.5}{3110.625} = 10.67 \text{ tons}$$

$$T_1 = 100 - 7.91 - 10.67 = 81.42 \text{ tons}$$

Example 16.2.3: *A copper zinc mill treats 1000 tons/day of ore assaying 2.7% Cu and 19.3% Zn and produced a Copper concentrate of grade 25.3% Cu, 5.1% Zn and zinc concentrate of 52.7% Zn, 1.2% Cu. The tailing assay is 0.15% Cu and 0.95% Zn. How much Cu and Zn concentrates are produced in the plant per day.*

Solution:

Notations as given in article 16.2 are followed

Weight of copper concentrate $= C_1 = F \times \dfrac{(f_1 - t_1)(c_{22} - t_2) - (f_2 - t_2)(c_{12} - t_1)}{(c_{11} - t_1)(c_{22} - t_2) - (c_{21} - t_2)(c_{12} - t_1)}$

$= 100 \times \dfrac{(2.7 - 0.15)(52.7 - 0.95) - (19.3 - 0.95(1.2 - 0.15)}{(25.3 - 0.15)(52.7 - 0.95) - (5.1 - 0.95)(1.2 - 0.15)} = 86.9 \text{ tons}$

Weight of zinc concentrate $= C_2 = F \times \dfrac{(f_2 - t_2)(c_{11} - t_1) - (f_1 - t_1)(c_{21} - t_2)}{(c_{22} - t_2)(c_{11} - t_1) - (c_{12} - t_1)(c_{21} - t_2)}$

$= 100 \times \dfrac{(19.3 - 0.95)(25.3 - 0.15) - (2.7 - 0.15)(5.1 - 0.95)}{(52.7 - 0.95)(25.3 - 0.15) - (1.2 - 0.15)(5.1 - 0.95)} = 347.6 \text{ tons}$

Weight of tailing $= T_1 = 1000 - 86.9 - 347.6 = 565.5$ tons

This problem can be worked out by using determinants.

Material balance equations for total material, copper and zinc are as follows:

$$C_1 + C_2 + T = 1000$$

$$25.3C_1 + 1.2C_2 + 0.15T = 1000 \times 2.7 = 2700$$

$$5.1C_1 + 52.7C_2 + 0.95T = 1000 \times 19.3 = 19300$$

$$\Delta = \begin{vmatrix} 1 & 1 & 1 \\ 25.3 & 1.2 & 0.15 \\ 5.1 & 52.7 & 0.95 \end{vmatrix} = \begin{vmatrix} 1 & 0 & 0 \\ 25.3 & 24.1 & 25.15 \\ 5.1 & -47.6 & 4.15 \end{vmatrix} = 1297.155$$

$$\Delta_1 = \begin{vmatrix} 1000 & 1 & 1 \\ 2700 & 1.2 & 0.15 \\ 19300 & 52.7 & 0.95 \end{vmatrix} = \begin{vmatrix} 1000 & 1 & 0 \\ 2700 & 1.2 & 1.05 \\ 19300 & 52.7 & 51.75 \end{vmatrix} = 112695$$

$$\Delta_2 = \begin{vmatrix} 1 & 1000 & 1 \\ 25.3 & 2700 & 0.15 \\ 5.1 & 19300 & 0.95 \end{vmatrix} = \begin{vmatrix} 1 & 1000 & 0 \\ 25.3 & 2700 & 25.15 \\ 5.1 & 19300 & 4.15 \end{vmatrix} = 450920$$

$$C_1 = \frac{\Delta_1}{\Delta} = \frac{112695}{1297.155} = 86.9 \text{ tons}$$

$$C_2 = \frac{\Delta_2}{\Delta} = \frac{450920}{1297.155} = 347.6 \text{ tons}$$

$$T_1 = 1000 - 86.9 - 347.6 = 565.5 \text{ tons}$$

Example 16.2.4: *Feed to a froth flotation plant assays 14% Pb and 10% Zn. The percent recovery of Pb in lead concentrate is 85% and percent recovery of Zn in zinc concentrate is 85%. If 10,000 tons of ore are treated per day to produce 1,500 tons of lead concentrate and 7,000 tons of tailings which assay 1% Pb and 1.5% Zn, what is the grade of the lead concentrate in terms of Pb and Zn and the grade of zinc concentrate in terms of Zn and Pb.*

Solution:

Notations as given in article 16.2 are followed

$F = 10,000$ tons/day; $\quad C_1 = 1,500$ tons/day; $\quad T = 7,000$ tons/day

$F = C_1 + C_2 + T \Rightarrow 10000 = 1500 + C_2 + 7000 \Rightarrow C_2 = 1500$ tons/day

$$\% \text{ Recovery of Pb} = \frac{C_1 c_{L1}}{F f_L} \times 100 = 85 \quad \Rightarrow \frac{1500 c_{L1}}{10000 \times 14} \times 100 = 85$$

$$\Rightarrow c_{L1} = 79.33\%$$

$$\% \text{ Recovery of Zn} = \frac{C_2 c_{z2}}{Ff_Z} \times 100 = 85 \quad \Rightarrow \quad \frac{1500 c_{z2}}{10000 \times 10} \times 100 = 85$$

$$\Rightarrow \quad c_{Z2} = 56.67\%$$

Lead balance $\qquad C_1 c_{L1} + C_2 c_{L2} + T t_L = Ff_L$

$$\Rightarrow \qquad 1500 \times 79.33 + 1500 \times c_{L2} + 7000 \times 1 = 10000 \times 14$$

$$\Rightarrow \qquad c_{L2} = 9.34\%$$

Zinc balance $\qquad C_1 c_{Z1} + C_2 c_{Z2} + T t_Z = Ff_Z$

$$\Rightarrow \qquad 1500 \times c_{Z1} + 1500 \times 56.67 + 7000 \times 1.5 = 10000 \times 10$$

$$\Rightarrow \qquad c_{Z1} = 3\%$$

	%Pb	%Zn
Lead concentrate	79.33	3.0
Zinc concentrate	9.34	56.67

Example 16.2.5: *A large copper concentrator has the following metallurgical data:*

Ratio of concentration = 37.2:1.0

Head assay: $Cu = 0.968\%$
                        $Au = 0.01 \ oz/t$
                        $Ag = 0.10 \ oz/t$

Concentrate assay: $Cu = 31.5\%$
                        $Au = 0.25 \ oz/t$
                        $Ag = 2.65 \ oz/t$

Tailing assay: $Cu = 0.124\%$
                        $Au = 0.0034 \ oz/t$
                        $Ag = 0.03 \ oz/t$

Calculate recovery of Cu, Au, Ag.

Solution:

Notations as given in article 16.2 are followed

$$\text{Ratio of concentration} = K = \frac{F}{C} = 37.2$$

$$\% \text{ Recovery of Cu} = \frac{Cc}{Ff} \times 100 = \frac{1}{37.2} \times \frac{31.5}{0.968} \times 100 = 87.5\%$$

$$\% \text{ Recovery of Au} = \frac{Cc}{Ff} \times 100 = \frac{1}{37.2} \times \frac{0.25}{0.01} \times 100 = 67.2\%$$

$$\% \text{ Recovery of Ag} = \frac{Cc}{Ff} \times 100 = \frac{1}{37.2} \times \frac{2.65}{0.1} \times 100 = 71.2\%$$

16.3 SEPARATION EFFICIENCY

Metallurgical performance of a beneficiation operation is assessed by most widely accepted measures of concentrate grade and recovery. The beneficiation operation can be conducted in different ways to yield different combinations of grade and recovery. If both the grade and recovery are greater for a particular case than the other, then the choice of process is simple. But if the results of one test show a higher grade but a lower recovery than the other, then the choice is no longer obvious. There were many attempts to combine recovery and concentrate grade into a single index to represent the metallurgical efficiency of separation. After reviewing such attempts, Schulz [29] proposed an index called separation efficiency. It is defined as

$$\text{Separation efficiency} = R_V - R_G$$

where R_V is % recovery of the valuable mineral, and R_G is % recovery of the gangue into the concentrate.

$$\text{\% Recovery of valuable mineral in to the concentrate} = R_V = \frac{Cc}{Ff} \times 100$$

% Recovery of valuable mineral is equal to metal recovery, assuming that all the valuable metal is contained in the same mineral.

$$\text{The gangue content of the concentrate} = \left(100 - 100\frac{c}{m}\right)\% = \frac{m - c}{m} \times 100\%$$

where m is the % metal of the valuable mineral

$$\text{Therefore, } R_G = \frac{C}{F} \times \frac{\text{Gangue content of concentrate}}{\text{Gangue content of feed}} \times 100 = \frac{C}{F}\frac{m - c}{m - f} \times 100$$

$$R_V - R_G = \frac{Cc}{Ff} \times 100 - \frac{C}{F}\frac{m - c}{m - f} \times 100 = \frac{C}{F}\frac{m(c - f)}{f(m - f)} \times 100 = \frac{Cc}{Ff}\frac{m(c - f)}{c(m - f)} \times 100$$

$$\text{Separation efficiency} = R_V - R_G = \frac{Cc}{Ff}\frac{m(c - f)}{c(m - f)} \times 100 \qquad (16.3.1)$$

Example 16.3.1: *A concentrator treating copper ore of 1.86% Cu yields 9.3% concentrate with a grade of 18.54% Cu. The tailing assays 0.15% Cu. Determine separation efficiency. The only copper mineral present in the ore is chalcopyrite. The atomic weights of Copper, Iron and Sulphur are 63.55, 55.85 and 32.06 respectively.*

Solution:

Given

Assay of copper ore $= f = 1.86\%$ Cu
Assay of concentrate $= c = 18.54\%$ Cu
Assay of tailing $= t = 0.15\%$ Cu
Weight of concentrate $= C = 9.3\%$ of feed

Let the basis of calculation is 100 tons of feed ore

Chemical formula for Chalcopyrite is $CuFeS_2$

Molecular weight of Chalcopyrite $= 63.55 + 55.85 + 32.06 \times 2 = 183.52$

$$\% \text{ Cu in Chalcopyrite} = m = \frac{63.55}{183.52} \times 100 = 34.63$$

$$\text{Separation efficiency} = \frac{Cc}{Ff} \frac{m(c-f)}{c(m-f)} \times 100$$

$$= \frac{9.3 \times 18.54 \times 34.63 \times (18.54 - 1.86)}{100 \times 1.86 \times 18.54 \times (34.63 - 1.86)} \times 100 = 88.1\%$$

Example 16.3.2: *Data obtained in an Iron ore concentration operation is given in Table 16.3.2.*

Table 16.3.2 Data of Iron ore concentration operation.

	Quantity, tons	Assay value, %Fe
Feed	1,44,000	54
Concentrate	96,000	61
Tailing	48,000	40

What is the separation efficiency. Assume that Hematite is the only Iron mineral present in the ore. Atomic weights of Iron and Oxygen are 55.85 and 16.00 respectively.

Solution:

Chemical formula for Hematite is Fe_2O_3

Molecular weight of Hematite $= 55.85 \times 2 + 16.00 \times 3 = 159.7$

$$\% \text{ Fe in Hematite} = m = \frac{55.85 \times 2}{159.7} \times 100 = 69.94$$

$$\text{Separation efficiency} = \frac{Cc}{Ff} \frac{m(c-f)}{c(m-f)} \times 100$$

$$= \frac{96000 \times 61 \times 69.94 \times (61 - 54)}{144000 \times 54 \times 61 \times (69.94 - 54)} \times 100 = 37.9\%$$

Example 16.3.3: *An Iron ore contains 20% Fe. The ore is enriched around 55% to 65% Fe by using different unit operations as indicated below:*

1. by HMS, the grade obtained is 55% Fe with 35% yield
2. by tabling, the grade obtained is 60% Fe with 30% yield
3. by hydrocyclone, the grade obtained is 65% Fe with 25% yield

Which of the above three operations performs better.
Iron is totally contained in Hematite. Atomic weights of Iron and Oxygen are 55.85 and 16.00 respectively.

Solution:

Chemical formula of Hematite $= Fe_2O_3$

Molecular weight of Hematite $= 55.85 \times 2 + 16.00 \times 3 = 159.7$

$$\% \text{ Fe in Hematite} = m = \frac{55.85 \times 2}{159.7} \times 100 = 69.94$$

$f = 20.0\%$ Fe

$$\text{Separation efficiency} = \frac{Cc}{Ff} \frac{m(c-f)}{c(m-f)} \times 100 = \frac{Cc}{Ff} \times 100 \times \frac{m(c-f)}{c(m-f)}$$

by HMS, Separation efficiency $= 35.00 \times \dfrac{69.94(55.00 - 20.00)}{55.0(69.94 - 20.00)} = 31.19\%$

by Tabling, Separation efficiency $= 30.00 \times \dfrac{69.94(60.00 - 20.00)}{60.0(69.94 - 20.00)} = 28.00\%$

by Hydrycyclone,

Separation efficiency $= 25.00 \times \dfrac{69.94(65.00 - 20.00)}{65.00(69.94 - 20.00)} = 24.24\%$

Therefore, HMS performs better.

16.4 ECONOMIC EFFICIENCY

Separation efficiency, though useful in comparing the performance of different operating conditions on selectivity, takes no account of economic factors. A high value of separation efficiency does not necessarily lead to the most economic return. Hence the most economic combination of recovery and concentrate grade which produces the greatest financial return per ton of ore treated in the plant is to be determined. This depends primarily on the current price of the valuable product, transportation costs to the smelter, refinery or other further treatment plant, and the cost of such further treatment.

A high grade concentrate incurs lower smelting costs, but the lower recovery means lower returns on final product. A low grade concentrate may achieve greater recovery of the values, but incurs greater smelting and transportation costs due to the included gangue minerals.

The net smelter return (NSR) for any recovery-grade combination can be calculated as follows:

$$NSR = \text{Payment for contained metal} - (\text{Smelter charges} + \text{Transport costs})$$

A smelter purchases concentrates and smelts them to salable metal. The price paid by the smelter depends primarily on the market price of the metal. Deductions are made based on all costs involved in smelting including smelting losses. In few ores, certain constituents such as phosphorus in manganese ore are detrimental to the smelting process. In such cases, penalties are deducted in terms of dollars per unit per ton.

The following is the simplified lead smelter schedule:

Payments
Lead: Deduct 1.5 units from net Pb assay, and pay for 90% of remainder
        at published price
Silver: Deduct 30 gm and pay for 95% of remainder at published price.

Deductions
Treatment charge: $ 148 per ton of concentrate
Freight: $ 15 per ton of concentrate

As per the lead smelter schedule, net smelter return can be calculated. Example 16.4.1 illustrates the calculation of net smelter return per ton of concentrate.

Example 16.4.1:   *A lead concentrate is produced from an ore assaying 10% PbS. The concentrate assays 80% PbS. In addition, the lead concentrate contains 2 gm of silver per kilogram of PbS. What is the return per ton of concentrate as per the lead smelter schedule given?*

   Payments
      Lead: Deduct 1.5 units from net Pb assay, and pay for 90% of remainder at published price
      Silver: Deduct 30 gm and pay for 95% of remainder at published price.

   Deductions
      Treatment charge: $ 148 per ton of concentrate
      Freight: $ 15 per ton of concentrate

   (Metal prices: Pb = $ 1.8/kg; Ag = $ 0.58/gm; %Pb in PbS = 86.6%)

Solution:

Basis: 1 ton (1000 kg) concentrate

Payment for lead = Mass of lead × return

$$= 1000 \times \left[\frac{80}{100} \frac{86.6}{100} - \frac{1.5}{100} \right] \times 1.8 \times \frac{90}{100} = \$ 1098$$

Payment for silver = Mass of silver × return

$$= \left[1000 \frac{80}{100} \times 2 - 30 \right] \times 0.58 \times \frac{95}{100} = \$ 865$$

Treatment charges = 148

Freight charges = 15

Net Smelter Return = 1098 + 865 − 148 − 15 = \$ 1,800

Example 16.4.2 illustrates the savings in processing of ore by calculating net smelter return for the ore directly shipping to smelter and for the same ore shipping to smelter after processing.

Example 16.4.2:   *A copper-gold ore assays Au = 0.45 oz per ton (1 oz = 28.35 gm)*
Ag = 4.50 oz per ton
Cu = 2.20%

This ore can either be shipped directly to a smelter for treatment or processed in a concentrator and the concentrate be shipped to a smelter for treatment. In the concentrator, 95% of the copper is recovered in the concentrate.

The concentrate assays: Au = 4.85 oz per ton
Ag = 45.75 oz per ton
Cu = 25.0%

The ore or concentrate shipped to a smelter pay for the metals according to the following smelter schedule.

Gold: *If 0.03 oz Au per dry ton or over,*
Pay for 96.75% at net mint price
Silver: *If 1.0 oz Ag per dry ton or over,*
Pay for 95% at the average price for silver during the week following delivery at the smelter. There will be a minimum deduction of 1 oz of silver per ton treated.
Copper: *Deduct from the wet copper assay 1.3%*
Pay for the rest of the copper at the daily net export price quoted by LME for the week following receipt at the plant, less a deduction of \$ 0.09 per kg of copper paid for.

Charges: *Base smelting charge is $ 150 per ton*
Freight: *$ 15 per ton from mine to smelter*
Processing: *$ 120 per ton of crude ore*
Prices: *Cu $ 5/kg; Au $ 1,163/oz; Ag $ 16.5/oz*

Calculate the savings due to processing of 100 tons of crude ore.

Solution:

a) When the crude ore is directly shipped to smelter:

Gold content	$= 100 \times 0.45$	$= 45.0$ oz Au
Smelter pays for	$= 0.9675 \times 45$	$= 43.5375$ oz Au
Silver content	$= 100 \times 4.50$	$= 450$ oz Ag
Smelter pays for	$= 0.95(4.5 - 1)100$	$= 332.5$ oz Ag
Copper content	$= 0.022 \times 100 \times 1000$	$= 2200$ kg Cu
Smelter pays for	$= \dfrac{2.2 - 1.3}{100} \times 100 \times 1000$	$= 900$ kg Cu

  Payments

Gold	$43.5375 \times 1,163 =$	$ 50,634.1
Silver	332.5×16.5 $=$	5,486.3
Copper	$900(5 - 0.09) =$	4,419.0
Total		**60,539.4**

  Charges

Smelting	$100 \times 150 =$	15,000
Freight	100×15 $=$	1,500
Total		**16,500**

   NSR $60,539.4 - 16,500$ $= \$ 44,039.4$

b) When the crude ore is processed and then shipped to smelter:

Copper recovered $= 95\%$ $= 0.95 \times \dfrac{2.2}{100} \times 1000 = 20.9$ kg

Copper recovered from 100 tons of ore $= 20.9 \times 100$ $= 2090$ kg

As the concentrate assays 25% Cu, 2090 kg of copper is equal to 25% of concentrate

Weight of concentrate $= \dfrac{2090}{0.25} = 8360$ kg $= 8.36$ tons

Smelter pays for $= 8.36 \times \dfrac{25 - 1.3}{100} \times 1000 = 1981.32$ kg Cu

Ag in concentrate $= 45.75 \times 8.36$ $= 382.47$ oz Ag
Smelter pays for $= 0.95(45.75 - 1)8.36 = 355.40$ oz Ag

Au in concentrate $= 4.85 \times 8.36$ $= 40.55$ oz Au
Smelter pays for $= 0.9675 \times 40.55$ $= 39.23$ oz Au

Payments

Gold	$39.23 \times 1,163$	$= \$ 45,624.5$	
Silver	355.4×16.5	$=$	$5,864.1$
Copper	$1981.32 \times (5 - 0.09)$	$=$	$9,728.3$
	Total		**61,216.9**

Charges

Processing	100×120	$=$	$12,000.0$
Smelting	8.36×150	$=$	$1,254.0$
Freight	8.36×15	$=$	125.4
	Total		**13,379.4**
NSR	$= 61,216.9 - 13,379.4$	$= \$ 47,837.5$	

Savings due to processing $= 47,837.5 - 44,039.4 = \$ 3,798.1$

Net smelter return is maximum if the perfect separation is assumed. In case of perfect separation, all the valuable mineral content is liberated, and complete separation of the valuable mineral into the concentrate with all the gangue reporting to tailings would be possible. The **economic efficiency** is the ratio of net smelter return in actual separation to that of the net smelter return assuming perfect separation. Using economic efficiency, plant efficiencies can be compared even during periods of fluctuating market conditions.

Example 16.4.3 illustrates the calculation of economic efficiency for a lead concentrate containing silver metal of example 16.4.1.

Example 16.4.3: *A concentrator treats a lead ore assaying 10% PbS. The concentrate assays 80% PbS. In addition, the lead concentrate contains 2 gm of silver per kilogram of PbS. Calculate the economic efficiency of lead concentrator as per the following simplified lead smelter schedule:*

Simplified lead smelter schedule

Payments

Lead: Deduct 1.5 units from net Pb assay, and pay for 90% of remainder at published price

Silver: Deduct 30 gm and pay for 95% of remainder at published price.

Deductions

Treatment charge: $ 148 per ton
Freight: $ 15 per ton

Metal Prices: Pb = $ 1.8/kg; Ag = $ 0.58/gm; %Pb in PbS = 86.6%

Solution:

Basis: 1 ton (1000 kg) concentrate

Payment for lead   = Mass of lead × return

$$= 1000 \times \left[\frac{80}{100} \frac{86.6}{100} - \frac{1.5}{100} \right] \times 1.8 \times \frac{90}{100} = \$ 1,098$$

Payment for silver   = Mass of silver × return

$$= \left[1000 \times \frac{80}{100} \times 2 - 30 \right] \times 0.58 \times \frac{95}{100} = \$ 865$$

Treatment charges   = 148

Freight charges   = 15

Net Smelter Return   = 1098 + 865 − 148 − 15 = \$ 1,800

Assuming perfect separation, 100% recovery of lead,
The concentrate grade is 86.6% lead (i.e. pure galena)

Payment for lead   = Mass of lead × return

$$= 1000 \times \left[\frac{100}{100} \frac{86.6}{100} - \frac{1.5}{100} \right] \times 1.8 \times \frac{90}{100} = \$ 1378.6$$

Payment for silver   = Mass of silver × return

$$= \left[1000 \times \frac{100}{100} \times 2 - 30 \right] \times 0.58 \times \frac{95}{100} = \$ 1085.5$$

Treatment charges   = 148

Freight charges   = 15

Net Smelter Return = 1378.6 + 1085.5 − 148 − 15 = \$ 2,301.1

Economic efficiency = $\dfrac{1800}{2301.1} \times 100 = 78.2\%$

16.5   PROBLEMS FOR PRACTICE

16.5.1:   *In an Iron ore concentration operation, the data obtained is shown in Table 16.5.1:*

Table 16.5.1 Data of Iron ore concentration operation for problem 16.5.1.

	Quantity tons	Assay value %Fe
Feed	1,50,000	54
Concentrate	1,00,000	61
Tailing	50,000	40

Calculate recovery, ratio of concentration, and ratio of enrichment.
[75.3%, 1.5, 1.13]

16.5.2:   From a 6.0% lead ore milled at the rate of 300 tons/day is produced a concentrate assaying 70% lead and a tailing with 0.5% lead. Calculate the concentrate recovered, percentage of recovery and lead lost in tailing.

[23.74 tons/day, 92.32%, 1.381 tons/day]

16.5.3:   A concentrator is fed 1000 tons/hr of ore assaying 10% PbS. It produces a concentrate assaying 80% PbS and a tailings assaying 0.19% PbS. What are the flow rates of the tailings and concentrate streams?

[877.09 tons/hr, 122.91 tons/hr]

16.5.4:   A concentrate plant treating copper ore with 1.86% metal operates to give 186 tons of concentrate with a concentration ratio of 10.753. The loss of metal in the tailing is 0.15%. Calculate how many tons of ore the plant was treating and the copper percent in the concentrate. Also calculate the recovery of metal in the concentrate.

[2000 tons, 18.54%, 92.7%]

16.5.5:   9% Zn ore produced a concentrate of 41% Zn and a tailing of 1% Zn when beneficiated at the rate of 100 tons/hr. What is the quantity of zinc lost per day?

[19.2 tons/day]

16.5.6:   A copper concentrator is milling 15,000 tons/day of a chalcopyrite ore assaying 1.15% copper. The concentrate and tailings produced average 32.7% and 0.18% copper, respectively. Calculate the tonnages of concentrate and tailing and copper recovery.

[447.4 tons/hr, 14,552.6 tons/hr, 84.8%]

16.5.7:   A PbS concentrate is produced by a rougher-cleaner flotation circuit. The cleaner tailings assay 15% PbS are recycled to the rougher cells, and the circulating load (recycle/fresh feed) is 0.3. The fresh feed assays 10% PbS and is delivered at the rate of 1000 tons/hr. The recovery (PbS in cleaner concentrate/PbS in fresh feed) and grade in the concentrate are 95% and 85% respectively. What are the flow rates and assays of the other streams?

[411.8 t/hr, 111.8 t/hr, 888.2 t/hr, 300 t/hr, 34%, 0.55%]

16.5.8:   A lead zinc mill treats 150 tons/day of ore assaying 8% Pb and 12% Zn and produced a lead concentrate of grade 75% Pb, 4% Zn and zinc concentrate of 55% Zn, 1% Pb. The tailing assay is 0.4% Pb and 0.8% Zn. How much Pb and Zn concentrates are produced in the plant per day.

[15.04 tons/day, 30.11 tons/day]

16.5.9:   A copper zinc ore containing 7.7% Cu and 11.9% Zn is treated to produce a copper concentrate assaying 50% Cu and 5% Zn, a zinc concentrate assaying 10% Cu and 50% Zn and a tailing containing 1% Cu and 2% Zn. Calculate recovery of copper and zinc in copper and zinc concentrates respectively and ratio of concentration of copper and zinc.

[64.94%, 84.03%, 10, 5]

16.5.10:   *6.0% lead ore milled at the rate of 300 tons/day produces a concentrate of 70% lead and tailing of 0.5% lead. Calculate the separation efficiency if the lead is totally contained in Galena. Atomic weights of Lead and Sulphur are 207.19 and 32.06 respectively.*

[90.7%]

16.5.11:   *A tin concentrator treats tin ore of 1% tin. Concentrate grade and recovery combinations possible are three as shown below:*

High grade	63% tin at 62% recovery
Medium grade	42% tin at 72% recovery
Low grade	21% tin at 78% recovery

Tin is totally contained in the mineral cassiterite (SnO_2). Atomic weights of Tin and Oxygen are 118.69 and 16.00. Determine which of these combinations of grade and recovery produce the highest separation efficiency.

[Low grade]

16.5.12:   *A tin concentrator treats an ore grading 1% tin and produces a concentrate assaying 42% tin at 72% recovery. Cassiterite is the tin mineral. % tin in cassiterite is 78.8%. Calculate the economic efficiency under the conditions of the smelter contract shown below.*

Tin smelter contract

Material: *Tin concentrates, assaying no less than 15% Sn, to be free from deleterious impurities not stated, and to contain sufficient moisture as to evolve no dust when unloaded at our works.*
Quantity: *Total production of concentrates*
Valuation: *Tin, less 1 unit per dry ton of concentrates*
Pricing: *Tin price of £ 8500 per ton*
Treatment charge: *£ 385 per dry ton of concentrates*
Transportation: *£ 20 per ton of concentrate*

[67.2%]

16.5.13:   *The assay data collected from a copper-zinc concentrator is shown below.*

Feed	0.7% Cu,	1.94% Zn
Cu concentrate	24.6% Cu,	3.40% Zn
Zn concentrate	0.4% Cu,	49.70% Zn

Calculate the overall economic efficiency under the following simplified smelter terms:

Copper: *Copper price: £ 1000/ton*
       Smelter payment: 90% of Cu content
       Smelter treatment charge: £ 30/ton of concentrate
       Transport cost: £ 20/ton of concentrate

Zinc: Zinc price: £ 400/ton
Smelter payment: 85% of Zn content
Smelter treatment charge: £ 100/ton of concentrate
Transport cost: £ 20/ton of concentrate

Copper is totally contained in chalcopyrite and zinc is totally contained in sphalerite. Tailings does not contain either copper or zinc. %Cu in Chalcopyrite is 34.63%; %Zn in Sphalerite is 67.1%.

[78.7%]

Chapter 17

Coal washing efficiency

In a laboratory float and sink test, the separation is clean and accurate as sufficient time is given to allow complete separation to take place. The data from such tests indicate what should be obtained under ideal conditions of operation. Such conditions do not exist in plant practice. In a continuously operating process, with constant discharge of float and sink, particles having specific gravity nearer to the medium may not have time to be separated and will be misplaced into the other product. Particles of high or low specific gravity are least effected. The difficulty, or ease, of separation depends on the amount of the material present having specific gravity nearer to the medium. Conversely, the efficiency of a particular separating process depends on its ability to separate material of specific gravity close to that of the medium.

Different expressions for coal washing efficiency have been evolved to accurately reflect the quantitative and/or qualitative aspects of a washing operation. Evaluation of performance of a washer is based on two types of criteria namely dependent and independent.

17.1 DEPENDENT CRITERIA

The criteria that depend both on the washability characteristics of the coal and on the characteristics of the washing equipment are usually called dependent criteria. They include Organic efficiency, Anderson efficiency, Ash error and Yield error. Separation efficiencies or dependent criteria are used to evaluate different equipment for application to different reserves.

17.1.1 Organic efficiency

Organic efficiency, also called recovery efficiency, is suggested by Fraser and Yancey [30] and is defined as the ratio between the actual yield of a desired product and the theoretically possible yield at the same ash content. Normally it is expressed as a percentage as follows:

$$\text{Organic efficiency} = \frac{\text{Actual yield of washed coal}}{\text{Theoretical yield of floats of same ash content}} \times 100$$

(17.1.1)

Organic Efficiency cannot be used to compare the efficiencies of different plants, as it is dependent criteria, and is much influenced by the washability of the coal. It is possible, for example, to obtain a high organic efficiency on a coal containing little near gravity material, even when the separating efficiency, as measured by partition data is quite inefficient.

17.1.2 Anderson efficiency

Anderson efficiency [31] indicates the correctly placed material and is defined as

Anderson efficiency

$$= 100 - (\text{Float in refuse} + \text{sink in clean coal at the effective density of cut})$$

$$(17.1.2)$$

17.1.3 Ash error

Ash error is closely related to recovery efficiency and it is the difference between the ash content of clean coal and the theoretical ash content obtained from the washability data at the same yield. Therefore, the smaller the ash error the larger the yield for a given ash content of the clean product.

17.1.4 Yield error (or) yield loss

Yield error is the difference between the actual yield of the clean coal obtained in the washing unit and the theoretical yield to be obtained from the washability data at the ash content of the actual yield. The smaller the yield error, the more accurate is the separation.

Example 17.1.1: *Heavy medium separation operation, fed with a raw coal at the rate of 85 tons/hr, yields 47 tons/hr of coal of 18% ash. The float and sink test of raw coal is given in Table 17.1.1.1. Determine organic efficiency and yield loss.*

Table 17.1.1.1 Float and sink test of raw coal.

Sp. gr.	Wt% of floated coal	Ash% in floated coal
1.3	04.62	06.64
1.4	27.22	14.88
1.5	22.66	22.61
1.6	19.58	30.34
1.7	10.34	41.50
1.8	12.48	50.92
1.8 (sink)	03.10	60.54
	100.00	

Solution:

Cumulative wt% of coal floated and cumulative ash% are calculated and shown in Table 17.1.1.2.

Table 17.1.1.2 Calculated values for example 17.1.1.

				Cumulative floats		
Sp. gr.	Wt%	Ash%	Ash product	Wt%	Ash product	Ash%
1.30	04.62	06.64	30.6768	04.62	30.6768	06.64
1.40	27.22	14.88	405.0336	31.84	435.7104	13.68
1.50	22.66	22.61	512.3426	54.50	948.0530	17.40
1.60	19.58	30.34	594.0572	74.08	1542.1102	20.82
1.70	10.34	41.50	429.1100	84.42	1971.2202	23.35
1.80	12.48	50.92	635.4816	96.90	2606.7018	26.90
1.80 (sink)	03.10	60.54	187.6740	100.00	2794.3758	27.94

Total floats ash curve is drawn as in Figure 17.1.1.

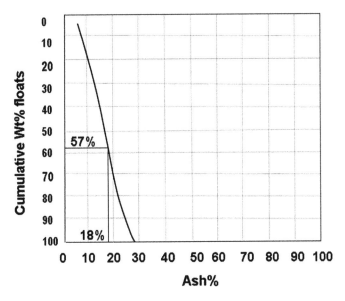

Figure 17.1.1 Total floats ash curve.

From the total floats ash curve, the yield at 18% ash = 57%.

The yield at 18% ash for feed tonnage of 85 tons/hr = 85 × 0.57

= 48.45 tons/hr

Actual yield in Heavy medium separation $= 47$ tons/hr

$$\text{Organic efficiency} = \frac{47}{48.45} \times 100 = 97\%$$

$$\text{Yield loss} = 48.45 - 47.0 = 1.45 \text{ tons/hr}$$

Example 17.1.2: *Raw coal with 30.4% ash has been washed by Heavy medium cyclone. It yielded a concentrate of 18% ash and rejects of 52% ash. If the theoretical recovery is 66.8% at the same ash level, determine the organic efficiency of the washery.*

Solution:

Given

Percent ash in feed $= f = 30.4\%$
Percent ash in clean coal $= c = 18\%$
Percent ash in tailing $= t = 52\%$

Let the feed to the washery be 100 tons.

Coal balance $F = C + T \Rightarrow \quad\quad 100 = C + T$
Ash balance $Ff = Cc + Tt \Rightarrow 100 \times 30.4 = 18C + 52T$

Solving these two equations gives $C = 63.5$ tons

The plant yield of clean coal $= 63.5$ tons $\Rightarrow 63.5\%$

$$\text{Theoretical recovery} = 66.8\%$$

$$\text{Organic efficiency} = \frac{63.5}{66.8} \times 100 = 95\%$$

Example 17.1.3: *A jigging plant of 350 TPD capacity treats the non coking coal of 45% ash to obtain clean coal of 34% ash. If the tonnage of the refuse material is 150 tons/day, what is the percent ash in refuge? Calculate the yield reduction factor.*

Solution:

Tonnage of coal fed $= F = 350$ tons/day
Tonnage of refuge obtained $= T = 150$ tons/day
Tonnage of clean coal obtained $= C = F - T = 350 - 150 = 200$ tons/day

Percent ash in feed $= f = 45\%$
Percent ash in clean coal $= c = 34\%$
Let percent ash in refuge $= t$

Ash balance $Ff = Cc + Tt$
$\Rightarrow \quad\quad\quad 350 \times 45 = 200 \times 34 + 150t$
$\Rightarrow \quad\quad\quad\quad\quad t = 59.7\%$

$\therefore$ Percent ash in refuge $= 59.7\%$

$$\text{Yield reduction factor} = \frac{F - C}{f - c} = \frac{350 - 200}{45 - 34} = 13.6$$

Example 17.1.4: *A coal washery cleans the coal stock by Dense medium separation using magnetite-water suspension as the medium. A sample of coal was tested using the float and sink experiment and the data obtained is shown in Table 17.1.4.1.*

Table 17.1.4.1 Float and sink experiment data for example 17.1.4.

Sp. gr.	Wt% of coal floated	Ash% in floated coal
1.287	5.52	0.91
1.314	19.16	3.17
1.341	16.22	8.39
1.377	20.58	12.75
1.423	19.12	14.67
1.461	9.97	21.44
1.461 (sink)	9.43	27.73

If the ultimate ash content of clean coal should not exceed 6%, determine the specific gravity of the medium to be used and the percent recovery.

Solution:

Cumulative wt% and cumulative ash% are calculated and shown in the Table 17.1.4.2.

Table 17.1.4.2 Calculated values for example 17.1.4.

Sp. gr.	Wt% of coal floated	Ash% in floated coal	Cum. wt% coal floated	Ash product	Cum. Ash product	Cum ash%
1	2	3	4	$5 = 2 \times 3$	6	$7 = 6/4$
1.287	5.52	0.91	05.52	05.02	05.02	0.91
1.314	19.16	3.17	24.68	60.74	65.76	2.67
1.341	16.22	8.39	40.90	136.09	201.85	4.94
1.377	20.58	12.75	61.48	262.40	464.25	7.55
1.423	19.12	14.67	80.60	280.49	744.74	9.24
1.461	9.97	21.44	90.57	213.76	958.50	10.58
1.461 (sink)	9.43	27.73	100.0	261.49	1219.99	12.19

Total floats ash curve and yield gravity curve are drawn and shown in Figure 17.1.4.

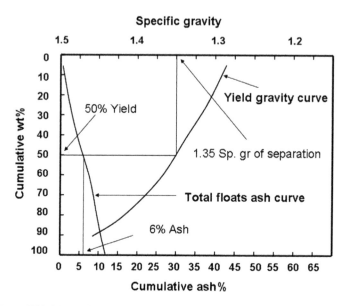

Figure 17.1.4 Total floats ash curve & Yield gravity curve for example 17.1.4.

From the graph, Yield is read corresponding to 6% ash and it is 50%. The specific gravity of separation for this 50% yield is read as 1.35.

∴   Specific gravity of the medium to be used is 1.35.

Percent recovery is 50%.

Example 17.1.5: *A coal having float and sink analysis shown in Table 17.1.5.1 has been washed in a Heavy Medium Separation Unit at 1.50 specific gravity.*

Table 17.1.5.1 Float and sink analysis data for example 17.1.5.

Sp. gr.	Wt% coal	Ash%
1.30 float	9.5	5.0
1.30–1.35	9.7	8.0
1.35–1.40	13.4	10.1
1.40–1.45	17.8	16.8
1.45–1.50	17.2	22.5
1.50–1.55	6.0	25.9
1.55–1.60	3.4	35.2
1.60–1.70	3.3	40.0
1.70–1.80	2.8	46.4
1.80–1.90	1.5	51.5
1.90–2.00	2.0	65.9
2.00 sink	13.4	79.8

Calculate the ash percent of feed coal. What is the yield of clean coal that can be expected from HMS? What would be the ash percent of clean coal and rejects?

Solution:

Cumulative wt% of floats for whole coal, Cumulative wt% of floats from 1.30 floats to 1.50 floats, from 1.55 floats to 2.00 sinks, Ash product values for whole coal, Cumulative ash product upto 1.50 floats and 1.50 floats to 2.00 sinks are also calculated and shown in the Table 17.1.5.2.

Table 17.1.5.2 Calculated values for example 17.1.5.

Specific gravity	Wt% of float coal	Ash %	Cum wt%	Cum wt%	Ash product	Cum. ash product
1.30	9.5	5.0	9.5		47.50	
1.35	9.7	8.0	19.2		77.60	
1.40	13.4	10.1	32.6		135.34	
1.45	17.8	16.8	50.4		299.04	
1.50	17.2	22.5	67.6	67.6	387.00	946.48
1.55	6.0	25.9	73.6		155.40	
1.60	3.4	35.2	77.0		119.68	
1.70	3.3	40.0	80.3		132.00	
1.80	2.8	46.4	83.1		129.92	
1.90	1.5	51.5	84.6		77.25	
2.00	2.0	65.9	86.6		131.80	
2.00 (sink)	13.4	79.8	100.0	32.4	1069.32	1815.37
	100.0				2761.85	

Ash percent in coal sample $= 2761.85/100 = 27.62\%$
Ash percent in clean coal $= 946.48/67.6 = 14.00\%$
Ash percent in rejects $= 1815.37/32.4 = 56.03\%$
Yield of clean coal $= 67.60\%$

17.2 INDEPENDENT CRITERIA

The criteria that depend on the characteristics of the washing unit and independent of the washability characteristics of the coal are called Independent criteria. Independent criteria are also referred as the sharpness of separation criteria or equipment performance measures. They include **Probabale Error, Error Area** and **Imperfection**. Sharpness of separation criteria may be used to evaluate different equipment for application to a specific coal reserve.

In commercial coal cleaning, coal of specific gravity well below the specific gravity of separation and impurities of specific gravity well above the specific gravity of separation report largely (or entirely) to their proper products, clean coal and refuse, respectively. As coal approaches the specific gravity of separation, however, more and more material tends to report to an improper product. Finally, an infinitesimal increment of coal, at the specific gravity of separation, is divided equally between the clean coal and refuse. The specific gravity of this increment is defined as the specific gravity of separation.

A bar chart showing the percentage of each specific gravity fraction of the feed coal that was recovered in clean coal was devised in 1912 by Hancock. In 1938 Tromp [32] and Fraser and Yancey has independently developed a curve which is plotted between percentage recoveries and mean specific gravities of the specific gravity fractions. This curve has several names such as **Tromp Curve, Partition curve, Distribution curve, Recovery curve,** and **Error curve.** The **Tromp Curve** is used to assess the sharpness of separation or to predict the performance of a washing unit. It is the curve drawn between **Partition coefficient** and the mean or nominal specific gravity for each specific gravity range. Partition coefficient is the percentage of feed coal of a certain nominal specific gravity which reports to floats.

To construct the Tromp curve, three sets of coal analytical data are required:

1 recovery of clean coal
2 float and sink washability analysis of the clean coal product
3 float and sink washability analysis of the refuse product

The recovery of clean coal can be determined either by direct weight or it can be predicted by using material balance equations, if the ash percent of feed coal, clean coal and refuse are known.

Let A_f, A_c and A_r are the ash percentages of feed coal, clean coal and refuse respectively. $Y_f, Y_c,$ and Y_r are the recovery or Yield percent of feed coal (100%), clean coal and refuse respectively. Then

$$\text{Yield balance equation} \quad Y_c + Y_r = Y_f \tag{17.2.1}$$
$$\text{Ash balance equation} \quad A_c \cdot Y_c + A_r \cdot Y_r = A_f \cdot Y_f \tag{17.2.2}$$

Solving these two equations

$$\Rightarrow \qquad Y_c\% = \frac{\left(A_r - A_f\right)}{\left(A_r - A_c\right)} \times 100 \tag{17.2.3}$$

i.e., Yield percentage of clean coal $= \dfrac{\text{Ash \% in refuse} - \text{Ash \% in feed}}{\text{Ash \% in refuse} - \text{Ash \% in clean coal}} \times 100$

Let an example be taken where ash percent of feed coal, clean coal and refuse are 27.60, 13.80 and 48.65 respectively. Then percent clean coal obtained is

$$Y_c\% = \frac{\left(A_r - A_f\right)}{\left(A_r - A_c\right)} \times 100 = \frac{(48.65 - 27.60)}{(48.65 - 13.80)} \times 100 = 60.4\%$$

Yield percentage of refuse coal $= Y_r\% = Y_f - Y_c = 100 - 60.4 = 39.6\%$

For obtaining float and sink washability analyses of clean coal and refuse product, representative samples of clean coal and refuse are collected separately from the washing unit after the machine has come to the normal operating state. The individual samples of clean coal and refuse are subjected to complete float and sink analysis in the same way as a sample of raw coal is tested for the evaluation of its washability

Table 17.2.1 Float and sink analyses of clean coal and refuse.

Sp. gr.	A Clean coal analysis Wt%	B Refuse analysis Wt%
1.30 floats	15.7	0
1.30–1.35	15.9	0.3
1.35–1.40	21.0	1.8
1.40–1.45	24.2	8.1
1.45–1.50	17.9	16.1
1.50–1.55	3.8	9.3
1.55–1.60	1.2	6.8
1.60–1.70	0.3	7.8
1.70–1.80	0	7.1
1.80–1.90	0	3.8
1.90–2.00	0	5.1
2.00 sink	0	33.8
	100.0	100.0

characteristics. Table 17.2.1 shows the float and sink data of clean coal and refuse products.

The percent weights of the fractions totaling 100 (in case of both clean coal and refuse) as shown in Table 17.2.1 are then expressed as percent weights of the feed coal by multiplying the individual weights by the actual yield percentages of clean coal (60.4%) and refuse (39.6%) and dividing the resultant figures by 100.

Reconstituted feed is then calculated by adding values of percent weights of feed coal for clean coal and refuse at each specific gravity range. Nominal specific gravity is determined by averaging lowest and highest specific gravity at each specific gravity range.

The partition coefficient is calculated as fraction of percent weights of feed coal for clean coal to the reconstituted feed at each nominal specific gravity and multiplied the resultant figures with 100. The partition coefficient can also be calculated as fraction of percent weights of feed coal for refuse to the reconstituted feed at each nominal specific gravity and multiplied the resultant figures with 100. Then the Tromp curve can be constructed by plotting the partition coefficient against the nominal specific gravity.

Table 17.2.2 shows calculated values for drawing Tromp Curve for 60.4% yield of clean coal and 39.6% yield of refuse.

While drawing Tromp curve, assumptions are required in plotting the lightest and heaviest fractions because they have no exact limiting specific gravities. If 1.30 is the lowest specific gravity used in the analysis, as frequently is the case, the point for the float should be plotted at a specific gravity that is midway between that of the lightest particle present and 1.30. Specific gravity of 1.26–1.28 is generally used. Any error involved in making this assumption generally has very little influence on the shape and position of the curve; it becomes important only when the specific

Table 17.2.2 Calculated values for drawing Tromp Curve.

	A	B	C	D	E	F	G	H
Sp. gr.	Clean coal analysis Wt%	Refuse analysis Wt%	Clean coal % of feed 0.604A	Refuse % of feed 0.396B	Reconstituted feed % C+D	Nominal Sp. gr.	Partition coefficient for clean coal $\frac{C}{E} \times 100$	Partition coefficient for refuse $\frac{D}{E} \times 100$
1.30 floats	15.7	0	9.5	0	9.5		100.0	0
1.30–1.35	15.9	0.3	9.6	0.1	9.7	1.325	99.0	1.0
1.35–1.40	21.0	1.8	12.7	0.7	13.4	1.375	94.8	5.2
1.40–1.45	24.2	8.1	14.6	3.2	17.8	1.425	82.0	18.0
1.45–1.50	17.9	16.1	10.8	6.4	17.2	1.475	62.8	37.2
1.50–1.55	3.8	9.3	2.3	3.7	6.0	1.525	38.3	61.7
1.55–1.60	1.2	6.8	0.7	2.7	3.4	1.575	20.6	79.4
1.60–1.70	0.3	7.8	0.2	3.1	3.3	1.65	6.1	93.9
1.70–1.80	0	7.1	0	2.8	2.8	1.75	0	100
1.80–1.90	0	3.8	0	1.5	1.5	1.85	0	100
1.90–2.00	0	5.1	0	2.0	2.0	1.95	0	100
2.00 sink	0	33.8	0	13.4	13.4		0	100
	100.0	100.0	60.4	39.6	100.0			

gravity of separation is unusually low. If the highest specific gravity is 1.80 in the analysis, the sink is usually plotted at 2.20–2.30, depending on what is known about its composition.

Figure 17.2.1.1 shows the Tromp Curve drawn for the partition coefficient relates to clean coal in which case it will decrease with increasing specific gravity. Similarly, Tromp Curve can also be drawn for the partition coefficient relates to refuse (Fig. 17.2.1.2) in which case it will increase with increasing specific gravity.

The Tromp curve of a particular washer remains unchanged with varying composition of raw coal provided that the feed coal having the same size grading are treated nearly at the same specific gravity of cut under more or less similar load conditions. The shape and gradient of the Tromp curve is a measure of the sharpness of separation of a washer. The steeper the slope, more sharp the separation. For an ideal separation, the misplaced materials in either of the products (clean coal and refuse) being zero, the curve is represented by a straight vertical line. It shows that all the particles lighter than separating specific gravity report to floats while all the heavier particles report to sinks. On the other hand, a Tromp curve tending to be horizontal throughout its length represents an extreme case where practically no separation is achieved. However, most of the Tromp curves remain in between these two extremes.

The Tromp curve for real separation shows that efficiency is highest for particles of specific gravity far from the operating specific gravity and decreases for particles approaching the operating specific gravity. The shape of the tails or loops at the extremities of the Tromp curve indicate the amount of high ash refuse particles and low ash coal particles misplaced in the clean coal and refuse products respectively. The shorter the length of the tails, the better is the efficiency of separation.

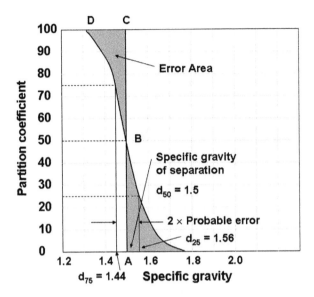

Figure 17.2.1.1 Tromp Curve relates to Clean coal.

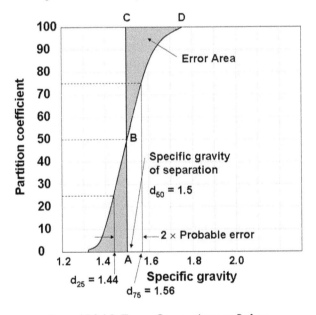

Figure 17.2.1.2 Tromp Curve relates to Refuse.

The value of the ordinate corresponding to any point on the abscissa gives the probability that a particle will be found in the sinks. 100 minus this value gives the probability that a particle will be found in floats.

The value of the specific gravity corresponding to partition coefficient of 50% is called the **partition specific gravity (partition density)** or **Effective specific gravity** of separation or Tromp cut point. At this specific gravity, a particle has an equal chance of going to either clean coal fraction or refuse fraction. Since the partition density is obtained from the curve, it need not coincide with any density in practice or with the operating density. All most all Tromp curves gives a reasonable straight line relationship between partition coefficient values of 25% and 75% and the slope of the line between these two values is used to indicate the accuracy of separation (or efficiency of separation) process.

It is found that partition curve is a function of particle size. Larger particles give closer separation. Hence coals of the same grading should be used when comparing the performance of different washers.

In Figure 17.2.1.1, the vertical straight line ABC represents the theoretical separating specific gravity. The actual separation is represented by the curve DBE.

It is common practice to use parameters which is derived from the Tromp curve, instead of using the whole Tromp curve for measuring the technical efficiency of a cleaning unit. The Ecart probable Moyen, Error area and Imperfection are widely used to describe the characteristics of the Tromp curve.

17.2.1 Probable error (or) Ecart Probable Moyen (E_p)

The Probable error of separation or the Ecart Probable Moyen (E_p) is a guide to the efficiency of separation of a washing unit. It represents the sharpness with which the coal and impurities are separated. It is a measure of the deviation of the Tromp curve from a perfect separation. It is defined as half the difference between the specific gravities corresponding to partition coefficient values of 25% and 75% and is calculated as:

$$E_p = \frac{1}{2}(d_{25} - d_{75}) \tag{17.2.4}$$

Generally, the steeper the Tromp curve, the lower the E_p value. The lower the E_p value, the smaller the difference in specific gravity between the 25% and 75% partition coefficient. The line is nearer to vertical between 25% and 75% partition coefficient, and hence the more efficient is the separation. Conversely a higher E_p value indicates a wider spread of specific gravity and a less efficient separation. In other words, a low E_p denotes a sharp separation, and a high E_p denotes a separation that is not sharp. An ideal separation has a vertical line with an $E_p = 0$. The usual range of E_p is from about 0.020 to 0.30 or more.

For the example considered, $d_{25} = 1.56$ and $d_{75} = 1.44$ from Figure 17.2.1.1.

Ecart Probable Moyen $= E_p = \frac{1}{2}(d_{25} - d_{75}) = \frac{1}{2}(1.56 - 1.44) = 0.06$

The E_p is not commonly used as a method of assessing the efficiency of separation in units such as tables, spirals, cones etc., due to many operating variables like wash water, table slope, speed etc. which can effect the separation efficiency. It is however, ideally suited to the relatively simple and reproducible Heavy Medium Separation process.

17.2.2 Error area or Tromp area

Error area is a measure of the sharpness of separation between clean coal and refuse. It is defined as the area between the actual Tromp curve and the theoretically perfect Tromp curve. In Figure 17.2.1.1, the area DBC represents the true floats that have reported to the sinks while the area ABE represents the true sinks that have reported to the floats. These two areas are not necessarily equal. These areas are called Error areas. These areas represent deviation from ideality. For a theoretically perfect separation, the error area is zero.

17.2.3 Imperfection

As Probable error depends on relative specific gravities, the imperfection is used as a further method of comparing separation processes. Imperfection involves the influence of the specific gravity of separation on the shape of the Tromp curve. It has been observed that there is a tendency for the curve to steepen as the specific gravity of separation decreases. In other words, separations at low specific gravity tend to be sharper than those at high specific gravity.

This concept contradicts the well-established principle that efficiency increases with increase in the specific gravity of separation. In an effort to develop criteria for sharpness of separation that would be independent of the specific gravity of separation, the term Imperfection (I) was originated and it is calculated as

$$I = \frac{E_p}{d_{50} - 1} \quad \text{for Jigs} \qquad (17.2.5)$$

$$I = \frac{E_p}{d_{50}} \quad \text{for Heavy medium baths} \qquad (17.2.6)$$

For the example considered, $E_p = 0.06$ and $d_{50} = 1.50$ from Figure 17.2.1.1.

$$\text{Imperfection} = I = \frac{E_p}{d_{50}} = \frac{0.06}{1.50} = 0.04$$

Imperfection is a numerical figure that characterizes a particular cleaning device regardless of the separating gravity. It is the coefficient to be preferred in expressing the performance of a washer.

All the three measures of separation, Ecart probable Moyen, Error area and Imperfection, have been developed due to cumbersomeness and inconvenience of using Tromp curve for evaluation of efficiencies. These measures of separation are defined as sharpness of separation criteria.

Table 17.2.3 shows the values of Effective density of separation (d_{50}), Ecart probable Moyen (E_p) and Imperfection (I) for different types of processes.

Table 17.2.3 Values of Independent criteria for different washing units.

Process Unit	d_{50}	E_p	I
Heavy medium bath	1.50	0.02	0.04
Heavy medium cyclone	1.60	0.03	0.05
Jig (25 mm × 10 mm)	1.65	0.06	0.09
(10 mm × 3 mm)	1.70	0.10	0.14
Water washing cyclone	1.65	0.15	0.23

From the data of Table 17.2.3, it can be seen that Heavy medium processes are more efficient than other coal cleaning methods.

Example 17.2.1: *Representative coking coal samples from the overflow and under-flow streams of a 600-mm dense-medium cyclone circuit were subjected to float and sink analysis and they are shown in Table 17.2.1.1.*

Table 17.2.1.1 Float and sink analyses of clean coal and refuse for example 17.2.1.

	Clean coal	Refuse
Sp. gr.	Wt%	Wt%
1.30 floats	13.02	1.15
1.30–1.40	57.21	4.28
1.40–1.50	26.09	21.43
1.50–1.60	3.22	20.72
1.60–1.70	0.35	9.02
1.70–1.80	0.12	6.22
1.80 sinks	0	37.18
	100.00	100.00

90% of the feed is reported to overflow. Draw the Tromp curve and determine specific gravity of separation, Ecart probable and Imperfection for this dense medium cyclone.

Solution:

Necessary calculations for drawing Tromp curve were done and shown in the Table 17.2.1.2. Tromp curve is shown in Figure 17.2.1.3.

Table 17.2.1.2 Calculated values for Tromp curve for example 17.2.1.

Sp. gr.	Cleans Wt %	Refuse Wt%	Cleans % feed B × 0.9	Refuse % feed C × 0.1	Calculated feed D + E	Mean Sp. gr.	Partition coefficient for cleans $\frac{D}{F} \times 100$
A	B	C	D	E	F	G	H
1.30 floats	13.02	1.15	11.71	0.12	11.83	1.26	99.0
1.30–1.40	57.21	4.28	51.49	0.43	51.92	1.35	99.2
1.40–1.50	26.09	21.43	23.48	2.14	25.62	1.45	91.7
1.50–1.60	3.22	20.72	2.90	2.07	4.97	1.55	58.4
1.60–1.70	0.35	9.02	0.31	0.90	1.21	1.65	25.6
1.70–1.80	0.12	6.22	0.11	0.62	0.73	1.75	15.1
1.80 sinks	0.00	37.18	0.00	3.72	3.72	2.20	0
			90.00	10.00			

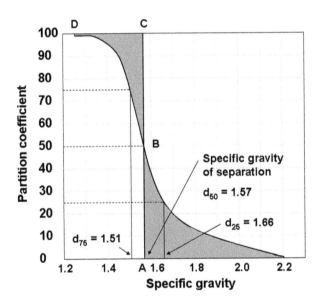

Figure 17.2.1.3 Tromp curve for example 17.2.1.

From the Tromp curve, specific gravity of separation, $d_{50} = 1.57$

$$d_{75} = 1.51$$

$$d_{25} = 1.66$$

$$\text{Ecart Probable Moyen} = E_p = \frac{1}{2}(d_{25} - d_{75}) = \frac{1}{2}(1.66 - 1.51) = 0.075$$

$$\text{Imperfection} = I = \frac{E_p}{d_{50}} = \frac{0.075}{1.57} = 0.048$$

Example 17.2.2: *A Coal with 49.9% ash was treated in a Baum Jig to obtain clean coal and refuse. Samples from clean coal and refuse were collected and analysed for their ash and found as 35.8% and 65.1% respectively. Float and sink analyses of clean coal and refuse are shown in Table 17.2.2.1.*

Table 17.2.2.1 Float and sink analyses of clean coal and refuse for example 17.2.2.

Sp. gr.	Floats of Clean coal	Floats of Refuse
1.40	12.50	0.50
1.45	9.40	0.60
1.50	9.50	0.80
1.55	12.40	1.00
1.60	12.00	1.50
1.65	9.00	2.50
1.70	6.90	4.00
1.80	10.70	11.30
1.90	4.90	8.40
2.00	5.50	12.40
2.10	3.40	11.50
2.20	3.80	45.50
	100.00	100.00

Draw the Tromp curve and determine specific gravity of separation, Ecart probable and Imperfection for this Baum jig operation.

Solution:

$$\text{Ash \% in feed } = 49.9$$
$$\text{Ash \% in cleans} = 35.8$$
$$\text{Ash \% in refuse} = 65.1$$

$$\text{Yield percentage of cleans} = \frac{\text{Ash \% in refuse} - \text{Ash \% in feed}}{\text{Ash \% in refuse} - \text{Ash \% in clean coal}} \times 100$$

$$= \frac{65.1 - 49.9}{65.1 - 35.8} \times 100 = 51.90$$

Percentage of refuse $= 100 - 51.90 = 48.10$

The necessary calculations for drawing Tromp curve were done and shown in the Table 17.2.2.2.

Table 17.2.2.2 Calculated values for Tromp curve for example 17.2.2.

Sp. gr.	Floats of Cleans	Floats of Refuse	Cleans %feed	Refuse %feed	Calculated feed	Mean Sp. gr.	Partition Coefficient for cleans
1.40	12.50	0.50	6.49	0.24	6.73	1.400	96.43
1.45	9.40	0.60	4.88	0.29	5.17	1.425	94.39
1.50	9.50	0.80	4.93	0.38	5.31	1.475	92.84
1.55	12.40	1.00	6.44	0.48	6.92	1.525	93.06
1.60	12.00	1.50	6.23	0.72	6.95	1.575	89.64
1.65	9.00	2.50	4.67	1.20	5.87	1.625	79.56
1.70	6.90	4.00	3.58	1.92	5.50	1.675	65.09
1.80	10.70	11.30	5.55	5.44	10.99	1.750	50.50
1.90	4.90	8.40	2.54	4.04	6.58	1.850	38.60
2.00	5.50	12.40	2.85	5.96	8.81	1.950	32.35
2.10	3.40	11.50	1.76	5.53	7.29	2.050	24.14
2.20	3.80	45.50	1.98	21.90	23.88	2.150	08.29
	100.00	100.00	51.90	48.10	100.00		

Figure 17.2.2 shows the Tromp curve drawn with the values of the Table 17.2.2.2.

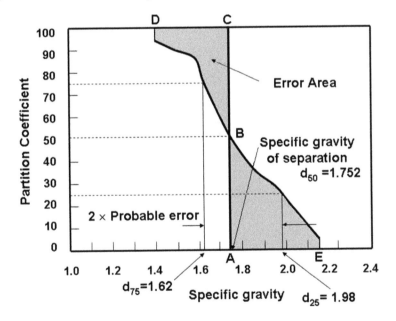

Figure 17.2.2 Tromp curve for example 17.2.2.

From the Tromp curve, specific gravity of separation, $d_{50} = 1.752$

$$d_{75} = 1.62$$
$$d_{25} = 1.98$$

Ecart Probabale Moyen $E_p = \frac{1}{2}(d_{25} - d_{75}) = \frac{1.98 - 1.62}{2} = 0.18$

Imperfection $I = \frac{E_p}{d_{50} - 1} = \frac{0.18}{1.752 - 1.0} = 0.239$

17.3 PROBLEMS FOR PRACTICE

17.3.1: *Raw coal feed to a washery has the float and sink analysis as shown in Table 17.3.1.*

Table 17.3.1 Float and sink analysis of raw coal for problem 17.3.1.

Sp. gr.	Yield%	Ash%
below 1.25	1.5	2.6
+1.25–1.35	25.9	7.3
+1.35–1.45	38.4	20.0
+1.45–1.55	15.2	26.7
+1.55–1.65	7.7	37.1
+1.65–1.75	2.2	43.2
above 1.75	9.1	57.1

On washing in a plant, a clean coal of 18% ash at an yield of 82% is obtained. Estimate the performance of the whashery.

[97.6%]

17.3.2: *Assuming 100% efficiency of a three product chance cone, determine yields of clean coal, middling and rejects if the coal, having the float and sink analysis shown in Table 17.3.2, is washed employing medium specific gravities of 1.5 and 1.7. Also calculate their ash percentages.*

Table 17.3.2 Float and sink analysis for problem 17.3.2.

Specific gravity	Wt% of coal floated	Ash%
1.40	12.35	15.28
1.50	34.96	22.08
1.60	25.04	33.17
1.70	3.45	42.37
1.80	13.06	48.69
1.90	2.61	53.15
1.90 (sink)	8.53	76.03

[47.31%, 28.49%, 24.20%, 20.31%, 34.28%, 58.81%]

17.3.3: *Calculate yield percent of clean coal, when a coal of 31% ash is treated in a froth flotation cell to obtain a froth of 15% ash. Ash percent of tailing is 55%.*

[60%]

17.3.4: *Coking coal of 22% ash is fed to a washery at the rate of 320 tons/day yielding 65% of clean coal with 15% ash. The tailings are further washed in order to obtain a middling for use in power plant. Equal tonnage of middling and final tailing is obtained. If the tailing contains 50% ash, what is the ash% of middling?*

[20%]

17.3.5: *When a coal of 18.65% ash is beneficiated in a heavy medium bath, it yields clean coal of 12.20% ash and refuse of 49.26% ash. Representative samples from clean coal and refuse are taken and subjected to float and sink analysis. The results are shown in Table 17.3.5.*

Table 17.3.5 Float and sink analyses of clean coal and refuse for problem 17.3.5.

Sp. gr.	Clean coal analysis Wt%	Refuse analysis Wt%
−1.30	83.34	18.15
1.30–1.40	10.50	10.82
1.40–1.50	03.35	09.64
1.50–1.60	01.79	13.33
1.60–1.70	00.30	08.37
1.70–1.80	00.16	05.85
1.80–1.90	00.09	05.05
1.90–2.00	00.07	40.34
+2.00	00.40	24.45

Draw Tromp curve and determine specific gravity of separation, Ecart probable and Imperfection for heavy medium bath.

[1.50, 0.11, 0.073]

Chapter 18

Process plant circuits

Varieties of circuits used in the plant have been considered in the previous chapters. However, water requirements in the circuits are not considered. Calculation of water requirements in the circuits is of vital importance. In this chapter, water requirements have been calculated in different circuits and complete material balance is illustrated for those circuits. Examples relating to contact time requirements for the flotation to complete, different examples on analysis and washability are also given.

First and foremost calculation essential in many process plants is the calculation of quantities of different ores obtained from different mines of varying qualities. In order to maintain consistent quality of the ore, this calculation is of prime importance for the process to complete as designed. For example, a process plant obtains iron ore from the mines A, B, C, D and E. The respective iron percents are shown as follows:

Mine	% Fe
A	61.11
B	54.89
C	57.09
D	56.76
E	59.82

The ore from five mines are blended according to the quantities mentioned in column 3 as follows:

Mine	% Fe	Tons	Iron content
A	61.11	250	15277.5
B	54.88	25	1372.0
C	57.08	50	2854.0
D	56.76	50	2838.0
E	59.82	25	1495.5
		450	23837.0

Then the average analysis of the blended ore is 23837/400 = 59.59% Fe.

Let us assume that at one time, supply of mines D and E are stopped. Then quantities of ore from mines A, B and C are to be adjusted in order to get blended ore more or less of 59.59% Fe. This can be done by trial and error. For an assumed value of 300 tons from A and 60 tons from C, about 66 tons of B is required to give 59.58% Fe for the blended ore.

Let us assume that after few days, ore from two new mines are supplied with 58.26% Fe and 57.44% Fe. Now the quantities of ore from five mines are to be calculated by trial and error to get the same 59.59% Fe to maintain the consistency of ore being fed to the process plant.

18.1 CIRCUITS WITH COMPLETE MATERIAL BALANCE

Example 18.1.1: *A two stage grinding circuit using a rod mill in open circuit and ball mill in closed circuit with a hydrocyclone classifier is used to grind 100 tons/hr of ore (sp. gr. 3.0). The circuit diagram is shown in Fig. 18.1.1.*

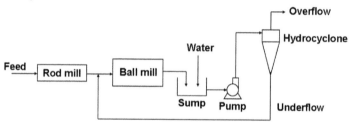

Figure 18.1.1 Closed circuit grinding diagram for example 18.1.1.

The circuit is sampled and analysed for −200 mesh material and percent solids. The results are given in Table 18.1.1.

Table 18.1.1 Results of tests for samples for example 18.1.1.

	% −200 mesh	% solids
Rod mill discharge	5	75.18
Ball mill discharge	20	78.33
Cyclone overflow	30	40.00
Cyclone underflow	15	80.00

(i) *Carryout a material balance and calculate the tons/hr of dilution water added to the sump*

(ii) *Calculate the flowrates of slurry, solids and water of all streams and specific gravities of all streams and illustrate the material balance of the whole circuit*

Solution:

Let F = Flow rate of feed solids to hydrocyclone

P = Flow rate of overflow solids from hydrocyclone

= Flow rate of solids fed to rod mill

U = Flow rate of underflow solids from hydrocyclone

Solids balance over hydrocyclone

$$F = P + U \quad \Rightarrow \quad F = 100 + U$$

-200 mesh material balance over hydrocyclone

$$Ff = Pp + Uu \quad \Rightarrow \quad (100 + U)20 = 100 \times 30 + U \times 15$$
$$\Rightarrow \qquad\qquad\qquad U = 200 \text{ tons/hr}; \quad F = 300 \text{ tons/hr}$$

Slurry balance over hydrocyclone

Flow rate of slurry fed to hydrocyclone
$$= \text{Flow rate of hydrocylone overflow slurry}$$
$$+ \text{Flowrate of hydrocyclone underflowslurry}$$

$$\Rightarrow \quad \frac{300}{\text{Fraction of solids in feed slurry}}$$

$$= \frac{100}{\text{Fraction of solids in overflow slurry}}$$

$$+ \frac{200}{\text{Fraction of solids in underflow slurry}}$$

$$= \frac{100}{0.40} + \frac{200}{0.80} = 250 + 250 = 500$$

$$\Rightarrow \quad \text{Fraction of solids in feed slurry to hydrocyclone} = \frac{300}{500} = 0.60$$

Flow rate of slurry fed to ball mill
$$= \text{Flow rate of slurry discharged from rod mill}$$
$$+ \text{Flow rate of hydrocyclone underflow slurry}$$

$$= \frac{100}{\text{Fraction of solids in rod mill slurry}}$$

$$+ \frac{200}{\text{Fraction of solids in underflow slurry}}$$

$$= \frac{100}{0.7518} + \frac{200}{0.80} = 133 + 250 = 383 \text{ tons/hr}$$
$$= \text{Flow rate of slurry discharged from ball mill}$$

Dilution water to be added to the sump
$$= \text{Flow rate of slurry fed to hydrocyclone}$$
$$- \text{Flow rate of slurry discharged from ball mill}$$
$$= 500 - 383 = 117 \text{ tons/hr}$$

Flow rate of solids in ball mill feed slurry
$$= \text{Flow rate of solids in rod mill discharge slurry}$$
$$+ \text{Flow rate of solids in hydrocyclone underflow slurry}$$
$$= 100 + 200 = 300 \text{ tons/hr}$$

% solids in ball mill feed slurry

$$= \% \text{ solids in ball mill discharge slurry}$$

$$= 78.33\%$$

Let the densities of slurries of rodmill feed and/(or) discharge, ball mill feed and/(or) discharge, hydrocyclone feed, overflow and underflow be ρ_{slr}, ρ_{slb}, ρ_{slh}, ρ_{slo} and ρ_{slu}.

By using equation $\dfrac{C_w}{\rho_p} + \dfrac{1 - C_w}{\rho_w} = \dfrac{1}{\rho_{sl}}$ all these density values are calculated as follows:

$$\frac{1}{\rho_{slr}} = \frac{0.7518}{3} + \frac{0.2482}{1} \quad \Rightarrow \quad \rho_{slr} = 2.00 \, \text{gm/cm}^3$$

$$\frac{1}{\rho_{slb}} = \frac{0.7833}{3} + \frac{0.2167}{1} \quad \Rightarrow \quad \rho_{slb} = 2.09 \, \text{gm/cm}^3$$

$$\frac{1}{\rho_{slh}} = \frac{0.60}{3} + \frac{0.40}{1} \quad \Rightarrow \quad \rho_{slh} = 1.67 \, \text{gm/cm}^3$$

$$\frac{1}{\rho_{slo}} = \frac{0.40}{3} + \frac{0.60}{1} \quad \Rightarrow \quad \rho_{slo} = 1.36 \, \text{gm/cm}^3$$

$$\frac{1}{\rho_{slu}} = \frac{0.80}{3} + \frac{0.20}{1} \quad \Rightarrow \quad \rho_{slu} = 2.14 \, \text{gm/cm}^3$$

Specific gravities of slurries of rodmill feed and/(or) discharge, ball mill feed and/(or) discharge, hydrocyclone feed, overflow and underflow are 2.00, 2.09, 1.67, 1.36 and 2.14 respectively.

Complete material balance for the circuit is illustrated in Fig. 18.1.1.1.

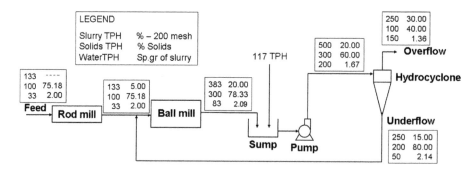

Figure 18.1.1.1 Circuit diagram with complete material balance for example 18.1.1.

Example 18.1.2: *A two stage grinding circuit using a rod mill in open circuit with a ball mill in closed circuit with a rake classifier is used to grind 40 tons per hour of ore (sp. gr. 3.0). The circuit diagram is given in Fig. 18.1.2.*

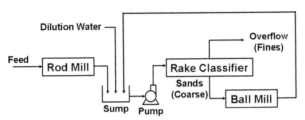

Figure 18.1.2 Circuit diagram for example 18.1.2.

The circuit is sampled and analyzed for percent −100 microns material and percent solids. The results are given in the Table 18.1.2.

Table 18.1.2 Circuit sampling results for example 18.1.2.

Stream	% −100 microns	% solids by wt
Circuit (Rod mill) Feed	5	80
Rod Mill Discharge	20	80
Classifier Sands (Ball Mill Feed)	20	75
Classifier Overflow	75	33.3
Ball Mill Discharge	47.5	75

Using the above data carry out a material balance and calculate:

(a) the tons/hr of dilution water added to the rake classifier
(b) the percent circulating load in the ball mill circuit
(c) the flow rates and specific gravities of all streams and illustrate the material balance of the whole circuit

Solution:

In closed circuit grinding operation feed solids to the grinding circuit will be equal to the overflow solids from the rake classifier at steady state. Therefore

Tons of solids in the rake classifier overflow $= 40$ tons/hr

Let the sands discharged from the rake classifier be S tons/hr

Balance of −100 microns material over the rake classifier

$$40 \times 20 + S \times 47.5 = 40 \times 75 + S \times 20$$

$$\Rightarrow \qquad\qquad S = 80 \text{ tons/hr}$$

Sands discharged from the rake classifier $= 80$ tons/hr

Dilution ratio of Rod mill discharge $= \dfrac{100 - 80}{80} = 0.25$

Dilution ratio of Ball mill discharge $= \dfrac{100 - 75}{75} = 0.333$

Dilution ratio of rake classifier sands discharge $= \dfrac{100 - 75}{75} = 0.333$

Dilution ratio of rake classifier fines discharge $= \dfrac{100 - 33.3}{33.3} = 2.0$

Let the dilution water added to the rake classifier be W

Water balance over the rake classifier

$$40 \times 0.25 + W + 80 \times 0.333 = 40 \times 2.0 + 80 \times 0.333$$

$\Rightarrow$ $\qquad\qquad W = 70\,\text{tons/hr}$

(a) Dilution water added to the rake classifier $= 70\,\text{tons/hr}$

(b) Percent circulating load in the ball mill circuit $= \dfrac{80}{40} \times 100 = 200\%$

(c) Slurry flow rates

Rod mill feed $=$ Rod mill discharge $= 40/0.80 = 50\,\text{tons/hr slurry}$
Ball mill feed $=$ Ball mill discharge $= 80/0.75 = 106.7\,\text{tons/hr slurry}$
Rake classifier overflow $= 40/0.333 = 120\,\text{tons/hr slurry}$

Percent -100 micron material in feed stream to rake classifier

$$= \dfrac{40 \times 20 + 80 \times 47.5}{120} = 38.3\%$$

% solids in feed stream to rake classifier

$$= \dfrac{40 + 80}{50 + 70 + 106.7} \times 100 = 52.9\%$$

Let the densities of slurries of rodmill feed and/(or) discharge, ball mill feed and/(or) discharge and/(or) rake classifier underflow, rake classifier feed and rake classifier overflow be ρ_{slr}, ρ_{slb}, ρ_{slc} and ρ_{slo}.

By using equation $\dfrac{C_w}{\rho_p} + \dfrac{1 - C_w}{\rho_w} = \dfrac{1}{\rho_{\text{sl}}}$ all these density values are calculated as follows:

$$\dfrac{1}{\rho_{\text{slr}}} = \dfrac{0.80}{3} + \dfrac{0.20}{1} \quad \Rightarrow \quad \rho_{\text{slr}} = 2.1\,\text{gm/cm}^3$$

$$\frac{1}{\rho_{slb}} = \frac{0.75}{3} + \frac{0.25}{1} \quad \Rightarrow \quad \rho_{slb} = 2.0\,\text{gm/cm}^3$$

$$\frac{1}{\rho_{slc}} = \frac{0.529}{3} + \frac{0.471}{1} \quad \Rightarrow \quad \rho_{slc} = 1.5\,\text{gm/cm}^3$$

$$\frac{1}{\rho_{slo}} = \frac{0.333}{3} + \frac{0.667}{1} \quad \Rightarrow \quad \rho_{slo} = 1.3\,\text{gm/cm}^3$$

Specific gravities of slurries of rodmill feed and/(or) discharge, ball mill feed and/(or) discharge and/or rake classifier underflow, rake classifier feed, and overflow are 2.1, 2.0, 1.5 and 1.3 respectively.

Complete material balance for the circuit is illustrated in Fig. 18.1.2.1.

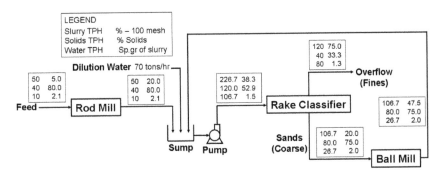

Figure 18.1.2.1 Circuit diagram with complete material balance for example 18.1.2.

Example 18.1.3: *A copper flotation circuit is used to concentrate 100 tons/hr of ore. The ore contains chalcopyrite as valuable mineral with siliceous gangue. Specific gravity of chalcopyrite and gangue are 4.2 and 2.6 respectively. %Cu in chalcopyrite is 34.6%. The circuit is shown in Fig. 18.1.3.*

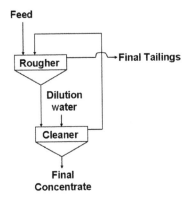

Figure 18.1.3 Flotation circuit for example 18.1.3.

The circuit is sampled and the results are shown in Table 18.1.3.

Table 18.1.3 Analyses of samples for example 18.1.3.

	% solids	%Cu
Circuit feed	33.30	0.50
Rougher concentrate	50.00	13.89
Final concentrate	40.00	25.00
Cleaner tailings	8.62	5.00
Final tailings	31.10	0.10

Using the above data, carry out a material balance and calculate the following:

(a) the copper recovery in   (i) the circuit
*   (ii) the cleaners*
*   (iii) the roughers*
(b) the tons/hr of dilution water added to the cleaners
(c) solids and slurry flowrates and specific gravities of all streams
(d) Illustrate the material balance of whole circuit

Solution:

Let F = Flow rate of ore to the circuit = 100 tons/hr
       C_R = Flow rate of rougher concentrate
       T = Flow rate of final tailing
       C = Flow rate of final concentrate
       T_C = Flow rate of cleaner tailing
       f = Assay value of ore feed
       c = Assay value of final concentrate
       t = Assay value of final tailing
       c_R = Assay value of rougher concentrate
       t_C = Assay value of cleaner tailing
       DR_f = Dilution ratio of feed stream
       DR_c = Dilution ratio of concentrate stream
       DR_t = Dilution ratio of tailing stream

Overall solids balance for the circuit

$$F = C + T \quad \Rightarrow \quad 100 = C + T \quad \Rightarrow \quad T = 100 - C$$

Overall copper balance for the circuit

$$Ff = Cc + Tt \quad \Rightarrow \quad 100 \times 0.5 = C \times 25 + T \times 0.1$$
$$\Rightarrow \quad 100 \times 0.5 = C \times 25 + (100 - C) \times 0.1$$
$$\Rightarrow \quad C = 1.6 \text{ tons/hr}; \quad T = 98.4 \text{ tons/hr}$$

Solids balance over cleaner

$$C_R = C + T_C \quad \Rightarrow \quad C_R = 1.6 + T_C$$

Copper balance over cleaner

$$C_R c_R = Cc + T_C t_C \quad \Rightarrow \quad (1.6 + T_C)13.89 = 1.6 \times 25 + T_C \times 5$$

$$\Rightarrow \quad T_C = 2 \, \text{tons/hr}; \quad C_R = 3.6 \, \text{tons/hr}$$

$$\text{Copper recovery in the circuit} = \frac{Cc}{Ff} \times 100 = \frac{1.6 \times 25}{100 \times 0.5} \times 100 = 80\%$$

$$\text{Copper recovery in cleaner} = \frac{Cc}{C_R c_R} \times 100 = \frac{1.6 \times 25}{3.6 \times 13.89} \times 100 = 80\%$$

$$\text{Copper recovery in rougher} = \frac{C_R c_R}{Ff + T_C t_C} \times 100$$

$$= \frac{3.6 \times 13.89}{100 \times 0.5 + 2 \times 5} \times 100 = 83.3\%$$

Overall water balance

$$F \times DR_f + \text{Dilution water} = C \times DR_c + T \times DR_t$$

$$\Rightarrow \quad 100 \times \frac{66.7}{33.3} + \text{Dilution water} = 1.6 \times \frac{60}{40} + 98.4 \times \frac{68.9}{31.1}$$

$$\Rightarrow \quad \text{Dilution water} = 20 \, \text{tons/hr}$$

$$\% \, \text{chalcopyrite in feed} = \frac{0.5}{34.6} \times 100 = 1.45\%$$

$$\% \, \text{gangue in feed} \quad = 100 - 1.45 = 98.55\%$$

Let the density of the feed solids $\quad = \rho_{fs}$

$$\frac{100}{\rho_{fs}} = \frac{1.45}{4.2} + \frac{98.55}{2.6} \quad \Rightarrow \quad \rho_{fs} = 2.61 \, \text{gm/cm}^3$$

$$\% \, \text{chalcopyrite in rougher concentrate} = \frac{13.89}{34.6} \times 100 = 40.14\%$$

$$\% \, \text{gangue in rougher concentrate} \quad = 100 - 40.14 = 59.86\%$$

Let the density of the rougher concentrate solids $= \rho_{rc}$

$$\frac{100}{\rho_{rc}} = \frac{40.14}{4.2} + \frac{59.86}{2.6} \quad \Rightarrow \quad \rho_{rc} = 3.07 \, \text{gm/cm}^3$$

$$\% \, \text{chalcopyrite in final concentrate} = \frac{25}{34.6} \times 100 = 72.25\%$$

$$\% \, \text{gangue in final concentrate} \quad = 100 - 72.25 = 27.75\%$$

Let the density of the final concentrate solids $= \rho_c$

$$\frac{100}{\rho_c} = \frac{72.25}{4.2} + \frac{27.75}{2.6} \quad \Rightarrow \quad \rho_c = 3.58 \, \text{gm/cm}^3$$

$$\% \, \text{chalcopyrite in cleaner tailings} = \frac{5}{34.6} \times 100 = 14.45\%$$

$$\% \, \text{gangue in cleaner tailings} \quad = 100 - 14.45 = 85.55\%$$

Let the density of the cleaner tailings solids $= \rho_{ct}$

$$\frac{100}{\rho_{ct}} = \frac{14.45}{4.2} + \frac{85.55}{2.6} \quad \Rightarrow \quad \rho_{ct} = 2.75 \,\text{gm/cm}^3$$

% chalcopyrite in final tailings $\quad = \dfrac{0.1}{34.6} \times 100 = 0.29\%$

% gangue in final tailings $\quad = 100 - 0.29 = 99.71\%$

Let the density of the final tailings $= \rho_t$

$$\frac{100}{\rho_t} = \frac{0.29}{4.2} + \frac{99.71}{2.6} \quad \Rightarrow \quad \rho_t = 2.60 \,\text{gm/cm}^3$$

Flow rate of slurry fed to the circuit $\quad = \dfrac{100}{0.333} = 300.3 \,\text{tons/hr}$

Flow rate of final tailing slurry $\quad = \dfrac{98.4}{0.311} = 316.3 \,\text{tons/hr}$

Flow rate of rougher concentrate slurry $\quad = \dfrac{3.6}{0.50} = 7.2 \,\text{tons/hr}$

Flow rate of final concentrate slurry $\quad = \dfrac{1.6}{0.40} = 4 \,\text{tons/hr}$

Flow rate of cleaner tailing slurry $\quad = \dfrac{2}{0.0862} = 23.2 \,\text{tons/hr}$

Complete material balance for the flotation circuit is illustrated in Fig. 18.1.3.1.

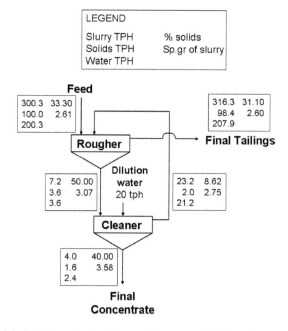

Figure 18.1.3.1 Flotation circuit with complete material balance for example 18.1.3.

Example 18.1.4:   *In an Iron Ore concentrator, Iron ore fines are fed to a cyclone at the rate of 719 dry tons/hr after grinding to the required size in a ball mill. The cyclone feed contains 40% solids by weight. The cyclone underflow is discharged at the rate of 469 dry tons/hr and found to contain 50% solids by weight. This underflow slurry is diluted to 24% solids and treated in a 2-start spiral concentrator from which a concentrate with 60% solids is obtained at the rate of 347 dry tons/hr. The overflow of the cyclone and the tailings of spiral concentrator are disposed off as tailing after recovering water in a thickener. The thickened slurry is found to contain 30% solids. Draw a neat flow diagram and determine whether the water recovered will be sufficient to dilute the cyclone underflow. If the assays of cyclone feed, cyclone underflow and the spiral concentrate are 64.77% Fe, 65.97% Fe and 67.30% Fe respectively, calculate the Iron lost in disposed slurry. Calculate flowrates of slurry and solids of all streams and illustrate the complete material balance of the circuit.*

Solution:

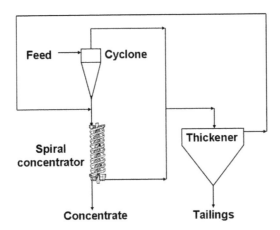

Figure 18.1.4 Flow diagram for example 18.1.4.

Given

	Flowrates tons/hr	% solids	% Fe
Feed to the cyclone	719	40	64.77
Underflow from the cyclone	469	50	65.97
Spiral concentrate	347	60	67.30
Feed to spiral concentrator	–	24	–
Thickened slurry	–	30	–

Flowrate of slurry fed to cyclone	$= \dfrac{719}{0.40} = 1797.5 \text{ tons/hr}$
Flowrate of cyclone underflow slurry	$= \dfrac{469}{0.50} = 938 \text{ tons/hr}$
Flowrate of cyclone overflow slurry	$= 1797.5 - 938.0 = 859.5 \text{ tons/hr}$
Flowrate of cyclone underflow slurry after adding water	$= \dfrac{469}{0.24} = 1954.2 \text{ tons/hr}$
Water added to cyclone underflow	$= 1954.2 - 938.0 = 1016.2 \text{ tons/hr}$
Flowrate of spiral concentrate slurry	$= \dfrac{347}{0.60} = 578.3 \text{ tons/hr}$
Flowrate of spiral tailing slurry	$= 1954.2 - 578.3 = 1375.9 \text{ tons/hr}$
Flowrate of thickener feed slurry	$= 859.5 + 1375.9 = 2235.4 \text{ tons/hr}$
Flowrate of dry solids in cyclone overflow	$= 719 - 469 = 250 \text{ tons/hr}$
% solids in cyclone overflow slurry	$= \dfrac{250}{859.5} \times 100 = 29.09\%$
Flowrate of dry solids in spiral tailings	$= 469 - 347 = 122 \text{ tons/hr}$
% solids in spiral tailing slurry	$= \dfrac{122}{1375.9} \times 100 = 8.87\%$
Flowrate of dry solids in thickener feed	$= 250 + 122 = 372 \text{ tons/hr}$
% solids in thickener feed slurry	$= \dfrac{372}{859.5 + 1375.9} \times 100 = 16.64\%$
Flowrate of thickener underflow slurry	$= \dfrac{372}{0.30} = 1240 \text{ tons/hr}$
Flowrate of water removed from thickener	$= 2235.4 - 1240.0 = 995.4 \text{ tons/hr}$
Flowrate of additional water required	$= 1016.2 - 995.4 = 20.8 \text{ tons/hr}$
Flowrate of water added to the cyclone feed	$= 1797.5 - 719.0 = 1078.5 \text{ tons/hr}$

Flowrate of total water required for the circuit $= 1078.5 + 20.8 = 1099.3 \text{ tons/hr}$

Volumetric flowrate of water required $= 1099.3 \text{ m}^3/\text{hr}$

$$= 1099.3 \times 219.3 = 24.1 \times 10^4 \text{ gallons/hr}$$

Let t_1 be the %Fe in cyclone overflow. Iron balance over cyclone

$$719 \times 64.77 = 469 \times 65.97 + 250t_1$$

$$\Rightarrow \qquad t_1 = 62.52\%$$

Let t_2 be the %Fe in spiral tailing. Iron balance over spiral concentrator

$$469 \times 65.97 = 347 \times 67.30 + 122t_2$$

$$\Rightarrow \qquad t_2 = 62.19\%$$

$$\text{Iron lost in disposed slurry} = 250 \times \frac{62.52}{100} + 122 \times \frac{62.19}{100} = 232.2 \text{ tons/hr}$$

$$\%\text{Fe in thickened slurry} = \frac{232.2}{250 + 122} \times 100 = 62.42\% \text{ Fe}$$

Complete material balance for the flow diagram is illustrated in Fig. 18.1.4.1.

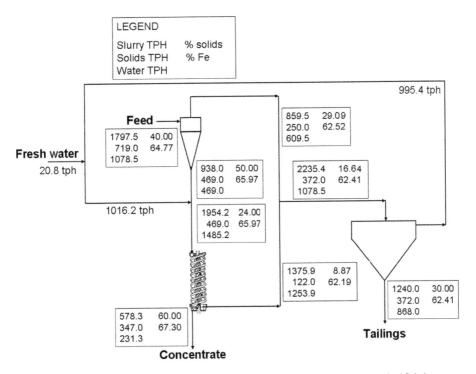

Figure 18.1.4.1 Flow diagram with complete material balance for example 18.1.4.

Example 18.1.5: *A two stage water only cyclone circuit as shown in Fig. 18.1.5 is used for cleaning 100 tons/hr of fine coal. The circuit was sampled and the results are given in Table 18.1.5.*

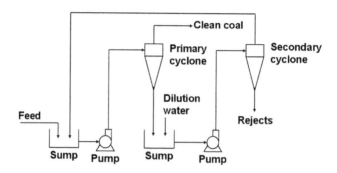

Figure 18.1.5 Two stage water only cyclone circuit for example 18.1.5.

Table 18.1.5 Results of tests for samples for example 18.1.5.

	% solids	Ash%
Circuit feed	12.50	25
Primary cyclone feed	13.15	25
Primary cyclone overflow (clean coal)	10.00	10
Primary cyclone underflow	25.00	47.5
Secondary cyclone overflow	16.67	25
Secondary cyclone underflow (rejects)	33.33	70

a)   Carry out a material balance for the circuit
b)   Calculate tons/hr of dilution water added to the sump
c)   Calculate the flowrates of slurry and solids of all streams and illustrate the comlete material balance of the circuit

Solution:

Let

F = Flow rate of circuit feed solids
C = Flow rate of clean coal solids
T = Flow rate of reject solids
P_f = Flow rate of primary cyclone feed solids
P_u = Flow rate of primary cyclone underflow solids
S_f = Flow rate of secondary cyclone feed solids
S_o = Flow rate of secondary cyclone overflow solids
W = Flow rate of dilution water

Balancing over primary cyclone

Solid balance $\qquad\qquad\qquad F + S_o = C + P_u$

$\Rightarrow \qquad\qquad\qquad\qquad\quad 100 + S_o = C + P_u$ $\qquad\qquad$ (I)

Slurry balance $\qquad \dfrac{100}{0.125} + \dfrac{S_o}{0.1667} = \dfrac{C}{0.1} + \dfrac{P_u}{0.25}$

$\Rightarrow \qquad\qquad\qquad 800 + 6S_o = 10C + 4P_u$ $\qquad\qquad$ (II)

Ash balance $100 \times 25 + S_o \times 25 = C \times 10 + P_u \times 47.5$

$\Rightarrow$ $2500 + 25S_o = 10C + 47.5P_u$ (III)

Balancing over secondary cyclone

Solid balance $P_u = S_o + T$ (IV)

Slurry balance $\dfrac{P_u}{0.25} + W = \dfrac{S_o}{0.1667} + \dfrac{T}{0.3333}$

$\Rightarrow$ $4P_u + W = 6S_o + 3T$ (V)

Ash balance $P_u \times 47.5 = S_o \times 25 + T \times 70$

$\Rightarrow$ $47.5P_u = 25S_o + 70T$ (VI)

On solving equations (I) to (VI), solids flowrates of all streams are obtained:

C = Flow rate of clean coal solids $= 75$ tons/hr
T = Flow rate of reject solids $= 25$ tons/hr
P_u = Flow rate of primary cyclone underflow solids $= 50$ tons/hr
S_o = Flow rate of secondary cyclone overflow solids $= 25$ tons/hr
W = Flow rate of dilution water $= 25$ tons/hr
P_f = Flow rate of primary cyclone feed solids $= 125$ tons/hr
S_f = Flow rate of secondary cyclone feed solids $= 50$ tons/hr

Slurry flow rates can be computed by knowing solids flowrates and % solids

Flow rate of circuit feed slurry $= 100/0.125$ $= 800$ tons/hr
Flow rate of secondary cyclone overflow slurry $= 25/0.1667$ $= 150$ tons/hr
Flow rate of primary cyclone feed slurry $= 125/0.1315$ $= 950$ tons/hr
Flow rate of clean coal slurry $= 75/0.10$ $= 750$ tons/hr
Flow rate of primary cyclone underflow slurry $= 50/0.25$ $= 200$ tons/hr
Flow rate of secondary cyclone feed slurry $= 200 + 25$ $= 225$ tons/hr
Flow rate of reject slurry $= 25/0.3333$ $= 75$ tons/hr

Circuit with complete material balance is shown in Fig. 18.1.5.1.

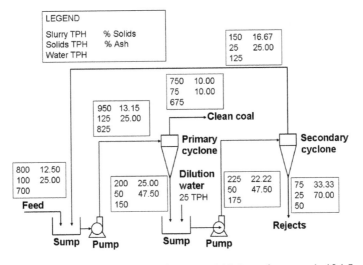

Figure 18.1.5.1 Circuit with complete material balance for example 18.1.5.

Example 18.1.6:   *90 dry tons/hr of lead ore with 10% moisture is fed to a rod mill. After grinding, the rod mill discharges a pulp of 65% solids by weight. It is diluted to 30% solids and then pumped to cyclone. The overflow of cyclone consists of 20% solids are sent to slime treatment plant. The cyclone underflow is fed to a concentration plant. On sampling and analysis it is found that cyclone underflow has 40% solids by weight and contains 6% lead. Concentration plant produced a lead concentrate and tailing assayed 45% and 0.2% respectively. The tailing slurry from concentration plant is found to have 38% solids by weight is dewatered to 65% solids in a thickener. The water recovered in a thickener is sent to mill header tank to supply the water to the rod mill feed as well as rod mill discharge.*

Draw the process flow diagram and calculate the volumetric flow rates of water required to add to the rod mill feed and discharge. Also calculate the flow rate of make-up water required to add to the mill header tank in order to supply the water without deficiency.

Calculate the flowrates of slurry and solids of all streams and illustrate the comlete material balance of the circuit.

Solution:

Process flow diagram for this example is shown in Fig. 18.1.6.

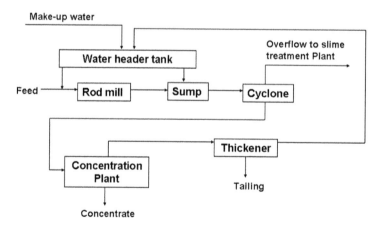

Figure 18.1.6 Process flow diagram for example 18.1.6.

Given

Flow rate of solids fed to rod mill	= 90 tons/hr
% moisture in feed to rod mill	= 10%
% solids in discharge from rod mill	= 65%
% solids in feed to cyclone	= 30%
% solids in cyclone overflow	= 20%

% solids in cyclone underflow   = 40%
% solids in tailing (thickener feed) = 38%
% solids in thickener underflow   = 65%
Assay value of cyclone underflow = 6% Pb
Assay value of concentrate   = 45% Pb
Assay value of tailing   = 0.2% Pb

Flow rate of lead ore fed to rod mill   = 90 tons/hr
Flow rate of water in the feed to the plant = $90 \times 10/90 = 10$ tons/hr
Flow rate of water in rod mill feed   = $90 \times 35/65 = 48.5$ tons/hr
$\therefore$ Water addition to rod mill feed   = $48.5 - 10.0 = 38.5$ tons/hr
$\Rightarrow 38.5 \, m^3/hr$

Flow rate of water in cyclone feed   = $90 \times 70/30 = 210$ tons/hr
$\therefore$ Water addition to cyclone feed   = $210 - 48.5 = 161.5$ tons/hr
= $161.5 \, m^3/hr$

Solids balance over cyclone

$$F = P + U \quad \Rightarrow \quad 90 = P + U \quad \Rightarrow \quad P = 90 - U$$

Water balance over cyclone

$$90 \times 70/30 = P \times 80/20 + U \times 60/40$$
$$\Rightarrow \quad 90 \times 70/30 = (90 - U) \times 80/20 + U \times 60/40$$
$$\Rightarrow \quad U = 60 \text{ tons/hr}$$
$$P = 30 \text{ tons/hr}$$

Solids balance over concentration plant

$$U = C + T \quad \Rightarrow \quad C = U - T = 60 - T$$

Lead balance over concentration plant

$$60 \times 6 = C45 + T \times 0.2$$
$$\Rightarrow \quad 60 \times 6 = (60 - T)45 + T \times 0.2$$
$$\Rightarrow \quad T = 52.2 \text{ tons/hr}$$
$$C = 7.8 \text{ tons/hr}$$

Water content in thickener feed = $52.2 \times 62/38 = 85.2$ tons/hr
= $85.2 \, m^3/hr$

Assuming no solids are lost in thickener overflow

Water content in thickener underflow = $52.2 \times 35/65 = 28.1$ tons/hr
= $28.1 \, m^3/hr$
$\therefore$ Thickener overflow water   = $85.2 - 28.1 = 57.1 \, m^3/hr$
Make-up water required to header tank = $38.5 + 161.5 - 57.1 = 142.9 \, m^3/hr$

Slurry flow rates can be computed by knowing solids flowrates and % solids

$$
\begin{aligned}
\text{Flow rate of slurry fed to rod mill} &= 90/0.90 = 100 \text{ tons/hr} \\
\text{Flow rate of slurry discharged from rod mill} &= 100 + 38.5 = 138.5 \text{ tons/hr} \\
\text{Flow rate of slurry fed to cyclone} &= 138.5 + 161.5 \\
&= 300 \text{ tons/hr} \\
\text{Flow rate of cyclone overflow slurry} &= 30/0.20 = 150 \text{ tons/hr} \\
\text{Flow rate of cyclone underflow slurry} &= 60/0.40 = 150 \text{ tons/hr} \\
\text{Flow rate of tailing slurry} &= 52.2/0.38 = 137.4 \text{ tons/hr} \\
\text{Flow rate of concentrate slurry} &= 150 - 137.4 = 12.6 \text{ tons/hr} \\
\text{Flow rate of thickener underflow slurry} &= 52.2/0.65 = 80.3 \text{ tons/hr} \\
\text{\% solids in concentrate} = \frac{7.8}{12.6} \times 100 \ &= 62\%
\end{aligned}
$$

Process flow diagram with complete material balance is shown in Fig. 18.1.6.1.

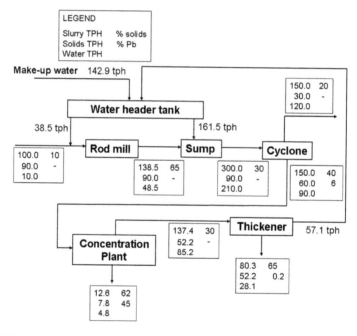

Figure 18.1.6.1 Process flow diagram with complete material balance for example 18.1.6.

18.2 FLOTATION CONTACT TIME

Example 18.2.1: *A pulp containing 40% solids by weight is conditioned for 6 minutes and pumped to flotation plant. If the rate of solids fed to flotation plant is 400 tons/hr and the density of solids is 2650 kg/m³, calculate the volume of the conditioning tank required.*

Solution:

$$\text{Volumetric flowrate of solids} = \frac{400 \times 1000}{2650} = 150.9 \, \text{m}^3/\text{hr}$$

$$\text{Mass flowrate of water} = \text{Mass flowrate of solids} \times \text{dilution ratio}$$

$$= 400 \times \frac{1 - 0.40}{0.40} = 600 \, \text{tons/hr}$$

$$\text{Volumetric flowrate of water} = \frac{600 \times 1000}{1000} = 600 \, \text{m}^3/\text{hr}$$

$$\text{Volumetric flowrate of slurry} = 600 + 150.9 = 750.9 \, \text{m}^3/\text{hr}$$

$$\text{Retention time} = 6 \, \text{minutes}$$

$$\text{Volume of conditioning tank} = 750.9 \times \frac{6}{60} = 75.1 \, \text{m}^3$$

Example 18.2.2: *An ore having specific gravity of 3.5 is to be floated at the rate of 3000 tons/day with a pulp of 25% solids and flotation time of 10 minutes. Find the number of flotation machines of mechanical type required, if each machine has a capacity of 0.75 m³ with 70% effective capacity.*

Solution:

Let the density of the ore and the pulp be ρ and ρ_{sl}.

$$\frac{1}{\rho_{sl}} = \frac{C_w}{\rho} + \frac{1 - C_w}{\rho_w} = \frac{0.25}{3.5} + 0.75 \quad \Rightarrow \quad \rho_{sl} = 1.217 \, \text{tons/m}^3$$

Weight of solids $= 3000 \, \text{tons/day}$

Weight of the pulp $= 3000/0.25 = 12000 \, \text{tons/day}$

$$= \frac{12000 \times 1000}{24 \times 60} = 8333.33 \, \text{kg/minute}$$

Volume of the pulp $= 8333.33/1217 = 6.847 \, \text{m}^3/\text{minute}$
Volume of the pulp in 10 minutes $= 6.847 \times 10 = 68.47 \, \text{m}^3$
Effective volume of each machine $= 0.75 \times 0.7 = 0.525 \, \text{m}^3$
Number of machines required $= 68.47/0.525 = 130.4 \approx 131$

Example 18.2.3: *Zinc ore containing sphalerite (ZnS) and Silica (SiO₂) is benefici-ated by froth flotation. The flotation circuit consists of Rougher, Scavenger and Cleaner bank of cells. The ore is fed to the rougher cells via conditioner at the rate of 1500 tons/day. Scavenger concentrate and cleaner tailings are re-fed to the rougher bank.*

Cleaner concentrate is the final concentrate and scavenger tailings is the final tailings. Tables 18.2.3.1 and 18.2.3.2 shows the composition of all streams and solid water ratio and contact time in each equipment.

Table 18.2.3.1 Composition of each stream of flotation circuit.

	Weight %	
	ZnS	SiO$_2$
Feed	4.0	96.0
Rougher concentrate	40.0	60.0
Rougher tailings	1.0	99.0
Scavenger concentrate	50.0	50.0
Scavenger tailings	0.3	99.7
Cleaner concentrate	98.0	2.0
Cleaner tailings	20.0	80.0

Table 18.2.3.2 Solid water ratio and contact time.

	Solid:Water (by weight)	Contact time (sec)
Rougher	1:3	450
Scavenger	1:5	700
Cleaner	1:7	600
Conditioner	1:3	450

Specific gravities of sphalerite and silica are 4.0 and 2.65 respectively.

a) *Illustrate the flotation circuit with neat diagram*
b) *Determine the flow rates of all streams*
c) *Determine the percentage yield*
d) *Using flotation cell of 2.5 m³ capacity, compute the number of flotation cells in each bank (rougher, scavenger, cleaner)*
e) *What should be the capacity of the conditioner?*

Solution:

The flotation circuit diagram is shown in Fig. 18.2.3.

Overall solids balance: $F = C + T$ $\Rightarrow$ $1500 = C + T$

Overall CuS balance: $Ff = Cc + Tt$

$\Rightarrow$ $1500 \times 4 = C \times 98 + T \times 0.3$

$\Rightarrow$ $C = 56.81$ tons/day

$\Rightarrow$ $T = 1443.19$ tons/day

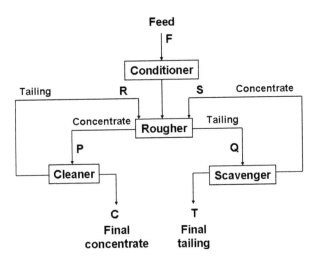

Figure 18.2.3 Flotation circuit diagram.

Solids balance for Scavenger: $Q = T + S$ $\Rightarrow$ $Q = 1443.19 + S$

CuS balance for Scavenger: $Qq = Tt + Ss$

$$Q \times 1 = 1443.19 \times 0.3 + S \times 50$$

$\Rightarrow$ $\qquad\qquad\qquad Q = 1463.81 \text{ tons/day}$

$\Rightarrow$ $\qquad\qquad\qquad S = 20.62 \text{ tons/day}$

Solids balance for Cleaner: $P = C + R$ $\Rightarrow$ $P = 56.81 + R$

CuS balance for Cleaner: $Pp = Cc + Rr$

$\Rightarrow$ $\qquad\qquad P \times 40 = 56.81 \times 98 + R \times 20$

$\Rightarrow$ $\qquad\qquad\qquad P = 221.55 \text{ tons/day}$

$\Rightarrow$ $\qquad\qquad\qquad R = 164.74 \text{ tons/day}$

$$\text{Percent Yield} = \frac{Cc}{Ff} \times 100 = \frac{56.81 \times 98}{1500 \times 4} \times 100 = 92.8\%$$

Calculation of cells required

Rougher feed contains F, R & S streams

Volume of solids in Rougher $= \dfrac{F}{\rho_f} + \dfrac{R}{\rho_r} + \dfrac{S}{\rho_s}$

Water to solids ratio in rougher $= 3$

Volume of water in F, R & S $= \dfrac{3(F + R + S)}{\rho_w}$

Total volume of streams F, R & $S = \dfrac{F}{\rho_f} + \dfrac{R}{\rho_r} + \dfrac{S}{\rho_s} + \dfrac{3(F + R + S)}{\rho_w}$

Density of solids in $F = \rho_f = \dfrac{1}{\dfrac{0.04}{4.0} + \dfrac{0.96}{2.65}} = 2.69 \text{ tons/m}^3$

Density of solids in $R = \rho_r = \dfrac{1}{\dfrac{0.2}{4.0} + \dfrac{0.8}{2.65}} = 2.84 \text{ tons/m}^3$

Density of solids in $S = \rho_s = \dfrac{1}{\dfrac{0.5}{4.0} + \dfrac{0.5}{2.65}} = 3.19 \text{ tons/m}^3$

Total volume of streams F, R & S

$$= \frac{1500}{2.69} + \frac{164.74}{2.84} + \frac{20.62}{3.19} + \frac{3(1500 + 164.74 + 20.62)}{1}$$

$$= 5678.17 \text{ m}^3/\text{day}$$

Volume of rougher = Volume of F, R & S × Contact time

$$= \frac{5678.17 \times 450}{3600 \times 24} = 29.57 \text{ m}^3$$

Number of cells for rougher $= \dfrac{29.57}{2.5} = 11.8 \approx 12$

$\therefore$ 12 cells are required in rougher

Q is the only feed to scavenger

Density of solids in $Q = \rho_q$ $\qquad = \dfrac{1}{\dfrac{0.01}{4.0} + \dfrac{0.99}{2.65}} = 2.66 \text{ tons/m}^3$

Water to solids ratio in scavenger $= 5$

Volume of water in Q $\qquad = \dfrac{5Q}{\rho_w}$

Total volume of $Q = \dfrac{Q}{\rho_q} + \dfrac{5Q}{\rho_w} = \dfrac{1463.81}{2.66} + \dfrac{5 \times 1463.81}{1} = 7869.4 \text{ m}^3/\text{day}$

Volume of scavenger $\qquad = \dfrac{7869.4 \times 700}{3600 \times 24} = 63.76 \text{ m}^3$

Number of cells for scavenger $= \dfrac{63.76}{2.5} = 25.5 \approx 26$

$\therefore$ 26 cells are required in scavenger

P is the only feed to cleaner

Density of solids in $P = \rho_p$ $\qquad = \dfrac{1}{\dfrac{0.4}{4.0} + \dfrac{0.6}{2.65}} = 3.06 \text{ tons/m}^3$

Water to solid ratio in P $\qquad = 7$

Volume of water in $P = \dfrac{7P}{\rho_w}$

Total volume of $P = \dfrac{P}{\rho_p} + \dfrac{7P}{\rho_w} = \dfrac{221.55}{3.06} + \dfrac{7 \times 221.55}{1} = 1623.25 \text{ m}^3/\text{day}$

Volume of cleaner $\qquad = \dfrac{1623.25 \times 600}{3600 \times 24} = 11.27 \text{ m}^3$

Number of cells in cleaner $= \dfrac{11.27}{2.5} = 4.5 \approx 5$

$\therefore$ 5 cells are required in cleaner

Total number of cells required $= 12 + 26 + 5 = 43$

Calculation of conditioner volume

Water to solid ratio in conditioner $= 3$

Volume of water in conditioner $\qquad = \dfrac{3F}{\rho_w}$

Volume of the pulp in conditioner $\quad = \dfrac{F}{\rho_f} + \dfrac{3F}{\rho_w} = \dfrac{1500}{2.69} + \dfrac{3 \times 1500}{1}$

$\qquad\qquad\qquad\qquad\qquad\qquad = 5057.62 \text{ m}^3/\text{day}$

As the contact time is 450 seconds,

Capacity of the conditioner $\qquad = \dfrac{5057.62 \times 450}{3600 \times 24} = 26.34 \text{ m}^3$

18.3 COAL ANALYSIS AND WASHABILITY

Example 18.3.1: *A group of students were given ROM Coal and asked to investigate. They crushed the coal to −38 mm and analyzed for its size using 38 mm, 25 mm, 13 mm, 4 mm, 2 mm, and 0.5 mm screens. Float and sink test was carried out for −38 + 25 mm fraction. −25 + 13 mm and −13 + 4 mm fractions were subjected to jigging operation separately. −4 + 2 mm fraction is analysed for its ash content. −2 + 0.5 mm fraction is separated by tabling operation. Froth flotation operation was conducted for −0.5 mm fraction. The following are the complete results of investigation:*

Size analysis:

Size, mm	−38 + 25	−25 + 13	−13 + 4	−4 + 2	−2 + 0.5	−0.5
Weight, mm	1780	5081	2166	400	447	217

Float and sink analysis of −38 + 25 mm fraction:

Specific gravity	Wt% of floats	Floats Ash%
1.40	12.35	18.95
1.50	34.96	27.89
1.60	25.04	34.56
1.70	3.45	45.67
1.80	13.06	54.32
1.90	2.61	65.43
1.90 (sink)	8.53	76.54

Jigging of −25 + 13 mm fraction:

Three fractions are obtained. Weight% of clean coal and mid-dling are 42% and 35%. The ash% of clean coal, middling and tailing are 18.91%, 33.51% and 67.89% respectively.

Jigging of −13 + 4 mm fraction:

Only two fractions are obtained. Weight% of clean coal is 62%. Ash% of clean coal and tailing are 21.48% and 52.69%.

Ash analysis of −4 + 2 mm fraction:

Size of the sample	$= -4 + 2\,mm$
Weight of empty crucible	$= 17.4567\,gm$
Weight of crucible with coal	$= 18.2491\,gm$
Weight of crucible with coal	
after heating at 750°C till constant weight	$= 17.6999\,gm$

Tabling of −2 + 0.5 mm fraction:

Weight% of cleans is 46%. Ash% of cleans and tailings are 13.45% and 36.24%.

Flotation of −0.5 mm fraction:

59% of cleans with 12.34% ash is obtained. Tailing ash% is 39.43%.

Answer the following questions:

a) Determine weight percent of each size fraction in ROM Coal
b) What is the ash% of ROM Coal?
c) Determine distribution of ash among various sizes of Coal

Solution:

Ash product is calculated for $-38 + 25$ mm fraction and shown in the Table 18.3.1.1.

Table 18.3.1.1 Ash product values.

Specific gravity	Wt% of floats	Floats Ash%	Ash product
1.40	12.35	18.95	234.033
1.50	34.96	27.89	975.034
1.60	25.04	34.56	865.382
1.70	3.45	45.67	157.562
1.80	13.06	54.32	709.419
1.90	2.61	65.43	170.772
1.90 (sink)	8.53	76.54	652.886
			3765.088

Ash% of $-38 + 25$ mm fraction $= 3765.088/100 = 37.65\%$

Jigging of $-25 + 13$ mm

Weight% of tailing $= 100 - 42 - 35 = 23\%$
Sum of ash products $= 42 \times 18.91 + 35 \times 33.51 + 23 \times 67.89$
$$= 3528.54$$
Ash% of $-25 + 13$ mm fraction $= 3528.54/100 = 35.29\%$

Jigging of $-13 + 4$ mm

Weight% of tailing $= 100 - 62 = 38\%$
Sum of ash products $= 62 \times 21.48 + 38 \times 52.69 = 3333.98$
Ash% of $-13 + 4$ mm fraction $= 3333.98/100 = 33.34\%$

Ash analysis of $-4 + 2$ mm fractions:

Weight of coal in crucible $= 18.2491 - 17.4567 = 0.7924$ gm
Weight of ash in crucible $= 17.6999 - 17.4567 = 0.2432$ gm
$$\text{Ash}\% = \frac{0.2432}{0.7924} \times 100 = 30.69\%$$
Ash% of $-4 + 2$ mm fraction $= 30.69\%$

Tabling of $-2 + 0.5$ mm

Weight% of tailing $= 100 - 46 = 54\%$
Sum of ash products $= 46 \times 13.45 + 54 \times 36.24 = 2575.66$
Ash% of $-2 + 0.5$ mm fraction $= 2575.66/100 = 25.76\%$

Flotation of -0.5 mm

Weight% of tailing $= 100 - 59 = 41\%$
Sum of ash products $= 59 \times 12.34 + 41 \times 39.43 = 2344.69$
Ash% of -0.5 mm fraction $= 2344.69/100 = 23.45\%$

Weight percentage of each size fraction and their ash percentages, Ash product and ash distribution for each size fraction are calculated and tabulated in Table 18.3.1.2.

Table 18.3.1.2 Calculated values for example 18.3.1.

Size mm	Weight gm	Wt%	Ash%	Ash Product	Ash distribution
−38 + 25 mm	1780	17.64	37.65	664.146	19.29
−25 + 13 mm	5081	50.35	35.29	1776.852	51.61
−13 + 4 mm	2166	21.46	33.34	715.476	20.78
−4 + 2 mm	400	3.96	30.69	121.532	3.53
−2 + 0.5 mm	447	4.43	25.76	114.117	3.32
−0.5 mm	217	2.15	23.45	50.418	1.47
	10091			3442.541	

From the above table,

$$\text{Ash\% of the coal sample} = 3442.541/100$$
$$= 34.43\%$$

Example 18.3.2: *It is required to wash the coal to yield clean coal of 15% ash. Two coals are available for the purpose. The float and sink analyses of these two coals are given in Table 18.3.2.1.*

Table 18.3.2.1 Float and sink analyses of two coals.

Sp. gr.	Coal A Wt%	Coal A Ash%	Coal B Wt%	Coal B Ash%
1.30 floats	9.6	4.97	10.1	5.44
1.30–1.40	23.0	9.18	18.5	14.32
1.40–1.50	35.0	19.64	17.2	21.58
1.50–1.60	9.4	29.27	20.0	29.75
1.60–1.70	3.3	40.00	13.9	39.56
1.70–1.80	2.8	46.42	3.8	46.30
1.80 sink	16.9	75.57	16.5	68.63

Determine the density of separation and yield percent of clean coal for these two coals. If the two coals are blended and separated, determine density of separation and yield percent of clean coal to get the same 15% ash. Find the difference in yield of clean coal if any.

Solution:

Cumulative weight percentages of floats and ash are determined for coal A and coal B and are given in Table 18.3.2.2 & 18.3.2.3.

Table 18.3.2.2 Cumulative percentages for coal A.

Sp. gr.	Wt% of floats	Cum Wt%	Ash%	Ash product	Cum. Ash product	Cum Ash%
1.30	9.6	9.6	4.97	47.712	47.712	4.97
1.40	23.0	32.6	9.18	211.140	258.852	7.94
1.50	35.0	67.6	19.64	687.400	946.252	14.00
1.60	9.4	77.0	29.27	275.138	1221.390	15.86
1.70	3.3	80.3	40.00	132.000	1353.390	16.85
1.80	2.8	83.1	46.42	129.976	1483.366	17.85
1.80 (sink)	16.9	100.0	75.57	1277.133	2760.499	27.60

Table 18.3.2.3 Cumulative percentages for coal B.

Sp. gr.	Wt% of floats	Cum Wt%	Ash%	Ash product	Cum. Ash product	Cum Ash%
1.30	10.1	10.1	5.44	54.944	54.944	5.44
1.40	18.5	28.6	14.32	264.920	319.864	11.18
1.50	17.2	45.8	21.58	371.176	691.040	15.09
1.60	20.0	65.8	29.75	595.000	1286.040	19.54
1.70	13.9	79.7	39.56	549.884	1835.924	23.04
1.80	3.8	83.5	46.30	175.940	2011.864	24.09
1.80 (sink)	16.5	100.0	68.63	1132.395	3144.259	31.44

Total floats ash curve and yield gravity curve are drawn for coal A and coal B as shown in Fig. 18.3.1 and 18.3.2.

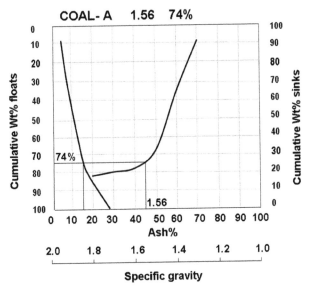

Figure 18.3.1 Total floats ash curve and yield gravity curve for coal A.

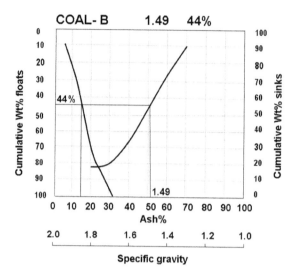

Figure 18.3.2 Total floats ash curve and yield gravity curve for coal B.

On the above curves, specific gravity of separation and yield percentages are read and shown for obtaining clean coal of 15% ash.

$$\text{Weight percentage of floats at 1.30 for blended coal} = \frac{9.6 + 10.1}{2} = 9.85\%$$

$$\text{Ash percentage of floats at 1.30 for blended coal} = \frac{9.6 \times 4.97 + 10.1 \times 5.44}{9.6 + 10.1} = 5.21\%$$

Similar calculations are done for all remaining floats and sink at 1.80 for blended coal and tablulated in 18.3.2.4. From these weight percent and ash percent values, cumulative weight percentages of floats and ash are determined and also shown in the same table.

Table 18.3.2.4 Cumulative percentages for blended coal.

Sp. gr.	Wt%	Cum Wt%	Ash%	Ash product	Cum. Ash product	Cum Ash%
1.30	9.85	9.85	5.21	51.3185	51.3185	5.21
1.40	20.75	30.60	11.47	238.0025	289.3210	9.45
1.50	26.10	56.70	20.28	529.3080	818.6290	14.44
1.60	14.70	71.40	29.60	435.1200	1253.7490	17.56
1.70	8.60	80.00	39.64	340.9040	1594.6530	19.93
1.80	3.30	83.30	46.35	152.9550	1747.6080	20.98
1.80 (sink)	16.70	100.00	72.14	1204.7380	2952.3460	29.52

Total floats ash curve and yield gravity curve are drawn for blended coal and shown in Fig. 18.3.3.

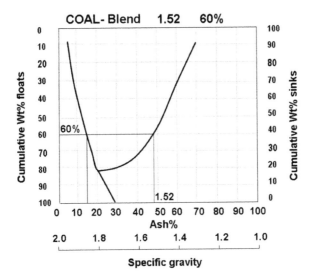

Figure 18.3.3 Total floats ash curve and yield gravity curve for Blend coal.

On the above curves, specific gravity of separation and yield percentage are read and shown for obtaining clean coal of 15% ash.

The results from the 6 curves are tabulated in Table 18.3.2.5.

Table 18.3.2.5 Results from 6 curves.

Coal	Specific gravity of separation	Yield of clean coal	Yield of clean coal
Coal A	1.56	74%	59%
Coal B	1.49	44%	
Blend Coal	1.52	60%	60%

It is to be noted that the specific gravity of separation to yield the clean coal of same 15% ash is different for two coals and blend coal. The yield of clean coal expected is increased by 1% (60 − 59) if two coals are blended and washed.

18.4 ADDITIONAL PROBLEMS FOR PRACTICE

18.4.1:   *A pulp of Iron ore having 30% solids is fed to a beneficiation plant at the rate of 100 tons/hr. Hematite is the only iron mineral in the ore and it contains 69.94% Fe. All other minerals present in the ore are gangue minerals. The assay value of the ore is 60% Fe. If 100% hematite is separated in beneficiation plant without any gangue, what is the rate of concentrate?*
[25.737 tons/hr]

18.4.2: *25 tons of dry ore per hour discharged from the cyclone overflow is fed to a flotation operation for upgradation. The cyclone is in closed circuit with ball mill and the ore is drawn from the fine ore bin as shown in Fig. 18.4.2. The feed from the fine ore bin is sampled, and is found to contain 5% moisture. The cyclone feed, underflow and overflow are sampled and found that they contain 33% solids, 65% solids and 15% solids.*

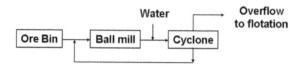

Figure 18.4.2 Closed circuit grinding diagram for problem 18.4.2.

Calculate the amount of water required to dilute the ball mill discharge, the circulating load on the circuit and percent circulating load.

[140.42 m^3/hr, 60.94 tons/hr, 243.8%]

18.4.3: *100 tons/hr of ore is grounded in a wet ball mill with 40% solids in closed circuit with a hydrocyclone. Calculate the weight of water required to be added in the ball mill if the %circulating load is 300 and %solids in hydrocyclone underflow is 50%. What is the % solids in hydrocyclone overflow? If the hydrocyclone overflow is thickened to 40% solids and the recovered water is reused in ball mill, how much more fresh water is required?*

[300 tons/hr, 25%, 150 tons/hr]

18.4.4: *In the circuit shown in Fig. 18.4.4, the rod mill is fed at the rate of 20 tons/hr of dry solids (density 2900 kg/m^3).*

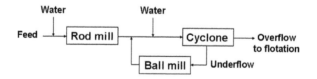

Figure 18.4.4 Circuit diagram for problem 18.4.4.

The cyclone feed contains 35% solids by weight, and size analysis on the rod mill discharge, ball mill discharge and cyclone feed gave

> *Rod mill discharge   26.9% + 250 microns*
> *Ball mill discharge   4.9% + 250 microns*
> *Cyclone feed   13.8% + 250 microns*

Calculate volumetric flow rate of feed to cyclone.

For the same circuit, if the rod mill is fed with 55 tons/hr of dry ore, and the percent solids in rod mill discharge, cyclone feed, cyclone overflow and underflow are 62%, 48%, 31% and 74% respectively, Calculate
(i) Amounts of water added to the rod mill and cyclone feed
(ii) Circulating load

$$[108.7 \ m^3/hr, \ 33.7 \ m^3/hr, \ 88.7 \ m^3/hr, \ 85.8 \ tons/hr]$$

18.4.5: *Iron ore beneficiation plant classify the grounded ore assaying 64.9% Fe in a hydrocyclone. The underflow of the hydrocyclone is treated in a bank rougher and cleaner spiral concentrators. The flow rate of the concentrate obtained is 323 dry tons/hr. The tailings of the rougher bank and cleaner bank has the same assay value of 62.0% Fe. The total flow rates of the tailings from both banks is 146 dry tons/hr. The flow rate of overflow of a hydrocyclone assays same as that of the spiral tailings is 250 dry tons/hr. Hydrocyclone overflow and the tailings of both the spirals are thickened in one thickener to recover water. Draw a neat flow diagram and indicate flow rates and assay values of all streams. What is the increase in the grade of the ore?*

$$[3.55\%]$$

18.4.6: *In a flotation plant, it is required to install a conditioner which can treat a pulp of 45% solids for 6 minutes with reagent. If the density of the solids the pulp contains is 5250 kg/m³ and throughput of the plant is 800 tph of dry solids, what is the volume of the conditioning tank required to install.*

$$[113 \ m^3]$$

18.4.7: *Tin ore is beneficiated by heavy medium separation. Representative samples from the float and sink streams are subjected to float and sink analysis and they are calculated as % feed and shown in Table 18.4.7.1.*

Table 18.4.7.1 Float and sink analyses of floats and sinks of HMS.

Sp.gr	Sinks %feed	Floats %feed
2.55 floats	0.0	1.57
2.55–2.60	0.55	8.67
2.60–2.65	3.52	22.59
2.65–2.70	5.31	14.35
2.70–2.75	5.24	6.67
2.75–2.80	6.88	4.04
2.80–2.85	6.26	1.61
2.85–2.90	2.32	0.24
2.90 sinks	10.18	0.0
	40.26	59.74

Draw Tromp curve considering sinks and calculate specific gravity of separation, Ecart probable and Imperfection.

$$[2.74, \ 0.07, \ 0.026]$$

18.4.8:   *70 gm of −200 mesh coal of 30% ash is subjected to oil agglomeration process at 10% pulp density. First 2% by weight of diesel oil is added and pulp is conditioned for 10 minutes at 700 rpm. Later 15% by weight of furnace oil is added and agitated for 15 minutes (agglomeration period) at 1200 rpm by adding sufficient amount of water. At the end, whole contents are screened on 30 mesh and agglomerates are removed and analysed. The agglomerates weighs 49 gm and their ash percent is 15%. Calculate the yield reduction factor.*

[1.4]

18.4.9:   *500 gm of graphite ore having the specific gravity 2.50 is added to 1.5 litres of water to prepare the feed to laboratory flotation cell. What will be the density of the pulp?*

This graphite pulp has been beneficiated in a 2 litre flotation cell with diesel oil as collector and pine oil as frother. Two samples from each feed, concentrate and tailing (totally six samples) are collected and subjected to analysis. The results are as follows (All weights are in grams):

	Feed	Concentrate	Tailing
Sample 1			
Wt. of empty volatile crucible	16.362	17.453	18.220
Wt. of crucible with graphite sample	17.162	18.253	19.020
Wt. of crucible with graphite sample after heating at 950°C for 7 minutes	17.122	18.197	18.972
Sample 2			
Wt. of empty crucible	17.354	18.368	16.650
Wt. of crucible with graphite sample	18.254	19.168	17.450
Wt. of crucible with graphite sample after heating at 800°C till constant weight	18.029	18.712	17.322

Calculate assay values of feed, concentrate and tailing in terms of percent fixed carbon. How much quantity of float fraction is obtained?

[1.2 gm/cm³, 20%, 50%, 10%, 125 gm]

Procedure for determination of bonds work index

EQUIPMENT AND MATERIALS REQUIRED

1. Standard 12″ × 12″ ball mill rotating at 70 rpm with a mechanical counter to register the number of revolutions.
2. Steel balls of following sizes:

Diameter, inch	Number of balls	Approximate weight, gm
1.50	43	8730
1.25	67	7197
1.00	10	705
0.75	71	2058
0.50	94	1441
Total	285	20131

Calculated surface area = $842 \, in^2$
(or)

Diameter, inch	Number of balls	Approximate weight, gm
1.50	36	9900
1.00	130	7175
0.75	119	3020
Total	285	20095

3. Standard nest of sieves ranging from 6 mesh to 200 mesh
4. Rotap sieve shaker
5. About 30 kg of the given ore

PREPARATION OF FEED MATERIAL

1. Crush about 5 kg of given ore to about ¼ inch in stages to avoid excess fines. Crush this in stages to all −6 mesh.
   Mix the sample well.

2. Carry out the sieve analysis on a representative sample of the crushed feed and determine the feed size F, i.e., the size in microns through which 80% of feed passes.

    Let this percent -200 mesh in the feed be 'x'

3. Place the feed sample in a 1,000 ml measuring cylinder.
    Compact by shaking to yield a volume of 700 ml.

    This unit volume of 700 ml is selected as the volume of the ore always present in the mill during closed circuit grinding tests.

    Let its weight be W gm.

EXPERIMENT

1. Load the mill with the balls and 700 ml of feed.
2. During first cycle, run the mill for N_1 (say 100) revolutions.
3. Unload the mill and sieve the ground material through 200 mesh sieve taking small portions at a time.

    Separate $+200$ mesh material and -200 mesh material.

    Weigh -200 mesh material. Let this weight be W_1 gm.
4. Discord -200 mesh material.

    Add new feed of W_1 gm to $+200$ mesh material and load this into the ball mill. ($+200$ mesh material is the circulating load)

    Weight of -200 mesh material in the feed $= \dfrac{Wx}{100}$

    Net weight of -200 mesh material formed during first cycle $= W_1 - \dfrac{Wx}{100}$

    Net weight of -200 mesh material produced per revolution

    (i.e., grindability in the first cycle) $G_1 = \dfrac{\left(W_1 - \dfrac{Wx}{100}\right)}{N_1}$

Closed circuit is achieved by screening through 200 mesh and replacing -200 mesh fraction by fresh addition of feed at the end of each cycle. For these tests, a circulating load of 250 percent is assumed. For this circulating load, the mill throughput will be 350 percent of the mill product and at equilibrium, mill product is equal to mill feed. So the weight of the product to be produced in each cycle will be $\dfrac{100}{3.5}$ or 28.6 percent of the total mill feed.

This is called **Intended Product Passing** or **IPP** (0.286 W)

For the second cycle, calculate the number of revolutions, N_2, for which the mill is to run.

Net weight of -200 mesh material to be produced in the second cycle
$$= IPP - \text{weight of } -200 \text{ mesh in new feed}$$
$$= 0.286W - \frac{W_1 x}{100}$$

Number of revolutions in second cycle, N_2

$$= \frac{\text{Net weight of } -200 \text{ mesh material to be produced}}{\text{Grindability of the first cycle}}$$

$$= \frac{\left(0.286 - \dfrac{W_1 x}{100}\right)}{G_1}$$

5. Now run the mill for N_2 revolutions.
   Unload the mill and sieve the ground material through -200 mesh.
     Separate $+200$ mesh material and -200 mesh material.
     Weigh -200 mesh material. Let this weight be W_2 gm.

6. Discord -200 mesh material
   Add new feed of W_2 gm to $+200$ mesh material and load this into the ball mill.
   ($+200$ mesh material is the circulating load)

Net weight of -200 mesh material produced during second cycle $= W_2 - \dfrac{W_1 x}{100}$

Grindability during second cycle $G_2 = \dfrac{\left(W_2 - \dfrac{W_1 x}{100}\right)}{N_2}$

For the third cycle

New feed to be added $= W_2$ gm

-200 mesh material in new feed added $= \dfrac{W_2 x}{100}$

Net weight of -200 to be produced in third cycle $= 0.286 W - \dfrac{W_2 x}{100}$

Number of revolutions in third cycle, $N_3 = \dfrac{\left(0.286 W - \dfrac{W_2 x}{100}\right)}{G_2}$

7. Now run the mill for N_3 revolutions.
   Unload the mill and sieve the ground material through -200 mesh.
     Separate $+200$ mesh material and -200 mesh material.
     Weigh -200 mesh material. Let this weight be W_3 gm.

Proceed as before.

Repeat the above procedure at each successive cycle. The net production of material finer than 200 mesh usually shows a definite trend either increasing or decreasing with each succeeding cycle. Continue the grinding cycles until a reversal or leveling off has occurred. From the last three cycles take the average value and find out the equilibrium grams/mill revolution or the ball mill grindability (G). At equilibrium, 28.6 percent of the mill discharge should be finer than the 200 mesh and make up feed must be equal to the product material formed during the cycle. The combined product from the final 3 cycles forms the equilibrium test product.

Carryout a sieve analysis on this product and plot the results in a manner similar to that used for the feed. From this plot, find the 80% passing size of the product (P).

OBSERVATIONS

1. Record the data as follows:
   Weight of 700 ml of feed $= W =$

   Percent -200 mesh material in feed $= x =$

   Weight of -200 mesh material in feed $= \dfrac{Wx}{100} =$

   Weight of the product to be produced for 250% circulating load $= 0.286\,W =$
   Record the experiment data in a tabular form in the following column-wise order:

   Cycle
   Number of revolutions
   -200 mesh material in the product (gm)
   -200 mesh material in the feed (gm)
   Net -200 mesh material produced (gm)
   Grindability (G) (gm/revolution)
   New feed to be added (gm)
   -200 mesh material in new feed (gm)
   Number of revolutions for next cycle
   Determine equilibrium grindability $= G =$
   (average value of last 3 cycles)

2. Calculate the work index of the ore by using the following formula:

$$W_i = \frac{4.45}{P_i^{0.22}\,G^{0.8}\left(P^{-0.5} - F^{-0.5}\right)}$$

   where

   W_i = work index
   P_i = mesh of grind in microns (74 microns in this case)
   G = ball mill grindability (gm/revolution)
   P = product size in microns
   F = feed size in microns

References

[1] Read, H.H.: Rutley's Elements of Mineralogy, *First Indian Edition, CBS Publishers & Distributors, Delhi*, 1984.

[2] Gaudin, A.M.: Principles of Mineral Dressing, *TMH Edition, Tata McGraw-Hill Publishing Company Ltd., New Delhi*, 2005.

[3] Taggart, Arthur F.: Handbook of Mineral Dressing, *A Wiley-Interscience publication, John Wiley & Sons, New York*, 1945.

[4] Schuhmann, R., Jr.: Principles of comminution, I – Size distribution and surface calculation, *Tech. Publs. AIME No. 1189*, 1940.

[5] Rosin, P., Rammler E.: The Laws Governing the Fineness of Powdered Coal, *J. Inst. Fuel*, 7, 1933.

[6] Tromp, K.F.: Collier Guardian, May 21, 1937.

[7] Atiq, S. et al.: Beneficiation Studies on the Low-Grade Chromite of Muslim Bagh, Balochistan, Pakistan, *Pakistan Journal of Scientific and Industrial Research*, Vol. 48, No. 2, March–April 2005.

[8] Chinnaiah: Beneficiation studies of Manganese ore of Kumsi, Shimoga, Southern India, *Indian Journal of Applied Research*, Vol. 4, Issue 2, Feb. 2004.

[9] Von Rittinger P. Ritter: Lehrbuch der Aufbereitungskunde, *Berlin*, 1867.

[10] Kick Friedrich: Das Gesetz der Proportionalem Widerstand und Seine Anwendung, *Leipzig*, 1885.

[11] Bond, F.C.: Third theory of comminution, *AIME Trans.*, 1952.

[12] Reynolds, Osborne: An Experimental Investigation of the Circumstances Which Determine Whether the Motion of Water Shall be Direct or Sinuous, and of the Law Resistance in Parallel Channels, Phil. Trans. Roy. Soc. London, 1883.

[13] Stokes, G.G.: Mathematical and Physical Paper III, *Cambridge University Press*, 1891.

[14] Rouse, Hunter: Nomogram for the Settling Velocity of Sphere, Report of the Committee on Sedimentation, 1936–1937, National Research Council 1937.

[15] Sir Isaac Newton: The Mathematical Principles of Natural Philosophy, Book II, *Translated into English by Andrew Motte*, 1729.

[16] Brown, G.G. & Associates: Unit Operations, *Wiley, New York*, 1950.

[17] Lynch, A.J. and Rao, T.C.: The operating characteristics of hydrocyclone classifiers, *Indian J. Technol.* 1967.

[18] Mishra, P.P., Mohapatra, B.K., Mahanta, K.: Upgradation of low grade siliceous manganese ore from Bonai-Keonjhar Belt, Orissa, India, *Journal of Minerals and Materials Chacterization and Engineering*, Vol. 8, No. 1, 2009.

[19] Vijaya Kumar, T.V., Rao, D.S., Gopalakrishna, S.J.: Mineralogical and separation characteristics of iron ore fines from Bellary-Hospet, India, with special emphasis on Beneficiation by Flotation, *Journal of Mining and Metallurgy*, 47A(I), 2011.

[20] Sunita Routray, Rao, D.S., Bhima Rao, R.: Preliminary studies on Recovery of Total Heavy Minerals from Konark-Ramchandi Beach, *Vistas in Geological Research U.U.Spl. Publ. in Geology (8)*, February 2009.

[21] Eltahir M. Moslim Magboul: Effective Processing of Low Grade Chromite Ore by Heavy Medium Separation Process, *University of Khartoum Engineering Journal*, Vol. 3, Issue 2, August 2013.

[22] Akbar Mehdilo et al.: Characterization and Beneficiation of Iranian Low grade Manganese Ore, *Physico chemical problems of Mineral Processing*, 49(2), 2013, www.minproc.pwr.wroc.pl/journal

[23] Richard O. Burt: Gravity Concentration Technology, *Elsevier Science Publishing Company Inc. New York*, 1984.

[24] Coe, G.D. (Glendale, D.): An explanation of washability curves for the interpretation of float-and-sink data on coal, *Information circular No. 7045, US Bureau of Mines*, 1938.

[25] Bird, B.M.: Interpretation of Float-and-sink Data, *Proceedings of the Third International Conference on Bituminous Coal, Pittsburgh*. Vol. 2. 1931.

[26] Sarkar, G.G., Bose, R.N., Mitra, S.K., Lahiri, A.: An Index for the comparison and correlation of washability characteristics of coal, *Presented to Fourth International Coal Preparation Congress, Harrogate (UK)*, May–June 1962.

[27] Subba Rao, D.V.: Evaluation of Washability Characteristics of Talcher coal, Orissa, *Indian Mineralogist, The Journal of the Mineralogical Society of India*, Vol. 36, No. 1, 2002.

[28] Mayer, F.W.: A new curve showing middling composition, *Gluckauf*, 86, 1950.

[29] Schultz, N.F.: Separation efficiency, *Trans. SME-AIME*, 1970.

[30] Fraser, T., Yancey, H.F.: Interpretation of Results of Coal Washing Tests, *Trans AIME*, Vol. 69, 1923.

[31] Anderson, W.W.: Quantitative efficiency of separation of coal cleaning equipment, *Trans AIME*, 187, 1950.

[32] Tromp, K.F.: New method for the evaluation of coal preparation, (German), *Gluckauf*, 73, 1937.

Further readings

1. Subba Rao, D.V.: Mineral Beneficiation – A Concise Basic Course, *A Balkema book, Taylor & Francis, Netherland, 2011.*
2. Subba Rao, D.V.: Coal – Its Beneficiation, *Em Kay Publications, Delhi, 2003.*
3. Subba Rao, D.V. & Gouricharan, T.: Coal Processing and Utilization, *A Balkema book, Taylor & Francis, Netherland, 2016.*
4. Barry A. Wills, Tim Napier-Munn: Mineral Processing Technology, *Elsevier Science & Technology Books, 2006.*
5. Narayanan, C.M. & Bhattacharyya, B.C.: Mechanical Operations for Chemical Engineers, *Khanna Publishers, Delhi, 1999.*
6. Sekhar, G.C.: Unit Operations in Chemical Engineering, Theory and Problems, *Pearson Education (Singapore) Pte. Ltd. 2005.*
7. Chopey Nicolas, P.: Handbook of Chemical Engineering Calculations, *McGraw Hill International.*
8. Christie John Greankoplis: Transport Processes and Separation Processes Principles, *Prentice Hall of India, 2003.*
9. Cytec Mining Chemicals Handbook, *Revised Edition, Cytec Industries Inc. 2002.*
10. Tsakalakis, K.: Use of simplified method to calculate closed crushing circuits, *Minerals Engineering, Vol. 13, No. 12, 2000, Elsevier Science Ltd.*

Subject index

Activators 195
Adventitious mineral matter 13
Aerated density 76
Angle of nip 106
Aperture 32
Ash 15
Ash error 268
Assay value 3

Bond's law 96
Bulk density 76
Buoyancy force 143

Calorific value 21
Characteristic curve 209
Choked crushing 113
Circulating load 118, 133
Clean coal 22
Cleaner 196, 239
Closed circuit crushing 118
Closed circuit grinding 132
Closed set 106
Coal measure 13
Coal seam 13
Coal substance 13
Coal washeries 22
Coefficient of friction 106
Coefficient of resistance 146
Coke 22
Coking coal 22
Collectors 195
Colloids 78
Combustibles 22
Complex ore 2
Compressibility 76
Concentrate 197
Conditioning 196
Consolidation trickling 193
Critical speed 130
Cumulative floats curve 209
Cumulative sinks curve 209

d_{50} 182
D80 38
Dead flux 182
Degree of liberation 93
Degree of washability 215
Density curve 203
Depressants 195
Dielectric separation 197
Dilution ratio 79
Disseminated ores 94
Distribution curve 274
Distribution factor 69
Drag coefficient 146
Drag force 144
Drift origin 13
Drum feeder 128
Dynamic density 76

Ecart Probable Moyen 278
Ecart probability 70
Ecart Terra 183
Economic efficiency 258, 263
Economic recovery 239
Efficiency
    Anderson 268
    Classifier 171
    Coal washing 267
    Economic 263
    Fraser & Yancey 267
    Grade 184
    Metallurgical 239
    Organic 267
    Screen 61–68
    Separation 256
Efficiency curve 68
Electrical separation 197
Electrodynamic separation 197
Electrostatic separation 197
Elementary assay curve 203
Elementary ash curve 209
Equal settling particles 157
Equivalent diameter 29

Equivalent size 29
Error area 279
Error curve 274
Extraneous mineral matter 13

F80 38
Fixed ash 205
Fixed carbon 14, 15
Float and sink 205
Flow ratio 182
Flowing film concentration 194
Free ash 205
Free crushing 113
Free particles 91
Free settling 155
Free settling ratio 158
Frother 195
Froth flotation 194

Gangue minerals 2
Grade 4
Gravity force 143
Grinding medium 127
Grinding mills 127
Gates-Gaudin-Schuhmann equation 42
Grain size 91
Gravity concentration 191

Heavy liquid separation 192
Heavy medium separation 192
High grade ore 4
High tension separation 197
Hindered settling 155
Hindered settling ratio 159
Humphrey spiral 194
Hydraulic water 165
Hydrocyclone 173
Hygroscopic moisture 14

Imperfection 183, 279
Industrial minerals 1
Inherent moisture 14
Insitu origin 13
Intergrown ores 94

Jigging 193

Kick's law 96

Laminar flow 144, 145
Law of conservation of mass 25
Lean ore 4
Liberation 91
Liberation size 91

Locked particles 91
Low grade ore 4

Magnetic Separation 196
Mass density 75
Massive ores 94
Material balance equation 25, 59
Maximum velocity 144
Mayer curve 227
Mesh of grind 132
Mesh number 32
Mesh size 32
Metal content 3
Metallic minerals 1, 3
Metallurgical coal 22
Metallurgical efficiency 239
Middling 22
Mineral characterization methods 12
Mineral matter 13, 20
Mineral moisture 14
Mining 10
Modifiers 195
Moisture 14

Near gravity material 211
Near mesh material 68
Near size material 68
Net smelter return 259
Non-coking coal 22
Non-metallic minerals 1, 2

Open circuit crushing 116
Open circuit grinding 132
Opening 32
Open set 106
Optimum ash 217
Optimum cut point 217
Optimum degree of washability 215
Optimum recovery 217
Optimum specific gravity 217
Ore 1, 2
Ore minerals 1
Overflow 55
Oversize 55

P80 38
Packed density 76
Particle size distribution 34, 37
Particle size distribution equations 42
Partition coefficient 69, 182
Partition curve 69, 274
Partition density 277
Partition specific gravity 277
Performance curve 68
pH regulators 195

Porosity 76
Primary grinding circuit 138
Probable error 70, 183, 278
Probability factor 69
Proximate analysis 14
Pulp 78
Pure coal 13

Rank 21
Ratio of concentration 238
Ratio of enrichment 239
Ratio of recovery 238
Reynolds number 144
Recovery curve 274
Reduced efficiency curve 184
Reduction ratio 105
Refuse 22
Regrinding circuit 138
Rejects 22
Retention time 86
Rich ore 4
Rittinger's law 96
Rock 1
Rougher 196, 239
Rosin-Rammler equation 43
Run-of-mine coal 22
Run-of-mine ore 8

Sample 27
Sampling 27
Sands 78
Scavenger 196, 239
Scoop feeder 128
Selectivity index 184
Separability curves 203
Set 106
Settling ratio 158
Sharpness index 184
Sieve analysis 34
Sieve cloth 31
Sieve scale 32
Simple ore 2
Sink and float 199
Size analysis 29
Sizing classifier 163
Slimes 78
Sludge 78
Slurry 78

Smelting 10
Sorting classifier 165
Specific gravity 75
Specific surface 30
Sphericity 30, 31
Spout feeder 128
Stratification 193
Surface diameter 29
Surface shape factor 31
Suspension 78

Tailing 197
Tenor 3
Terminal velocity 144, 146
Test sieve 31
Throughput 115
Total carbon 15
Total floats ash curve 209
Total sinks ash curve 209
Tromp area 279
Tromp curve 68, 274
Tromp cut point 278
Tumbling mills 127
Turbulent flow 144, 145
Turbulent resistance 147

Underflow 55
Undersize 55

Valuable minerals 1
Viscous resistance 146
Voidage 76
Void fraction 76
Volatile matter 14
Volume diameter 29
Volume shape factor 31

Washability index 213
Washability number 216
Washability curves 210
Washed coal 22
Weight density 75
Working density 76

Yield error 268
Yield gravity curve 209
Yield loss 268
Yield reduction factor 212

Index for calculations

80% passing size 39, 40

Angle of nip 109
Ash 15–17
Average size 37, 113

Capacity of rolls 111
Circulating load 120–123, 134–137
Closed circuit crushing 120–123
Closed circuit grinding 134–137,
    139–141
Complete material balance 288–304
Compound in mineral 4–7
Compressibility 77
Critical speed 130, 131
Cumulative plots 39–41

Density 77–86
Dilution ratio 83, 84, 181, 185

Ecart probability 72, 73
Ecart Probable Moyen 278–284
Efficiency of classifier 172
Efficiency curve 180–187
Economic efficiency 258–263

Feed size analysis 174
Float and sink analysis 218–227
Flow rates 175, 176, 288–304
Frequency plots 40

Gangue in ore 6–9
Gates-Gaudin-Schumann plot 42–48
    equation 42–48

Head assay, grade 51, 52

Imperfection 68–73, 273–284

Material from graph 41
    from equation 48

Maximum grade of concentrate 8, 9
Mayer curve 227–234
Metal distribution 49, 50
Metal in mineral 3, 5, 6, 7
Metallurgical efficiency 241
Minerals from chemical analysis 6, 7
Mineral in ore 6–9

Near gravity material, curve 212
Net smelter return 259–263

Open circuit crushing 116, 117
Operating speed 130, 131
Optimum degree of washability 216,
    221
Organic efficiency 270

Partition coefficient 273–284
Partition curve 273–284
Power 98–103
Proximate analysis 15–17
    on different bases 21

Ratio of concentration 240
Ratio of enrichment 240
Recovery 180, 240–250
Reduction ratio 114, 115
Reynolds number 148–154
Roll size 108, 110
Rosin-Rammler plot 45–47
    equation 45–47

Screen efficiency 62–68
Separation by classification 166–171
Separation efficiency 256–258
Settling ratio 161, 167–171
Sharpness index 184, 186
Sink and float analysis 202
Size analysis 174
Specific surface 37
Speed of ball mill 130, 131

Terminal velocity 148–154

Volume of conditioner 305–309
Volume of flotation cells 305–309
Volume of liquids for solution 88
Volume of tank 86, 87

Washability curves 219–226
Washability index 214, 221

Washability number 216, 221
Water required 288–304
Work index 102
Working density 77

Yield loss 270
Yield reduction factor 222–227, 270

Milton Keynes UK
Ingram Content Group UK Ltd.
UKHW051852071024
449327UK00025B/1927